T0250518

CRC Series in Chromatography

Editors-in-Chief

Gunter Zweig, Ph.D. and Joseph Sherma, Ph.D.

General Data and Principles
Gunter Zweig, Ph.D. and
Joseph Sherma, Ph.D.

Lipids
Helmut K. Mangold, Dr. rer. nat.

Hydrocarbons
Walter L. Zielinski, Jr., Ph.D.

Carbohydrates
Shirley C. Churms, Ph.D.

Inorganics
M. Qureshi, Ph.D.

Drugs
Ram Gupta, Ph.D.

Phenols and Organic Acids
Toshihiko Hanai, Ph.D.

Terpenoids
Carmine J. Coscia, Ph.D.

Amino Acids and Amines
S. Blackburn, Ph.D.

Steroids
)seph C. Touchstone, Ph.D.

Polymers
Charles G. Smith,
Norman E. Skelly, Ph.D.,
Carl D. Chow, and Richard A. Solomon

**Pesticides and Related
Organic Chemicals**
Joseph Sherma, Ph.D. and
Joanne Follweiler, Ph.D.

Plant Pigments
Hans-Peter Köst, Ph.D.

**Nucleic Acids and
Related Compounds**
Ante M. Krstulovic, Ph.D.

CRC
Handbook
of
Chromatography
Plant Pigments

Volume I
Fat-Soluble Pigments

Editor

Hans-Peter Köst, Dr. rer. nat.
Botanic Institute
University of Munich
F.R.G.

Editors-in-Chief

Gunter Zweig, Ph.D.
President
Zweig Associates
Arlington, Virginia
(Deceased)

Joseph Sherma, Ph.D.
Professor of Chemistry
Lafayette College
Easton, Pennsylvania

CRC Press
Taylor & Francis Group
Boca Raton London New York

CRC Press is an imprint of the
Taylor & Francis Group, an **informa** business

First published 1988 by CRC Press
Taylor & Francis Group
6000 Broken Sound Parkway NW, Suite
300 Boca Raton, FL 33487-2742

Reissued 2018 by CRC Press

© 1988 by Taylor & Francis
CRC Press is an imprint of Taylor & Francis Group, an Informa business

No claim to original U.S. Government works

This book contains information obtained from authentic and highly regarded sources. Reasonable efforts have been made to publish reliable data and information, but the author and publisher cannot assume responsibility for the validity of all materials or the consequences of their use. The authors and publishers have attempted to trace the copyright holders of all material reproduced in this publication and apologize to copyright holders if permission to publish in this form has not been obtained. If any copyright material has not been acknowledged please write and let us know so we may rectify in any future reprint.

Except as permitted under U.S. Copyright Law, no part of this book may be reprinted, reproduced, transmitted, or utilized in any form by any electronic, mechanical, or other means, now known or hereafter invented, including photocopying, microfilming, and recording, or in any information storage or retrieval system, without written permission from the publishers.

For permission to photocopy or use material electronically from this work, please access www.copyright.com (http://www.copyright.com/) or contact the Copyright Clearance Center, Inc. (CCC), 222 Rosewood Drive, Danvers, MA 01923, 978-750-8400. CCC is a not-for-profit organiza-tion that provides licenses and registration for a variety of users. For organizations that have been granted a photocopy license by the CCC, a separate system of payment has been arranged.

Trademark Notice: Product or corporate names may be trademarks or registered trademarks, and are used only for identification and explanation without intent to infringe.

A Library of Congress record exists under LC control number: 87021821

Publisher's Note
The publisher has gone to great lengths to ensure the quality of this reprint but points out that some imperfections in the original copies may be apparent.

Disclaimer
The publisher has made every effort to trace copyright holders and welcomes correspondence from those they have been unable to contact.

ISBN 13: 978-1-138-10504-1 (hbk)
ISBN 13: 978-1-138-55830-4 (pbk)
ISBN 13: 978-1-315-15056-7 (ebk)

Visit the Taylor & Francis Web site at http://www.taylorandfrancis.com and the CRC Press Web site at http://www.crcpress.com

CRC SERIES IN CHROMATOGRAPHY

SERIES PREFACE

The fat-soluble photosynthetic pigments present in plants and algae, including chlorophylls, carotenoids, and related pigments, comprise an important class of compounds with an extensive literature. Dr. Köst and his co-authors have done an admirable job in searching out and organizing much of the critical chromatographic data and methodology in the present volume.

Because of the chemical nature of these prenyllipid compounds, liquid chromatography is preferred for their isolation, separation, and determination. The most widely used methods include low pressure column LC, paper chromatography, TLC, and, most recently, HPLC. All of these methods are covered by Dr. Köst.

Chromatography was "invented" in the early 1900s by Michael Tswett, a Russian botanist and plant physiologist who first applied liquid-solid chromatography on a column of chalk to resolution of the complex natural mixture of yellow and green chloroplast pigments in the extracts of leaves he was studying. On a personal note, I was fortunate to work with Dr. Harold Strain for five summers at the Argonne National Laboratory when I first began to teach. Dr. Strain was one of the first important American chromatography experts and used all variations of liquid chromatography extensively in his studies of photosynthetic pigments. My experience with Dr. Strain set the foundation for my lifelong career of research and writing in chromatography.

Readers of this Handbook are asked to contact the Series Editor if they find errors or omissions in coverage as well as with suggestions for future volumes and authors within the Handbook of Chromatography series.

Joseph Sherma

THE EDITORS-IN-CHIEF

Gunter Zweig, Ph.D., received his undergraduate training at the University of Maryland, College Park, where he was awarded the Ph.D. in biochemistry in 1952. Two years following his graduation, Dr. Zweig was affiliated with the late R. J. Block, pioneer in paper chromatography of amino acids. Zweig, Block, and Le Strange wrote one of the first books on paper chromatography, which was published in 1952 by Academic Press and went into three editions, the last one authored by Gunter Zweig and Dr. Joe Sherma, the co-Editor-in-Chief of this series. *Paper Chromatography* (1952) was also translated into Russian.

From 1953 to 1957, Dr. Zweig was research biochemist at the C. F. Kettering Foundation, Antioch College, Yellow Springs, Ohio, where he pursued research on the path of carbon and sulfur in plants, using the then newly developed techniques of autoradiography and paper chromatography. From 1957 to 1965, Dr. Zweig served as lecturer and chemist, University of California, Davis and worked on analytical methods for pesticide residues, mainly by chromatographic techniques. In 1965, Dr. Zweig became Director of Life Sciences, Syracuse University Research Corporation, New York (research on environmental pollution), and in 1973 he became Chief, Environmental Fate Branch, Environmental Protection Agency (EPA) in Washington, D.C. From 1980 to 1984 Dr. Zweig was Visiting Research Chemist in the School of Public Health, University of California, Berkeley, where he was doing research on farmworker safety as related to pesticide exposure.

During his government career, Dr. Zweig continued his scientific writing and editing. Among his works are (many in collaboration with Dr. Sherma) the now 11-volume series on *Analytical Methods for Pesticides and Plant Growth Regulators* (published by Academic Press); the pesticide book series for CRC Press; co-editor of *Journal of Toxicology and Environmental Health;* co-author of basic review on paper and thin-layer chromatography for *Analytical Chemistry* from 1968 to 1980; co-author of applied chromatography review on pesticide analysis for *Analytical Chemistry*, beginning in 1981.

Among the scientific honors awarded to Dr. Zweig during his distinguished career were the Wiley Award in 1977, the Rothschild Fellowship to the Weizmann Institute in 1963/64; and the Bronze Medal by the EPA in 1980.

Dr. Zweig authored or co-authored over 80 scientific papers on diverse subjects in chromatography and biochemistry, besides being the holder of three U.S. patents. In 1985, Dr. Zweig became president of Zweig Associates, Consultants in Arlington, Va.

Following his death on January 27, 1987, the Agrochemicals Section of the American Chemical Society posthumously elected him a Fellow and established the Gunther Zweig Award for Young Chemists in his honor.

Joseph Sherma, Ph.D., received a B.S. in Chemistry from Upsala College, East Orange, N.J., in 1955 and a Ph.D. in Analytical Chemistry from Rutgers University in 1958, carrying on his thesis research in ion exchange chromatography under the direction of the late William Rieman III. Dr. Sherma joined the faculty of Lafayette College in September, 1958, and is presently Charles A. Dana Professor and Head of the Chemistry Department.

Dr. Sherma, independently and with others, has written over 300 research papers, chapters, books, and reviews involving chromatography and other analytical methodology. He is editor for residues and trace elements of the *Journal of the Association of Official Analytical Chemists* and a member of the advisory board of the *Journal of Planar Chromatography*. He is a consultant on analytical methodology for many companies and government agencies.

Dr. Sherma has received two awards for superior teaching at Lafayette College and the 1979 Distinguished Alumnus Award from Upsala College for outstanding achievements as an educator, researcher, author, and editor. He is a member of the ACS, Sigma Xi, Phi Lambda Upsilon, SAS, AIC, and AOAC. Dr. Sherma's current interests are in quantitative TLC, mainly applied to clinical analysis, pesticide residues, and food additives.

INTRODUCTION

While doing scientific work, many people regularly come across colored compounds that are either contained within plant or animal tissues or that perhaps represent an excreted component of the culture medium. Often pigments, because of their light-absorbing properties, have well-defined tasks, e.g., protection from light or as sensor and antenna pigments. Needless to say, when involved with the analysis of these pigments, one should have suitable literature available. Not only in scientific research, but also in routine analysis the exact determination of plant pigments plays a more and more pronounced role. For example, carotenoids may be used as food dyes, enhancing an unappetizing color. Their exact application necessitates exact analysis.

For this purpose, the *Handbook of Chromatography: Plant Pigments*, Volume I: *Fat-Soluble Pigments* has been compiled, in its essence an assemblage of tables where, besides data obtained by modern separation methods, older sources, often difficult to access, have also been included to give maximum possible information. It is a simple truth that if a pigment is unambiguously identified and described, it will keep the same chromatographic properties, the same absorption maxima, and the same molar extinction coefficient forever! Especially in older books, there are many valuable data that may easily be overlooked and "buried" by the nearly logarithmically growing flood of data published today. On the other hand, scientific methods of analysis and identification at present are developing more rapidly than ever: for example, modern high performance liquid chromatography (HPLC). Often a sample with little or no prior preparation can be injected directly onto tiny columns. The setup often is so sensitive that literally traces of pigments, not noticeable to the naked eye, are sufficient to obtain qualitative as well as quantitative data. Unfortunately, however, the new methods are rather expensive in terms of apparatus, equipment, and supply of suitable chemicals and solvents. This creates a clear limit of availability, especially for small laboratories and individual researchers. Most scientists, when dealing with a simple problem of separation and identification, just do not need sophisticated equipment, but reliable and cheap methods that are nevertheless reasonably quick and easy to handle. Also, even with the best possible instrumental equipment, it is indispensible to carry out some preliminary separation and identification steps before committing to the use of an expensive column that might easily be rendered inoperative by incompatible compounds.

The present *Handbook* is intended to give information on not only the most recent but also the proven older techniques.

In this sense, I wish the users of the book good success.

The Editor

THE EDITOR

Dr. Hans-Peter Köst, Dr. rer. nat. is at present Privatdozent at the Botanic Institute of the University of Munich, Munich, F.R.G. Dr. Köst received his chemistry diploma from the University of Saarbrücken in 1970 and subsequently was awarded the degree of Dr. rer. nat. in Natural Science, specializing in phytochemistry, from Munich University in 1974. From 1975 to 1977 he was a postdoctoral fellow at the University of California, Los Angeles.

In 1981 he was promoted to Dr. rer. nat. habil. and was thereafter lecturer at the Munich Ludwigs-Maximilian-University. From 1984 to 1986 he substituted for different professors, including a period of one semester in Saarbrücken.

Dr. Köst is a member of the German Chemical Society (Gesellschaft Deutscher Chemiker, GDCh) and the American Chemical Society. He is also a member of the International Society for the Study of the Origin Of Life and the International Electrophoresis Society, as well as an elected member of the New York Academy of Sciences.

Dr. Köst has published about 50 articles in scientific journals, chiefly dealing with the chemistry and physiology of tetrapyrrolic pigments of plants and animals.

ACKNOWLEDGMENTS

H.-P. Köst wishes to thank all these many people who have helped with advice and practical support to finish the present volume:

I was given very substantial assistance by my former technician G. Widerer, in industry.

Dr. E. Schropp has helped a great deal in planning the conception of the carotenoid part of the volume.

I would like to gratefully acknowledge the competent support of Dr. E. Benedikt in preparing the ''Porphyrins'' part of the present volume and for much graphic work, especially concerning the ''Carotenoids'' and the ''Porphyrins'' sections of the book.

My special thanks are devoted to all my friends and colleagues who helped me in all thinkable ways. I don't want to fail to thank CRC press for help, advice, and often patience. My special thanks are devoted to Ms. Amy Skallerup.

Finally, I want to emphasize the invaluable support in compiling data given by my wife, Dr. E. Köst-Reyes. Equally valuable, however, is the moral support she has given me over all these last years.

ADVISORY BOARD

Bruce Burnham, Ph.D.
President
Porphyrin Products
Logan, Utah

Brian H. Davies, Ph.D.
Professor
Department of Biochemistry and
  Agricultural Biochemistry
University College of Wales
Aberystwyth
Wales

Eliana Köst-Reyes, Ph.D.
Jesenwang
Federal Republic of Germany

Hugo Scheer, Dr.habil.rer.nat.
Professor
Botanical Institute
University of Munich
Munich
Federal Republic of Germany

CONTRIBUTORS

Eva Benedikt, Ph.D.
Botanical Institute
University of Munich
Munich
Federal Republic of Germany

Bruce Burnham, Ph.D.
President
Porphyrin Products
Logan, Utah

Brian H. Davies, Ph.D.
Professor
Department of Biochemistry and
  Agricultural Biochemistry
University College of Wales
Aberystwyth
Wales

Hans-Peter Köst, Ph.D.
Research and Development
Serva-Technik, GmbH
Heidelberg
Federal Republic of Germany

Eliana Köst-Reyes, Ph.D.
Jesenwang
Federal Republic of Germany

Hugo Scheer, Dr.habil.rer.nat.
Professor
Botanical Institute
University of Munich
Munich
Federal Republic of Germany

Eva Schropp, Ph.D.
Botanical Institute
University of Munich
Munich
Federal Republic of Germany

TABLE OF CONTENTS

PART II: PORPHYRINS (EXCLUSIVE OF CHLOROPHYLLS)
Bruce F. Burnham and Hans-P. Köst

CHROMATOGRAPHIC METHODS FOR THE SEPARATION OF PORPHYRINS AND METALLOPORPHYRINS

TABLES FOR THE ESTIMATION AND SEPARATION OF PORPHYRINS AND METALLOPORPHYRINS

PART III: CHLOROPHYLLS
Hugo Scheer

CHROMATOGRAPHIC METHODS FOR THE SEPARATION OF CHLOROPHYLLS

High Performance Liquid Chromatography (HPLC) 292

Part I: Carotenoids

B. H. Davies and Hans-P. Köst

Chromatographic Methods for the Separation of Carotenoids

INTRODUCTION

Many of the vividly red, orange, or yellow flowers and fruits, as well as a number of animals, owe their appearance to the presence of a class of more-or-less unsaturated tetra-terpenoids called "carotenoids".[1] The name comes from the carrot, *Daucus carota*, from which the prechromatographic "carotene" was isolated (Wackenroder, 1831); only by chromatography could it be shown that there are α-, β-, γ-, and δ-carotenes in the carrot.[1-3]

Since then, the carotenoids have been extensively studied in many branches of natural science.[1,3-5] The greater lability of the chlorophylls during autumnal necrosis reveals the carotenoids in the "fall colors" of deciduous trees. Carotenoids are present in the thylakoid membranes of higher plants, algae, and photosynthetic bacteria; here one part of their function is to serve with lesser or greater efficiency as accessory pigments for light-harvesting in photosynthesis. They are not confined to the photosynthetic organelles; however, their presence and synthesis in so many fungi, yeasts, and bacteria which sometimes exhibit intense colors suggest another, wider function: the universal function of carotenoids as photoprotectants (compare References 3 to 5, 7 to 9).

Carotenoids are also contained in the display apparatus of a variety of flowers;[10,11] here they help to attract insects. Many animals contain carotenoids, also, but only via their diet, for they cannot synthesize them as plants can. Chemically, carotenoids are hydrocarbons with numerous conjugated double bonds. The first carotenoid whose structure was elucidated (by Karrer in 1930) was lycopene, the red pigment of tomatoes and other fruits (for historic background, see References 1 and 12).

SOME REMARKS ON CAROTENOID FORMATION AND SOURCES[3-5,7-10,13-15]

The first "typical" intermediate in carotenoid biosynthesis is isopentenyldiphosphate, which is formed via hydroxymethylglutaryl CoA and then mevalonic acid from the condensation of three molecules of acetyl CoA which arise from the intermediary metabolism of carbohydrates and lipids. By the action of an isomerase, isopentenyldiphosphate is converted to dimethylallyldiphosphate, which easily splits off a diphosphate anion while leaving a carbonium ion. The carbonium ion may now condense head-to-tail with one molecule of isopentenyldiphosphate to form geranyldiphosphate. By the addition of a further molecule of isopentenyldiphosphate, farnesyldiphosphate, a C-15 intermediate, is synthesized. After their formation from this compound, two tetraisoprenoid geranyl-geranyldiphosphates are condensed to molecules of the phytoene, a colorless compound which contains only three conjugated double bonds.

Via stepwise dehydrogenation, phytoene is converted to phytofluene, ζ-carotene, and neurosporene. Eventually, the intensely red-colored lycopene is formed. Lycopene may be converted via ring formation — the α- or β-ionone rings of α- or β-carotene, respectively.

Animal pigments may have undergone considerable modifications; an example for such a modified pigment is the conversion of β-carotene into canthaxanthin by the brine shrimp, *Artemia salina* L. (Crustacea, Branchiopoda).[15] Some animals accumulate carotenoids in quantity (e.g., goldfish, flamingo), while others retain just as much as they require to convert via intestinal bisectioning of the molecule into retinal, which is the basis of all animal vision. Further conversion products are its primary alcohol retinol (vitamin A) and the corresponding acid (retinoic acid).

Since all of these "retinoids" are derived from dietary plant carotenoids, it is clear that the analysis of plant carotenoids has considerable nutritional significance.

Today, another source of carotenoids is chemical synthesis; its route is often according to published procedures (compare References 1 and 16). A number of synthetic carotenoids are used as food dyes. Since treating the principles of synthesis would exceed the scope of this book, the reader is referred to the respective literature.[16] Several hundreds of carotenoids are known, and their number is increasing logarithmically.[1,17] (See Table I.3.)

CAROTENOID CHARACTERIZATION[1,19]

Carotenoid Handling and Storage

Carotenoids, once extracted, are labile and require protection from heat, light, and oxygen. No more heat than absolutely necessary should be used in their manipulation. Low boiling solvents should be employed wherever possible and the concentration of carotenoid solutions should be carried out under reduced pressure using a rotatory evaporator. Carotenoids should not be exposed to bright light; the laboratory window(s) should not face the sun, and vessels, chromatography columns, and TLC tanks should be covered with black cloth, aluminum foil, etc., where appropriate. It is good practice to develop thin-layer chromatograms (TLC) in an inert atmosphere like nitrogen, although it may not always be necessary. As some carotenoids are acid labile (isomerization; furanoid epoxide formation), contact with acids must be avoided. Alkali is used for saponification, but it must be borne in mind that some carotenoids (e.g., astaxanthin) are also alkali labile.[1]

Carotenoids should be stored in aliphatic or aromatic hydrocarbon solvents (under nitrogen or argon), in the dark, and in the deep freeze ($-20°C$).

Carotenoid Crystallization and Melting Points

Around 1950, when many carotenoid syntheses were carried out (for review see Reference 16), an important criterion of purity was the determination of melting points of the crystalline synthetic compounds; for that purpose enough crystalline material was available. Although chromatographic isolation procedures provide enough material for carrying out a variety of spectroscopic and chemical identification procedures, there is often not sufficient material for the production of crystals, and achievement of crystallization requires much experience. For example, carotenoids tend to form mixed crystals, thereby sometimes barring the possibility of purification by fractionated crystallization. Another drawback of carotenoid crystallization is the rapid decomposition of "crystalline" but semipure material, which participants of phytochemical courses can confirm. This decomposition is most pronounced if carotenoids are obtained as a liquid, noncrystalline film. Only where pure carotenoids are crystallized can they be conserved in the solid state and kept for a long time, especially when stored under an inert gas, like argon or nitrogen, and in the dark.

A quick, general description of how to carry out carotenoid crystallization is given in the following: the method is based on the fact that many carotenoids dissolve poorly in methanol, but very well in aromatic solvents like benzene or toluene. In a centrifuge tube, the carotenoid to be crystallized (e.g., lycopene) is dissolved and hot methanol is added to begin crystallization. The sample is allowed to stand for 2 hr and then centrifuged. The crystals obtained are washed with boiling methanol and dried over phosphorus pentoxide.

Spectroscopic Methods[20]

The spectroscopic methods usually applied comprise IR spectroscopy, UV-vis spectroscopy, nuclear magnetic resonance spectroscopy (NMR), and mass spectroscopy. The latter two methods are of especially great value in characterizing carotenoids. Unless the structure of a carotenoid has been confirmed by ^{1}H NMR and mass spectroscopy (and, in the case of a carotenoid with one or more chiral centers, by CD), it cannot be considered as unambiguously identified. This does not, of course, apply to the known carotenoids of some plants.

UV-Vis Spectroscopy of Carotenoids

A key feature of carotenoids is their long system of conjugated double bonds on their hydrocarbon "backbone". These conjugated double bonds give rise to the yellow or red carotenoid color. Mostly, carotenoids exhibit three distinct absorption maxima which have been listed in Table I.5. Usually, the carotenoid double bonds are in all-*trans* positions (see figures for Table I.5); however, *cis*-isomers of individual carotenoids are also known. The absorption spectra (= electron spectra) of the carotenoids are very much dependent on the solvent (solvent effect[20]). A few examples are given: the wavelength of absorption of some carotenoids (e.g., lycopene) exhibits a bathochromic shift when homologous, primary, aliphatic alcohols with increasing chain lengths are used as solvents. Table I.1 gives the absorption maxima of lycopene in aliphatic *n*-alcohols. If we compare aliphatic and aromatic solvents, a bathochromic shift may be observed in aromatic solvents (Table I.2).

Recording of Absorption Spectra

Many spectrophotometers do not measure precisely in terms of reproducibility of wavelength. It is, therefore, best to first scan the spectrum of the carotenoid to be measured and then to overlay the spectra with the spectrum of a standard dydimium or a holmium oxide filter in order to enable correlations to the correct wavelength (filter available, e.g., from E. Zeiss, Oberkochen, F.R.G.). Although it is advisable not to use too concentrated solutions, absorption values A of up to 0.8 are preferable in order to exploit fully the measuring range of the spectrophotometer. The cuvettes used (preferably quartz) should be freshly cleaned, matching pairs. Solvents have to be redistilled according to the procedures recommended; they should be free of contaminants, especially from those absorbing in the UV region. For quantitative determinations, the volume of the carotenoid solutions should be known. Since carotenoids tend to bleach, absorption measurements should be carried out immediately following pigment elution. It should always be borne in mind that the values measured might be too low because of pigment bleaching due to light, oxydation, etc. If the absorption coefficient is not known, it is a good practice to use the values combined in Table I.6 as a guide. In most spectrophotometers 1-cm light-path standard cuvettes are used. Other cuvettes may be used as well in order to achieve a sufficient absorbance of very dilute solutions without the need to concentrate.

Quantitative Determination of Carotenoids

After their separation by a chromatographic procedure, the amount of carotenoid(s) obtained may be determined, for example, for further work or yield calculations of enzymic synthesis, etc. Usually, there will not be enough carotenoid to be determined by direct weighing. Therefore, quantitative determinations of carotenoids in most cases are carried out spectrophotometrically via the extinction coefficients A (1%, 1 cm) or molar extinction (absorbance) coefficients ϵ. Electronic spectra (= absorption spectra) may invariably be recorded, to enable the facile determination of concentration and/or absolute amount of carotenoid present. We have compiled data on molar extinction coefficients in Table I.6 listing names in strictly alphabetical order.

Infrared Spectroscopy of Carotenoids[20]

The technique of infrared spectroscopy of carotenoids is mostly used in connection with synthetic work. IR spectra indicate qualitatively the functional groups present. So it is of value to detect acetylenic (2170 cm^{-1}), allenic, or hydroxy groups and keto-functions, especially those inaccessible to chemical reagents in some pigments like capsanthin and fucoxanthin.[20] Since a conjugated polyene system usually causes weak peaks only, the technique of infrared spectroscopy has not been considered as major in carotenoid chemistry. Infrared spectra of carotenoids may either be taken from a KBr-pellet, in solution (for example, chloroform), from another suitable solvent, or from thin films. Infrared spectroscopy has been considered as being useful for the detection of colorless contaminants. For more information, the reader is referred to the literature (for review, see Reference 20).

¹H NMR Spectroscopy of Carotenoids[20]

One of the most prominent features of proton magnetic resonance spectroscopy is that the chemical shifts and coupling constants in a good approximation generally depend only on the immediate environment of the different protons. The interpretation of proton magnetic resonance spectra will, therefore, give detailed information on structural subunits that are present in an unknown molecule. This feature is termed additivity; it simplifies the interpretation of spectra considerably. Coupling constants and chemical shifts of some carotenoids have been tabulated.[20] ¹H NMR spectra usually are recorded in deuteriochloroform (purity >99.5%) using tetramethylsilane (TMS) as an internal standard. Much of the information gained applies to the end groups of the carotenoid. Additional information can be obtained from the shifts of methylene and methine protons. For example, the appearance of CH signals between about 3 and 5.5 ppm indicates the presence of an –OR-substituted end group, whereas the presence of –CH$_2$– multiplets at about 2 and 2.5 ppm indicates a neighboring double bond or carbonyl function; 2, 3, and 4-protons of a β end group usually give rise to characteristic multiplets that will help to identify the compound. For more information, the reader is referred to the literature.[17,20]

Separation of Carotenoids by Chromatography

Introductory Remarks

To achieve a satisfactory reproducibility of R$_f$ values and retention volumes or retention times, a number of parameters have to be controlled. A few of these parameters are the following:

1. A constant physical and chemical composition of the sorbents used, including particle size, water content, and respective degree of activation, is imperative.
2. The solvent systems employed should be freshly prepared. Only analytical grade solvents should be used. Otherwise, impurities like peroxides, etc., may decompose sensitive compounds. Also, minute amounts of impurities acting as solvent system components (like water) may greatly influence R$_f$ values.
3. Constant temperature must be maintained.
4. R$_f$ values vary to some degree with the applied amount; overloading has to be strictly avoided, especially since it usually results in separations of poor quality.
5. The degree of chamber saturation is very important.
6. Always exclude light when colored compounds are being separated, especially carotenoids.
7. When working with carotenoids it is good practice to work in an atmosphere of nitrogen.
8. One has always to bear in mind that plants do not possess a time-constant pigment composition. For example, pigments of flowers are limited in their accessibility to the reproductive period of the plant. Drying in the oven should be avoided altogether, since it destroys many carotenoids and other pigments. A number of plants (e.g., yellow pansies) may be freeze-dried, however, without considerable losses.[24] Generally, special care has to be taken in the handling of the individual plant material and the preparation of extracts.
9. In order to achieve real reproducibility, the experiments should be repeated a number of times.

The goal of the present work is to give quick and precise information on the chromatographic separation of plant pigments, including a certain number of related synthetic pigments or pigments from sources other than plants. More detailed information is given in the notes to the respective chromatographic tables.

Alternative Procedures for the Analysis of Carotenoids

The rapid development of small computers in recent years (e.g., the IBM AT®, the Apple®, and others) facilitated extensive mathematical treatment of recorded absorption spectra. For carotenoids, a method has recently been proposed for the "deconvolution" of spectra of biochemical mixtures into the amounts of its individual constituents.[20a] This method has been named "multicomponent analysis", and since it might circumvent the necessity of carrying out an actual chromatographic separation, a short description follows: the problem solved by multicomponent analysis (MCA) is the determination of the unknown relative concentrations c_i in an overdetermined system of equations. Due to systematical or statistical errors, this system does not provide exact solutions. To determine solutions with the smallest possible errors, the criterion of least squares is chosen. The solution is calculated with the aid of a suitable program on a computer which is directly linked to a precise spectrophotometer. The computer will then determine the amount of each individual pigment. Concentrations or relative amounts are printed out together with the applicable standard deviation. One must, of course, bear in mind that only mixtures with known qualitative composition may be analyzed. Prior to the actual analysis, it is necessary to record a "matrix" of absorption spectra of the components. There is no way to avoid the separation and identification of the individual components of mixtures of unknown qualitative composition. The main potential application of this relatively new method seems to lie in standard determinations, e.g., in the fruit juice industry.

REFERENCES

1. **Isler, O., Ed.,** *Carotenoids,* Birkhäuser-Verlag, Basel, 1971.
2. **Heftmann, E., Ed.,** *Chromatography,* 3rd ed., Van Nostrand, Reinhold, New York, 1975.
2a. **Stahl, E.,** *Dünnschichtschromatographie,* Springer-Verlag, Berlin, 1967.
3. **Goodwin, T. W., Ed.,** *The Comparative Biochemistry of Carotenoids,* Chapman and Hall, London, 1952.
4. **Goodwin, T. W.,** *The Biochemistry of the Carotenoids,* Vol. 1, Plants, Chapman and Hall, London, 1980.
5. **Britton, G.,** *The Biochemistry of Natural Pigments,* Cambridge University Press, Cambridge, 1983.
6. **Pfander, H., Ed., with Gerspacher, M., Rychinir, M., and Schwabe, R.,** *Key to Carotenoids: Lists of Natural Carotenoids,* rev. ed., Birkhäuser, Basel, 1987.
7. **Goodwin, T. W., Ed.,** *Chemistry and Biochemistry of Plant Pigments,* Vol. 2, 2nd ed., Academic Press, London, 1976.
8. **Davies, B. H.,** Carotenoids, in *Chemistry and Biochemistry of Plant Pigments,* Vol. 2, 2nd ed., Goodwin, T. W., Ed., Academic Press, London, 1976, 38.
9. **Britton, G.,** *The Biochemistry of Natural Pigments,* Cambridge University Press, Cambridge, 1983.
10. **Czygan, F.-C., Ed.,** *Pigments in Plants,* 2nd ed., G. Fischer Verlag, Stuttgart, 1980.
11. **Sitte, P., Falk, H., and Liedvogel, B.,** *Carotinoide,* Springer-Verlag, Berlin, 1934, 117.
12. **Zechmeister, L.,** *Carotinoide,* Springer, Berlin, 1934.
13. **Davies, B. H.,** *Pigments in Plants,* 2nd ed., Czygan, F.-C., Ed., G. Fischer Verlag, Stuttgart, 1980, 31.
14. **Davies, B. H.,** C-30 Carotenoids, in *International Review of Biochemistry,* Vol. 14., *Biochemistry of Lipids II,* Goodwin, T. W., Ed., University Park Press, Baltimore, 1977, 51.
15. **Davies, B. H., Hsu, W.-J., and Chichester, C. O.,** *Comp. Biochem. Physiol.,* 33, 601, 1970.
16. **Mayr, H. and Isler, O.,** in *Carotenoids,* Isler, O., Ed., Birkhäuser-Verlag, Basel, 1971, 328.
17. 7th Int. Symp. on Carotenoids, Book of Abstracts, University of München, Munich, 1984.
18. **Weast, R. C. and Astle, M. J., Eds.,** *Handbook of Chemistry and Physics,* 62nd ed., CRC Press, Boca Raton, Fla., 1981—82.
19. **Liaaen-Jensen, S.,** in *Carotenoids,* Iler, O., Ed., Birkhäuser-Verlag, Basel, 1971, 61.
20. **Vetter, W., Englert, G., Rigassi, N., and Schwieter, U.,** in *Carotenoids,* Isler, O., Ed., Birkhäuser-Verlag, Basel, 1971, 189.
20a. **Jochum, P. and Schrott, E. L.,** *Anal. Chim. Acta,* 157, 211, 1984.
21. **Karrer, P. and Juncker, E.,** *Carotinoide,* Birkhäuser-Verlag, Basel, 1948.

22. **Frye, A. H.,** *J. Org. Chem.,* 16, 914, 1951.
23. **Zweig, G. and Sherma, J.,** *Handbook Series in Chromatography,* Vols. 1 and 2, 4th ed., CRC Press, Boca Raton, Fla., 1978.
24. **Wanner, G. and Köst, H.-P.,** unpublished.

Tables for the Estimation and Separation of Carotenoids

All carotenoid tables have been compiled by G. Widerer, E. Schropp, and H.-P. Köst. Structures were drawn by E. Benedikt and H.-P. Köst.

GENERAL TABLES

Table I.1
MAIN ABSORPTION MAXIMA OF
LYCOPENE IN VARIOUS ALIPHATIC
n-ALCOHOLS

Alcohol	Max 1 (nm)	Max 2 (nm)	Max 3 (nm)
Methanol	443	470	501
Ethanol	446	473	503.5
n-Butanol	449	475	507
n-Octanol	451	479	511

Table I.2
MAIN ABSORPTION MAXIMA OF
LYCOPENE IN VARIOUS SOLVENTS

Solvent	Max 1 (nm)	Max 2 (nm)	Max 3 (nm)
Acetone	446	472	503
n-Hexane	445	470	505
Methylene chloride	455	483	516
Chloroform	457	485	518
Carbon tetra-chloride	455	483	516
Cyclohexane	446	474	506
Benzene	456	484	519
Toluene	456	484	517
Pyrrole	462	490	525

Table based on author's data.

Table I.3

NAME LIST FOR CAROTENOIDS (TABLES I.5, I.6, I. PC, I. TLC, I. LC)

NOTES

The present chapter of the Handbook deals with the chromatographic separation of a vast number of carotenoids. We, therefore, thought it practical to unify all carotenoids dealt with in Tables I.5 through I. LC 1 in the form of a "name list". Table I.3 contains names, synonyms, formulas, molecular weights, and structures of carotenoids mainly of plant origin; however, other naturally occurring carotenoids have been included as well, if advisable. *First vertical row:* this row contains alphabetically arranged pigment names appearing in Tables I.5 through I. LC 1; also, synonyms, molecular ("sum") formula $C_xH_yO_z$, molecular weights, and structures.[1,6,18] The molecular weights are rounded. The values are calculated from the molecular formula using a programmable desk-top computer employing the following molecular weights:[18]

C = 12.011 g mol^{-1}
H = 1.0079 g mol^{-1}
O = 15.9994 g mol^{-1}

Where the name of a pigment is marked with an asterisk (*), different structures have been described in the literature. *Second vertical row:* names which are actually included in Tables I.5 through I. LC 1. *Third vertical row:* completes list of names of each pigment; arrow indicates name under which further detailed information can be found in this table. Consult Reference 6 in the "Methods" section for recent clarification of this subject.

Pigment	Names mentioned in Tables I.5 through I. LC	All synonyms	Sum formula	Mol wt	Structure
Actinioerythrin	Actinioerythrin Actinioerythrol (free diol)	Actinioerythrol (free diol) 3,3'-Dihydroxy-2,2'-dinor-β,β-carotene-4,4'-dione-3,3'-diacylate	$C_{38}H_{48}O_4$	568.79	
Actinioerythrin-bis-α-ketol (free diol)	Actinioerythrin-bis-α-ketol Actinioerythrin Actinioerythrol	→ Actinioerythrin			
Adonirubin	4,4'-Diketo-3-hydroxy-β-carotene 3-Hydroxy-4,4'-diketo-β-carotene	→ 3-Hydroxy-4,4'-diketo-β-carotene			
Aleuriaxanthin	Aleuriaxanthin	$(2'R)$-1',16'-Didehydro-1',2'-dihydro-β,ψ-caroten-2'-ol	$C_{40}H_{56}O$	552.88	
Alloxanthin	Alloxanthin	Cryptomonaxanthin	$C_{40}H_{52}O_2$	564.85	
Cynthiaxanthin	Cynthiaxanthin	Cynthiaxanthin 7,8,7',8'-Dehydrozeaxanthin 3,3'-Dihydroxy-7,8,7',8'-dehydro-β-carotene Pectenoxanthin $(3R,3'R)$-7,8,7',8'-Tetradehydro-β,β-carotene-3,3'-diol 7,8,7',8'-Tetradehydrozeaxanthin			
Anchovyxanthin	Zeaxanthin	→ Zeaxanthin			
Anhydrodeoxy-flexixanthin	Anhydrodeoxy-flexixanthin 4-Ketotorulene	3',4'-Didehydro-β,ψ-caroten-4-one 3',4'-Didehydro-4-keto-γ-carotene 4-Ketotorulene	$C_{40}H_{52}O$	548.85	

Table I.3 (continued)

NAME LIST FOR CAROTENOIDS (TABLES I.5, I.6, I. PC, I. TLC, I. LC)

Pigment	Names mentioned in Tables I.5 through I. LC	All synonyms	Sum formula	Mol wt	Structure
Anhydroeschschscholtzxanthin	Anhydroeschschscholtzxanthin	Dianhydroeschschscholtzxanthin 2,3,2′,3′,4′,5′-Hexadehydro-4,5′-retro-β,β-carotene	$C_{40}H_{50}$	530.84	
Anhydrorhodovibrin	Anhydrorhodovibrin "P-481"	1-Methoxy-3,4-didehydro-1,2-dihydro-ψ,ψ-carotene 1-Methoxy-1,2-dihydro-3,4-dehydrolycopene "P-481"	$C_{41}H_{58}O$	566.91	
Anhydrosaproxanthin	Anhydrosaproxanthin Celaxanthin	Celaxanthin 3′,4′-Dehydrorubixanthin 3′,4′-Didehydro-β,ψ-caroten-3-ol 3-Hydroxy-3′,4′-dehydro-γ-carotene 3-hydroxytorulene	$C_{40}H_{54}O$	550.87	
Antheraxanthin	Antheraxanthin	3,3′-Dihydroxy-5,6-epoxy-β-carotene 5,6-Epoxy-5,6-dihydro-β,β-carotene-3,3′-diol 5,6-Epoxyzeaxanthin Zeaxanthin-5,6-epoxide	$C_{40}H_{56}O_{3}$	584.88	
Aphanicin	Aphanicin Canthaxanthin 4,4′-Diketo-β-carotene Euglenanone	→ Canthaxanthin			
"Aphanicol"	4,4′-Dihydroxy-β-carotene Isozeaxanthin	→ 4,4′-Dihydroxy-β-carotene			
Aphanin	Aphanin Echinenone 4-Keto-β-carotene Myxoxanthin	→ Echinenone			

Name	Synonyms	Semisystematic name / cross-reference	Formula	MW
Aphanizophyll	Aphanizophyll	4-Hydroxymyxoxanthophyll 2'-(β,L-Rhamnopyranosyloxy)-3',4'-didehydro-1',2'-dihydro-β,ψ-carotene-3,4,1'-triol	$C_{46}H_{66}O_8$	747.02
"Aphanol"	4-Hydroxy-β-carotene Isocryptoxanthin "Myxoxanthol"	→ 4-Hydroxy-β-carotene		
"β-Apo-2-carotenal"	β-Apo-8'-carotenal	→ β-Apo-8'-carotenal		
"β-Apo-3-carotenal"	β-Apo-10'-carotenal	→ β-Apo-10'-carotenal		
"β-Apo-4-carotenal"	β-Apo-4-carotenal	→ β-Apo-12'-carotenal		
	β-Apo-12'-carotenal			
β-Apo-2'-carotenal	β-Apo-2'-carotenal	3',4'-Didehydro-2'-apo-β-caroten-2'-al	$C_{37}H_{48}O$	508.79
β-Apo-8'-carotenal	β-Apo-8'-carotenal	"β-Apo-2-carotenal" 8'-Apo-β-caroten-8'-al "β-Carotenal"	$C_{30}H_{40}O$	416.65
β-Apo-10'-carotenal	β-Apo-10'-carotenal	"β-Apo-3-carotenal" 10'-Apo-β-caroten-10'-al	$C_{27}H_{36}O$	376.58
β-Apo-12'-carotenal	β-Apo-4-carotenal β-Apo-12'-carotenal	"β-Apo-4-carotenal" 12'-Apo-β-caroten-12'-al	$C_{25}H_{34}O$	350.54
6'-Apo-ψ-caroten-6'-al	Apo-2-lycopenal Apo-6'-lycopenal	→ Apo-6'-lycopenal		
8'-Apo-β-caroten-8'-al	β-Apo-8'-carotenal	→ β-Apo-8'-carotenal		
10'-Apo-β-caroten-10'-al	β-Apo-10'-carotenal	→ β-Apo-10'-carotenal		
12'-Apo-β-caroten-12'-al	β-Apo-4-carotenal β-Apo-12'-carotenal	→ β-Apo-12'-carotenal		
8'-Apo-ψ-caroten-8'-al	Apo-3-lycopenal Apo-8'-lycopenal	→ Apo-3-lycopenal		
Apo-8,8'-carotendial	Crocetindialdehyde	→ Crocetindialdehyde		
β-Apo-2'-carotenoic acid	β-Apo-8'-carotenoic acid	β-Apo-8'-carotenoic acid		
β-Apo-4'-carotenoic acid	β-Apo-4'-carotenoic acid Neurosporaxanthin	→ Neurosporaxanthin		

Table I.3 (continued)
NAME LIST FOR CAROTENOIDS (TABLES I.5, I.6, I. PC, I. TLC, I. LC)

Pigment	Names mentioned in Tables I.5 through I. LC	All synonyms	Sum formula	Mol wt	Structure
β-Apo-8'-carotenoic acid	β-apo-8'-carotenoic acid	8'-Apo-β-caroten-8'-oic acid "β-Apo-2'-carotenoic acid"	$C_{30}H_{40}O_2$	432.64	
β-Apo-10'-carotenoic acid	β-Apo-10'-carotenoic acid	10'-Apo-β-caroten-10'-oic acid	$C_{27}H_{36}6O_2$	392.58	
β-Apo-12'-carotenoic acid	β-Apo-12'-carotenoic acid	12'-Apo-β-caroten-12'-oic acid Retinylidenetiglic acid	$C_{25}H_{34}O_2$	366.54	
4'-Apo-β-caroten-4-oic acid	β-Apo-4'-carotenoic acid Neurosporaxanthin	→ Neurosporaxanthin			
8'-Apo-β-caroten-8'-oic acid 10'-Apo-β-caroten-10'-oic acid 12'-Apo-β-caroten-12'-oic acid	β-Apo-8'-carotenoic acid β-Apo-10'-carotenoic acid β-Apo-12'-carotenoic acid	→ β-Apo-8'-carotenoic acid → β-Apo-10'-carotenoic acid → β-Apo-12'-carotenoic acid			
β-Apo-2'-carotenol	β-Apo-2'-carotenol	3',4'-Didehydro-2'-apo-β-caroten-2'-ol	$C_{37}H_{50}O$	510.80	
β-Apo-8'-carotenol	β-Apo-8'-carotenol	8'-Apo-β-caroten-3-ol β-Citraurinene	$C_{30}H_{42}O$	418.66	
β-Apo-10'-carotenol 8'-Apo-β-caroten-3-ol "Apo-2-lycopenal"	β-Apo-10'-carotenol β-Apo-8'-carotenol "Apo-2-lycopenal" Apo-6'-lycopenal "Lycopenal"	— → β-Apo-8'-carotenol → Apo-6'-lycopenal			
Apo-3-lycopenal	Apo-3-lycopenal Apo-8'-lycopenal	Apo-8'-lycopenal 8'-Apo-ψ-caroten-8'-al	$C_{30}H_{40}O$	416.65	

Name	Alternative/systematic names	Formula	MW
Apo-6'-lycopenal	"Apo-2-lycopenal" Apo-6'-lycopenal "Lycopenal" Apo-3-lycopenal		
Apo-8'-Lycopenal	"Apo-2-lycopenal" 6'-Apo-ψ-caroten-6'-al "Lycopenal" → Apo-3-lycopenal	$C_{32}H_{42}O$	442.68
Apo-12'-violaxanthal	5,6-Epoxy-3-hydroxy-5,6-dihydro-12'-apo-β-caroten-12'-al	$C_{25}H_{34}O_3$	382.54
Apo-10'-violaxanthin	5,6-Epoxy-3-hydroxy-5,6-dihydro-10'-al → 5,6-Epoxy-3-hydroxy-5,6-dihydro-10'-apo-β-caroten-10'-al		
"Apoviolaxanthinal"	5,6-Epoxy-3-hydroxy-5,6-dihydro-10'-al → 5,6-Epoxy-3-hydroxy-5,6-dihydro-10'-apo-β-caroten-10'-al		
β-Citraurin 3-Hydroxy-10'-apo-β-caroten-10'-al	"Apo-2-zeaxanthinal" → β-Citraurin "Apo-3-zeaxanthinal" → 3-Hydroxy-10'-apo-β-caroten-10'-al		
"Apo-4-zeaxanthinal"	3-Hydroxy-12'-apo-β-caroten-12'-al	$C_{25}H_{34}O_2$	366.54
Astacene	"Euglenarhodon" "Salmon acid" 3,4,3',4'-Tetraketo-β-carotene "4,5,4',5'-Tetraketo-β-carotene"	$C_{40}H_{48}O_4$	592.82
Astaxanthin	3,3'-Dihydroxycanthaxanthin (3S,3'S)-3,3'-Dihydroxy-β,β-carotene-4,4'-dione 3,3'-Dihydroxy-4,4'-diketo-β-carotene "Ovoester"	$C_{40}H_{52}O_4$	596.85
Astaxanthindiacetate	—		
Astaxanthindiester	—		
Astaxanthinmonoester	—		

Table I.3 (continued)
NAME LIST FOR CAROTENOIDS (TABLES I.5, I.6, I. PC, I. TLC, I. LC)

Pigment	Names mentioned in Tables I.5 through I. LC	All synonyms	Sum formula	Mol wt	Structure
Asterinic acid	Asterinic acid	Mixture of 7,8,7′,8′-tetra-dehydroastaxanthin and 7,8-didedehydroastaxanthin			
Asteroidenone	3′-Hydroxyechinenone 4-Keto-3′-hydroxy-β-carotene	→ 3′-Hydroxyechinenone			
Aurochrome	Aurochrome	"ξ-Carotene"* β-Carotene-5,8,5′,8′-diepoxide 5,8,5′,8′-Diepoxy-5,8,5′,8′-te-trahydro-β,β-carotene	$C_{40}H_{56}O_2$	568.88	
Auroxanthin	Auroxanthin	5,8,5′,8′-Diepoxy-5,8,5′,8′-te-trahydro-β,β-carotene-3,3′-diol "5,8,5′,8′-Diepoxyzeaxanthin" "3,3′-Dihydroxy-5,8,5′,8′-die-poxy-β-carotene"	$C_{40}H_{56}O_4$	600.88	
Azafrin	Azafrin	(5R,6R)-5,6-Dihydroxy-5,6-dihydro-10′-apo-β-caroten-10′-oic acid	$C_{27}H_{38}O_4$	426.59	
"Bacterial phytoene"	4,4′-Diapophytoene	→ 4,4′-Diapophytoene			
Bacterioerythrin α	Rhodoviolascin Spirilloxanthin	→ Rhodoviolascin			
"Bacterioerythrin β"	OH-Lycopene Rhodopin	→ OH-Lycopene			
Bacteriopurpurin α	Rhodoviolascin Spirilloxanthin	→ Rhodoviolascin			

Name		Conversion	Formula	MW
Bacterioruberin	Bacterioruberin			
	Bacterioruberin α	→ Bacterioruberin α		
Bacterioruberin α	Bacterioruberin	Bacterioruberin	$C_{50}H_{76}O_4$	741.15
	Bacterioruberin α	2,2'-Bis-(3-hydroxy-3-methyl-butyl)-3,4,3',4'-tetradehydro-1,2,1',2'-tetrahydro-ψ,ψ-caro-tene-1,1'-diol		
		"Didemethylated spirilloxanthin"		
"α-Bacterioruberin mono-methyl ether"	Hydroxyspirilloxanthin	→ OH-Spirilloxanthin		
"Bisdehydro-β-carotene"	OH-Spirilloxanthin			
	3,4,3',4'-Bisdehydro-β-carotene	→ 3,4,3',4'-Bisdehydro-β-carotene		
3,4,3',4'-Bisdehydro-β-carotene	3,4,3',4'-Bisdehydro-β-carotene	"Bisdehydro-β-carotene" "Dehydrocarotene III" Retrobisdehydro-β-carotene 3,4,3',4'-Tetradehydro-β,β-carotene	$C_{40}H_{52}$	532.85
Bisdehydrolycopene (2R,6S,2'R,6'S)-2,2'-Bis-(4-hydroxy-3-methyl-2-bu-tenyl)-γ,γ-carotene	"Dehydrolycopene" Sarcinaxanthin	→ "Dehydrolycopene" → Sarcinaxanthin*		
(2R,6R,2'R,6'R)-2,2'-Bis-(4-hydroxy-3-methyl-2-bu-tenyl)-ε,ε-carotene	Sarcinaxanthin	→ Sarcinaxanthin*		
2,2'-Bis-(3-hydroxy-3-meth-ylbutyl)-3,4,3',4'-tetradehy-dro-1,2,1',2'-tetrahydro-ψ,ψ-carotene-1,1'-diol	Bacterioruberin Bacterioruberin α	→ Bacterioruberin α		
2,2'-Bis(O-methyl-5-C-meth-ylpentosyloxy)-3,4,3',4'-te-tradehydro-1,2,1',2'-tetrahydro-ψ,ψ-carotene-1,1'-diol	Oscillol-2,2'-di(O-methyl-methyl-pentoside)	→ Oscillol-2,2'-di(O-methyl-methyl-pentoside)		

Table I.3 (continued)
NAME LIST FOR CAROTENOIDS (TABLES I.5, I.6, I. PC, I. TLC, I. LC)

Pigment	Names mentioned in Tables I.5 through I. LC	All synonyms	Sum formula	Mol wt	Structure
2,2'-Bis-(β-L-rhamnopyrano-syloxy)-3,4,3',4'-tetradehy-dro-1,2,1',2'-tetrahydro-ψ,ψ-carotene-1,1'-diol	Oscillaxanthin	→ Oscilloxanthin			
Bixin	Bixin	Bixin natural Bixin *cis* Bixin lower melting "α-Bixin" "Bixin II"	$C_{25}H_{30}O_4$	394.51	
"α-Bixin" "Bixin II"	Bixin Bixin	→ Bixin → Bixin			
Caloxanthin	Caloxanthin	6,7-Didehydro-5,6-dihydro-β,β-carotene-3,3'-diol	$C_{40}H_{56}O_2$	568.88	
Canthaxanthin	Aphanicin Canthaxanthin 4,4'-Diketo-β-carotene Euglenanone	Aphanicin β,β-Carotene-4,4'-dione Chlorellaxanthin 4,4'-Diketo-β-carotene Euglenanone	$C_{40}H_{52}O_2$	564.85	
Capsanthin	Capsanthin	(3R,3'S,5'R)-3,3'-Dihydroxy-β,κ-caroten-6'-one	$C_{40}H_{56}O_3$	584.88	
Capsanthin-diester	Capsanthin-diester	—			
Capsanthin-5,6-epoxide	Capsanthin-5,6-epoxide	Capsanthin-monoepoxide 5,6-Epoxy-3,3'-dihydroxy-5,6-dihydro-β,κ-caroten-6'-one	$C_{40}H_{56}O_4$	600.88	

Name	Cross-reference / Systematic name	Formula	Mol wt
Capsanthin-5,6-epoxidediester	—		
Capsanthinmonoepoxide	→ Capsanthin-5,6-epoxide		
Capsanthinmonoester	→ Capsanthin-5,6-epoxide		
Capsochrome	5,8-Epoxy-3,3'-dihydroxy-5,8-dihydro-β,κ-caroten-6'-one	$C_{40}H_{56}O_4$	600.88
Capsorubin	(3S,5R,3'S,5'R)-3,3'-Dihydroxy-κ,κ-caroten-6,6'-dione	$C_{40}H_{56}O_4$	600.88
Capsorubin-diester	—		
16'-Carboxyl-3',4'-dehydro-γ-carotene	→ Torularhodin		
"Carcinoxanthin"	→ Chrysanthemaxanthin		
Chrysanthemaxanthin			
Flavoxanthin			
Caricaxanthin	→ β-Cryptoxanthin		
Cryptoxanthin			
β-Cryptoxanthin			
"β-Carotenal"	→ β-Apo-8'-Carotenal		
β-Apo-8'-Carotenal			
"Carotene"	→ β-Carotene		
β-Carotene			
β,β-Carotene			
Flavacin	→ Flavacin		
Mutatochrome			
β-Zeacarotene	→ β-Zeacarotene		
β_1-Zeacarotene			
α-Carotene	β,ε-Carotene	$C_{40}H_{56}$	536.88
β,ε-Carotene	(6'R)-β,ε-Carotene		
β-Carotene	"Carotene"	$C_{40}H_{56}$	536.88
β,β-Carotene	β,β-Carotene		
	ψ,α-Carotene		
	Neo-β-carotene		
	Pseudo-α-carotene		
β,β-Carotene	→ β-Carotene		
β,ε-Carotene	→ α-Carotene		

Table I.3 (continued)
NAME LIST FOR CAROTENOIDS (TABLES I.5, I.6, I. PC, I. TLC, I. LC)

Pigment	Names mentioned in Tables I.5 through I. LC	All synonyms	Sum formula	Mol wt	Structure
(6′R)-β,ε-Carotene	α-Carotene	→ α-Carotene			
β,φ-Carotene	β,ε-Carotene	→ β-Isorenieratene			
β,ψ-Carotene	β-Isorenieratene	→ γ-Carotene			
	β,ψ-Carotene				
	γ-Carotene				
	Pro-γ-carotene				
γ-Carotene	β,ψ-Carotene	β,ψ-Carotene	$C_{40}H_{56}$	536.88	
	γ-Carotene	Poly-*cis*-γ-carotene			
	Pro-γ-carotene	Pro-γ-carotene			
		"Sphaerobolin"			
δ-Carotene	δ-Carotene	(6R)-ε,ψ-carotene	$C_{40}H_{56}$	536.88	
ε-Carotene	ε-Carotene	ε₁-Carotene	$C_{40}H_{56}$	536.88	
	ε₁-Carotene	(6R,6′R) -ε,ε-Carotene			
ε₁-Carotene	ε-Carotene	→ ε-Carotene			
	ε₁-Carotene				
(6R,6′R)-ε,ε-Carotene	ε-Carotene	→ ε-Carotene			
	ε₁-Carotene				
(6R)-ε,ψ-Carotene	δ-Carotene	→ δ-Carotene			
ζ-Carotene	ζ-Carotene	"ζ-Carotene"	$C_{40}H_{60}$	556.91	
	7,8,7′,8′-Tetrahydro-ψ,ψ-carotene	5,6,7,8,5′,6′,7′,8′-Octahydrolycopene			
		7,8,7′,8′-Tetrahydro-ψ,ψ-carotene			
		7,8,7′,8′-Tetrahydrolycopene			

ζ-Carotene (asym.)	7,8,11,12-Tetrahydro-ψ,ψ-carotene	Iso-ζ-carotene	C$_{40}$H$_{60}$
	7,8,11,12-Tetrahydrolycopene	7,8,11,12-Tetrahydro-ψ,ψ-carotene	
		7,8,11,12-Tetrahydrolycopene	556.91
		7,8,11',12'-Tetrahydrolycopene	
ϑ-Carotene	ϑ-Carotene	ϑ-Carotene	C$_{40}$H$_{60}$
"ξ-Carotene"*	Aurochrome	Mixture of → ζ-carotene and → ζ-carotene asym.	556.91
	ζ-Carotene	→ Aurochrome	
	7,8,7',8'-Tetrahydro-ψ,ψ-Carotene	→ ζ-Carotene	
φ-Carotene	Flavorhodin	→ Neurosporene	
	Neurosporene		
	Proneurosporene		
	"Proietrahydrolycopene"		
	"Tetrahydrolycopene"		
	"5,6,5',6'-Tetrahydrolycopene"		
φ,φ-Carotene	Isorenieratene	→ Isorenieratene	
	Leprotene		
φ,χ-Carotene	Renieratene	→ Renieratene	
φ,ψ-Carotene	Chlorobactene	→ Chlorobactene	
χ,χ-Carotene	Renierapurpurin	→ Renierapurpurin	
"ψ-Carotene"	Flavorhodin	→ Neurosporene	
	Neurosporene		
	Proneurosporene		
	"Proietrahydrolycopene"		
	"Tetrahydrolycopene"		
	"5,6,5',6'-Tetrahydrolycopene"		
ψ,α-Carotene	β-Carotene	→ β-Carotene	
	β,β-Carotene		
ψ,ψ-Carotene	Lycopene	→ Lycopene	
	Polycislycopene		
	Prolycopene		
	Rhodopurpurin		

Table I.3 (continued)
NAME LIST FOR CAROTENOIDS (TABLES I.5, I.6, I. PC, I. TLC, I. LC)

Pigment	Names mentioned in Tables I.5 through I. LC	All synonyms	Sum formula	Mol wt	Structure
β-Carotene-diepoxide	β-Carotene-diepoxide β-Carotene-5,6,5′,6′-diepoxide 5,6,5′,6′-Diepoxy-β-carotene "Diepoxy-β-carotene"	β-Carotene-5,6,5′,6′-diepoxide "Diepoxy-β-carotene" 5,6,5′,6′-Diepoxy-β-carotene 5,6,5′,6′-Diepoxy-5,6,5′,6′-tetrahydro-β,β-carotene	$C_{40}H_{56}O_2$	568.88	
β-Carotene-5,6,5′,6′-diepoxide	β-Carotene-diepoxide β-Carotene-5,6,5′,6′-diepoxide 5,6,5′,6′-Diepoxy-β-carotene	→ β-Carotene-diepoxide			
β-Carotene-5,8,5′,8′-diepoxide	Aurochrome	→ Aurochrome			
(3R,3′R)-β,β-Carotene-3,3′-diol	Zeaxanthin	→ Zeaxanthin			
β,β-Carotene-4,4′-diol	4,4′-Dihydroxy-β-carotene Isozeaxanthin	→ 4,4′-Dihydroxy-β-carotene			
(3R,3′R,6′R)-β,ε-Carotene-3,3′-diol	3-Hydroxy-3′-hydroxy-α-carotene Lutein "Xanthophyll"	→ Lutein			
ψ,ψ-Carotene-16,16′-diol	Lycophyll Physalien	→ Lycophyll → Physalien			
(3R,3′R)-β,β-Carotene-3,3′-diol dipalmitate					
(3R,3′R,6′R)-β,ε-Carotene-3,3′-diol dipalmitate	Helenien Lutein dipalmitate Aphanicin	→ Helenien			
β,β-Carotene-4,4′-dione	Canthaxanthin 4,4′-Diketo-β-carotene Euglenanone	→ Canthaxanthin			
β,κ-Carotene-3′,6′-dione	Cryptocapsone	→ Cryptocapsone			
β,κ-Carotene-3′,6′-dione	Cryptocapsone	→ Cryptocapsone			

Name	Synonyms	Cross-references	Formula	MW
α-Carotene-epoxide	α-Carotene-5,6-epoxide; "5,6-Monoepoxy-α-carotene"	α-Carotene-5,6-epoxide; 5,6-Epoxy-5,6-dihydro-β-carotene; "5,6-Monoepoxy-α-carotene"; → α-Carotene-epoxide	$C_{40}H_{56}O$	552.88
α-Carotene-5,6-epoxide	α-Carotene-5,6-epoxide; "5,6-Monoepoxy-α-carotene"		$C_{40}H_{56}O$	552.88
β-Carotene-monoepoxide	β-Carotene-monoepoxide; 5,6-Monoepoxy-β-carotene	β-Caroten-5,6-epoxide; β-Carotene-monoepoxide; 5,6-Epoxy-5,6-dihydro-β,β-carotene; 5,6-Monoepoxy-β-carotene	$C_{40}H_{56}O_3$	584.88
β,β-Carotene-2,3,3'-triol	β,β-Carotene-2,3,3'-triol	—		
β,ε-Carotene-3,19,3'-triol	Pyrenoxanthin; "Trollein"	→ "Trollein"		
β,ε-Carotene-3,20,3'-triol	Pyrenoxanthin; Cryptoxanthin; β-Cryptoxanthin	→ Pyrenoxanthin; → β-Cryptoxanthin		
(3R-)β,β-Caroten-3-ol				
β,β-Caroten-4-ol	4-Hydroxy-β-carotene; Isocryptoxanthin; Myxoxanthol	→ 4-Hydroxy-β-carotene		
β,ε-Caroten-3'-ol	α-Cryptoxanthin; α-Cryptoxanthin; Zeinoxanthin	→ α-Cryptoxanthin; → α-Cryptoxanthin		
(3R,6'R-)β,ε-Caroten-3-ol				
β,ε-Caroten-4-ol	4-Hydroxy-α-carotene; Rubixanthin; Gazaniaxanthin; Lycoxanthin	→ 4-Hydroxy-α-carotene; → Rubixanthin; → Gazaniaxanthin; → Lycoxanthin		
(3R-)β,ψ-Caroten-3-ol				
(3R-)5'-cis-β,ψ-Caroten-3-ol				
ψ,ψ-Caroten-16-ol				

Table I.3 (continued)
NAME LIST FOR CAROTENOIDS (TABLES I.5, I.6, I. PC, I. TLC, I. LC)

Pigment	Names mentioned in Tables I.5 through I. LC	All synonyms	Sum formula	Mol wt	Structure
(β-)Carotenonaldehyde	(β-)Carotenonaldehyde	5,6-Dioxo-10′-apo-5,6-seco-β-caroten-10′-al	$C_{27}H_{36}O_3$	408.58	
β-Carotenone	β-Carotenone	5,6,5′,6′-Diseco-β,β-carotene-5,6,5′,6′-tetrone	$C_{40}H_{56}O_4$	600.88	
β,β-Caroten-4-one	Aphanin Echinenone 4-Keto-β-carotene Myxoxanthin	→ Echinenone			
β,ε-Caroten-4-one β,ψ-Caroten-4-one Celaxanthin	4-Keto-α-carotene 4-Keto-γ-carotene Anhydrosaproxanthin Celaxanthin	→ 4-Keto-α-carotene → 4-Keto-γ-carotene → Anhydrosaproxanthin			
Chlorellaxanthin	Aphanicin Canthaxanthin 4,4′-Diketo-β-carotene Euglenanone	→ Canthaxanthin			
Chlorobactene	φ,ψ-Carotene	φ,ψ-Carotene	$C_{40}H_{52}$	532.85	
Chloroxanthin	Chloroxanthin OH-Neurosporene	"Dihydroxyneurosporene" "1-Hydroxy-1,2-dihydroneurosporene" 1′-Hydroxy-1′,2′-dihydroneurosporene Hydroxyneurosporene 1-Hydroxy-1,2,7′,8′-tetrahydrolycopene 1,2,7′,8′-Tetrahydro-ψ,ψ-caroten-1-ol	$C_{40}H_{60}O$	556.91	

Name	Synonyms	Formula	MW	Structure
Chrysanthemaxanthin	Chrysanthemaxanthin Flavoxanthin "Carcinoxanthin" "3,3'-Dihydroxy-5,8-epoxy-α-carotene" 5,8-Epoxy-5,8-dihydro-β,ε-carotene-3,3'-diol "5,8-Epoxy-3,3'-dihydroxy-α-carotene" "5,8-Epoxylutein" Flavoxanthin	$C_{40}H_{56}O_3$	584.88	
Citranaxanthin	Citranaxanthin 5',6'-Dihydro-5'-apo-18'-nor-β-caroten-6'-one	$C_{33}H_{44}O$	456.71	
α-Citraurin	α-Citraurin 3-Hydroxy-8'-apo-ε-caroten-8'-al	$C_{30}H_{40}O_2$	432.64	
β-Citraurin	β-Citraurin "Apo-2-zeaxanthinal" "3-Hydroxy-β-apo-2-carotenal" (3R)-3-Hydroxy-8'-apo-β-caroten-8'-al	$C_{30}H_{40}O_2$	432.64	
β-Citraurinene Citroxanthin "Compound X"	→ β-Apo-8'-carotenol → Flavacin → 4,4'-Diapophytoene			
β-Apo-8'-carotenol Flavacin Mutatochrome 4,4'-Diapophytoene				
Corynexanthin	Decaprenoxanthin monoglucoside 2-(4-(β-D-Glucopyranosyloxy)-3-methyl-2-butenyl)-2'-(4-hydroxy-3-methyl-2-butenyl)-ε,ε-carotene	$C_{56}H_{82}O_7$	867.26	
"C₃₀-phytoene"	4,4-Diapophytoene → 4,4'-Diapophytoene			

Table I.3 (continued)
NAME LIST FOR CAROTENOIDS (TABLES I.5, I.6, I. PC, I. TLC, I. LC)

Pigment	Names mentioned in Tables I.5 through I. LC	All synonyms	Sum formula	Mol wt	Structure
Crocetin	Crocetin Crocetin *trans*	Crocetin *trans* Crocetin stable "α-Crocetin" "Crocetin I" 8,8'-Diapocarotene-8,8'-dioic acid	$C_{20}H_{24}O_4$	328.41	
"Crocetin I"	Crocetin Crocetin *trans*	→ Crocetin			
"α-Crocetin"	Crocetin Crocetin *trans*	→ Crocetin			
"γ-Crocetin" Crocetin trans	Crocetindimethylester Crocetin Crocetin *trans*	→ Crocetindimethylester → Crocetin			
Crocetindial	Crocetindialdehyde	→ Crocetindialdehyde			
Crocetindialdehyde	Crocetindialdehyde	Apo-8,8'-carotenedial Crocetindial 8,8'-Diapocarotene-8,8'-dial	$C_{20}H_{24}O_2$	296.41	
Crocetin-di-(β,D-glucosyl)-ester	Crocetin-di-(β,D-glucosyl)-ester	Di(β,D-glucosyl)-8,8'-diapocarotene-8,8'-dioate	$C_{32}H_{44}O_{14}$	652.69	
Crocetindimethylester	Crocetindimethylester	"γ-Crocetin" Dimethylcrocetin Dimethyl-8,8'-diapocarotene-8,8'-dioate	$C_{22}H_{28}O_4$	356.46	
Crocetin-(β,D-gentiobiosyl-β,D-glucosyl)-ester	Crocetin-(β,D-gentiobiosyl-β,D-glucosyl)-ester	β,D-Gentiobiosyl-β,D-glucosyl-8,8'-diapocarotene-8,8'-dioate	$C_{38}H_{54}O_{19}$	814.83	

Crocoxanthin	Crocoxanthin	7,8-Dehydrozeinoxanthin (3R,6'R)-7,8-Didehydro-β,ε-caroten-3-ol 3-Hydroxy-7,8-dehydro-α-carotene	$C_{40}H_{54}O$	550.87
Cryptocapsin	Cryptocapsin	(3'S,5'R)-3'-Hydroxy-β,κ-caroten-6'-one	$C_{40}H_{56}O_2$	568.88
Cryptocapsone	Cryptocapsone	β,κ-Carotene-3',6'-dione	$C_{40}H_{54}O_2$	566.87
Cryptochrome	Cryptochrome	Cryptoxanthin-5,8,5',8'-diepoxide 5,8,5',8'-Diepoxy-5,8,5',8'-tetrahydro-β,β-caroten-3-ol "3-Hydroxy-5,8,5',8'-diepoxy-β-carotene"	$C_{40}H_{56}O_3$	584.88
Cryptoflavin	Cryptoflavin Cryptoxanthin-5,8-epoxide	Cryptoxanthin-5,8-epoxide 5,8-Epoxycryptoxanthin 5,8-Epoxy-5,8-dihydro-β,β-caroten-3-ol 3-Hydroxy-5,8-epoxy-β-carotene	$C_{40}H_{56}O_2$	568.88
Cryptomonaxanthin	Alloxanthin Cynthiaxanthin	→ Alloxanthin		
"Cryptoxanthene"	Cryptoxanthin β-Cryptoxanthin	→ β-Cryptoxanthin		
Cryptoxanthin	Cryptoxanthin β-Cryptoxanthin	→ β-Cryptoxanthin		

Table I.3 (continued)
NAME LIST FOR CAROTENOIDS (TABLES I.5, I.6, I. PC, I. TLC, I. LC)

Pigment	Names mentioned in Tables I.5 through I. LC	All synonyms	Sum formula	Mol wt	Structure
α-Cryptoxanthin*	α-Cryptoxanthin Zeinoxanthin	β,ε-Caroten-3'-ol (3R,6'R)-β,ε-Caroten-3-ol 3-Hydroxy-α-carotene 3'-Hydroxy-α-carotene Physoxanthin Zeinoxanthin	$C_{40}H_{56}O$	552.88	
β-Cryptoxanthin	Cryptoxanthin β-Cryptoxanthin	"Caricaxanthin" (3R)-β,β-Caroten-3-ol "Cryptoxanthene" Cryptoxanthin "Cryptoxanthol" 3-Hydroxy-β-carotene Physoxanthin	$C_{40}H_{56}O$	552.88	
Cryptoxanthin diepoxide	Cryptoxanthin diepoxide β-Cryptoxanthin 5,6,5',6'-diepoxide	β-Cryptoxanthin-5,6,5',6'-diepoxide 5,6,5',6'-Diepoxy-5,6,5',6'-tetrahydro-β,β-caroten-3-ol	$C_{40}H_{56}O_3$	584.88	
Cryptoxanthin 5,8,5',8'-diepoxide	Cryptochrome	→ Cryptochrome			
β-Cryptoxanthin 5,6,5',6'-diepoxide	Cryptoxanthin diepoxide β-Cryptoxanthin 5,6,5',6'-diepoxide	→ Cryptoxanthindiepoxide			
Cryptoxanthinepoxide	Cryptoxanthinepoxide Cryptoxanthin 5,6-epoxide	Cryptoxanthin-5,6-epoxide β-Cryptoxanthin-5,6-monoepoxide 5,6-Epoxy-5,6-dihydro-β,β-caroten-3-ol	$C_{40}H_{56}O_2$	568.88	

Name	Name / synonym	→ Reference	Formula	M.W.
Cryptoxanthin 5,6-epoxide	Cryptoxanthinepoxide	→ Cryptoxanthinepoxide		
	Cryptoxanthin 5,6-epoxide			
Cryptoxanthin 5,8-epoxide	Cryptoxanthin 5,8-epoxide	→ Cryptoflavin		
	Cryptoflavin			
β-Cryptoxanthin 5,6-monoepoxide	Cryptoxanthinepoxide	→ Cryptoxanthinepoxide		
"Cryptoxanthol"	Cryptoxanthin 5,6-epoxide			
	Cryptoxanthin			
	β-Cryptoxanthin	→ β-Cryptoxanthin		
"Cucurbitaxanthin"	3-Hydroxy-3'-hydroxy-α-carotene			
	Lutein	→ Lutein		
	"Xanthophyll"			
Cynthiaxanthin	Alloxanthin	→ Alloxanthin		
	Cynthiaxanthin			
"Decahydro-β-carotene" 7,8,11,12,15,7',8',11',12',15'-Decahydro-ψ,ψ-carotene	Phytofluene	→ Phytofluene		
Decahydrolycopene	Lycopersene	→ Lycopersene		
	Lycopersene	→ Lycopersene		
Decaprenoxanthin	Sarcinaxanthin	→ Sarcinaxanthin		
Decaprenoxanthin monoglucoside	Corynexanthin	→ Corynexanthin		
Deepoxyneoxanthin	"Trollein"	→ "Trollein"*		
Dehydroadonirubin	Dehydroadonirubin	3-Oxocanthaxanthin Phoeniconone Phoenicoxanthin* 3,4,4'-Triketo-β-carotene	$C_{40}H_{52}O_3$	580.85
Dehydroadonixanthin	Dehydroadonixanthin	β-Doradecin 3'-Hydroxy-3,4-diketo-β-carotene "3'-Hydroxyeuglenanone"	$C_{40}H_{52}O_3$	580.85
"Dehydrocarotene II"	3,4-Dehydro-β-carotene			
	3,4-Monodehydro-β-carotene	→ 3,4-Monodehydro-β-carotene		
Dehydrocarotene III	3,4,3',4'-Bisdehydro-β-carotene	→ 3,4,3',4'-Bisdehydro-β-carotene		

Table I.3 (continued)
NAME LIST FOR CAROTENOIDS (TABLES I.5, I.6, I. PC, I. TLC, I. LC)

Pigment	Names mentioned in Tables I.5 through I. LC	All synonyms	Sum formula	Mol wt	Structure
"Dehydro-β-carotene"	"Dehydro-β-carotene" Dehydroretrocarotene Dehydroretro-β-carotene Retrodehydrocarotene Retrodehydro-β-carotene	Dehydroretrocarotene Dehydroretro-β-carotene 4,4'-Didehydro-β-carotene 4,5'-Didehydro-4,5'-retro-β,β-carotene Isocarotene Retrodehydrocarotene Retrodehydro-β-carotene	C$_{40}$H$_{54}$	534.87	
3,4-Dehydro-ψ-carotene	3,4-Dehydro-β-carotene 3,4-Monodehydro-β-carotene	→ 3,4-Monodehydro-β-carotene			
3',4'-Dehydro-γ-carotene Dehydrogenans-P-439 Dehydrogenans-phytoene Dehydrogenans-phytofluene	Torulene Sarcinaxanthin Phytoene Phytofluene	→ Torulene → Sarcinaxanthin → Phytoene → Phytofluene			
Dehydrohydroxyechinenone	Dehydrohydroxyechinenone "Euglenanone"	3,4-Diketo-β-carotene "Euglenanone" 3-Hydroxy-4-oxo-2,3-dehydro-β-carotene 3-Oxoechinenone	C$_{40}$H$_{52}$O$_{2}$	564.85	
3'-Dehydrolutein	3-Hydroxy-3'-keto-α-carotene Philosamiaxanthin	→ 3-Hydroxy-3'-keto-α-carotene			
"Dehydrolycopene"	Bisdehydrolycopene "Dehydrolycopene"	Bisdehydrolycopene 3,4,3',4'-Tetradehydro-ψ,ψ-carotene 3,4,3',4'-Tetradehydrolycopene	C$_{40}$H$_{52}$	532.85	

Name	Synonyms	Semi-systematic name / derivatives	Formula	MW
3,4-Dehydrolycopene	3,4-Dehydrolycopene	3,4-Didehydro-ψ,ψ-carotene Monodehydrolycopene	$C_{40}H_{54}$	534.87
15,15'-Dehydrolycopersene	Phytoene Torularhodin-aldehyde	→ Phytoene → Torularhodin-aldehyde		
"3',4'-Dehydro-17-oxo-γ-carotene"				
"3,4-Dehydro-18-oxo-γ-carotene"	Torularhodin-aldehyde	→ Torularhodin-aldehyde		
11,12-Dehydrophytoene	Phytofluene	→ Phytofluene		
2'-Dehydroplectaniaxanthin	2'-Dehydroplectaniaxanthin	1'-Hydroxy-3',4'-didehydro-1',2'-dihydro-β,ψ-caroten-2'-one 1'-Hydroxy-2'-keto-1',2'-dihydrotorulene	$C_{40}H_{54}O_2$	566.87
Dehydroretrocarotene	"Dehydro-β-carotene" Dehydroretrocarotene Dehydroretro-β-carotene Retrodehydrocarotene Retrodehydro-β-carotene	→ "Dehydro-β-carotene"		
Dehydroretro-β-carotene	"Dehydro-β-carotene" Dehydroretrocarotene Dehydroretro-β-carotene Retrodehydrocarotene Retrodehydro-β-carotene	→ "Dehydro-β-carotene"		
3,4-Dehydrorhodopin	3,4-Dehydrorhodopin "OH-P 481"	3,4-Didehydro-1,2-dihydro-ψ,ψ-caroten-1-ol 1,2-Dihydro-3,4-dehydro-1-OH-lycopene "OH-P 481" "OH-P 482"	$C_{40}H_{56}O$	552.88
Dehydrorhodovibrin	Hydroxyspirilloxanthin OH-Spirilloxanthin	→ OH-Spirilloxanthin		
3',4'-Dehydrorubixanthin	Anhydrosaproxanthin Celaxanthin	→ Anhydrosaproxanthin		
Dehydrosqualene	4,4'-Diapophytoene	→ 4,4'-Diapophytoene		
3,4-Dehydrotorulene	3,4-Dehydrotorulene	—		
7,8,7',8'-Dehydrozeaxanthin	Alloxanthin Cynthiaxanthin	→ Alloxanthin		
7,8-Dehydrozeinoxanthin	Crocoxanthin	→ Crocoxanthin		

Table I.3 (continued)
NAME LIST FOR CAROTENOIDS (TABLES I.5, I.6, I. PC, I. TLC, I. LC)

Pigment	Names mentioned in Tables I.5 through I. LC	All synonyms	Sum formula	Mol wt	Structure
Deoxyflexixanthin	Deoxyflexixanthin 1',2'-Dihydro-1'-hydroxy-4-ketotorulene 4-Keto-1',2'-dihydro-1'-hydroxytorulene	1',2'-Dihydro-1'-hydroxy-4-ketotorulene 1'-Hydroxy-3',4'-didehydro-1',2'-dihydro-β,ψ-caroten-4-one 4-Keto-1',2'-dihydro-1'-hydroxytorulene	$C_{40}H_{54}O_2$	556.87	
Diadinoxanthin	Diadinoxanthin	3,3'-Dihydroxy-7,8-dehydro-β-caroten-5',6'-epoxide 5,6-Epoxy-7',8'-didehydro-5,6-dihydro-β,β-carotene,3,3'-diol	$C_{40}H_{54}O_3$	582.86	
Dianhydroeschscholtzxanthin 8,8'-Diapocarotene-8,8'-dial 8,8'-Diapocarotene-8,8'-dioic acid	Anhydroeschscholtzxanthin Crocetindialdehyde Crocetin Crocetin *trans*	→ Anhydroeschscholtzxanthin → Crocetindialdehyde → Crocetin			
4,4'-Diapo-ζ-carotene	4,4'-Diapo-ζ-carotene	7,8,7',8'-Tetrahydro-4,4'-diapocarotene	$C_{30}H_{44}$	404.68	
4,4'-Diapo-lycopen-4-al	4,4'-Diapo-Lycopen-4-al	—			
4,4'-Diaponeurosporene	4,4'-Diaponeurosporene	7,8-Dihydro-4,4'-diapocarotene	$C_{30}H_{42}$	402.66	
4,4'-Diapo-7,8,11,12,7',8',11',12'-octahydro-ψ,ψ-carotene	4,4'-Diapophytoene	→ 4,4'-Diapophytoene			

Name	Synonyms	Formula	MW
4,4'-Diapophytoene	"Bacterial phytoene", "Compound X", C_{30}-Phytoene, Dehydrosqualene, 4,4'-Diapo-7,8,11,12,7',8',11',12'-octahydro-ψ,ψ-carotene, 7,8,11,12,7',8',11',12'-Octahydro-4,4'-diapocarotene	$C_{30}H_{48}$	408.71
4,4'-Diapophytofluene	7,8,11,12,7',8'-Hexahydro-4,4'-diapocarotene	$C_{30}H_{46}$	406.69
4,4'-Diapo-7,8,11,12-tetrahydrolycopene	7,8,11,12-Tetrahydro-4,4'-diapocarotene	$C_{30}H_{44}$	404.68
Diatoxanthin	$(3R,3'R$-$)$7,8-Didehydro-β,β-caroten-3,3'-diol, 7,8-Didehydrozeaxanthin	$C_{40}H_{54}O_2$	566.87
3',4'-Didehydro-2'-apo-β-caroten-2'-al	→ β-Apo-2'-carotenal		
3',4'-Didehydro-2'-apo-β-caroten-2'-ol	→ β-Apo-2'-carotenol		
7,8-Didehydroastaxanthin	Part of asterinic acid, → Asterinic acid, 3,3'-Dihydroxy-7,8-didehydro-β,β-carotene-4,4'-dione	$C_{40}H_{50}O_4$	
3',4'-Didehydro-β,ψ-carotene-16'-al	Torularhodin-aldehyde, → Torularhodin-aldehyde		
4,4'-Didehydro-β-carotene	"Dehydro-β-carotene", Dehydroretrocarotene, Dehydroretro-β-carotene, Retrodehydrocarotene, Retrodehydro-β-carotene, → "Dehydro-β-carotene"		
3,4-Didehydro-β,β-carotene 3,4-Monodehydro-β,β-carotene	3,4-Dehydro-β-carotene, 3,4-Monodehydro-β-carotene, → 3,4-Monodehydro-β-carotene		

Table I.3 (continued)
NAME LIST FOR CAROTENOIDS (TABLES I.5, I.6, I. PC, I. TLC, I. LC)

Pigment	Names mentioned in Tables I.5 through I. LC	All synonyms	Sum formula	Mol wt	Structure
3′,4′-Didehydro-β,ψ-carotene	Torulene	→ Torulene			
3,4-Didehydro-ψ,ψ-carotene	3,4-Dehydrolycopene	→ 3,4-Dehydrolycopene			
(3R,3′R)-7,8-Didehydro-β,β-carotene-3,3′-diol	Diatoxanthin	→ Diatoxanthin			
7,8-Didehydro-β,ε-carotene-3,3′-diol	Monadoxanthin	→ Monadoxanthin			
3′,4′-Didehydro-β,ψ-caroten-16′-oic acid	Torularhodin	→ Torularhodin			
(3R,6′R)-7,8-Didehydro-β,ε-caroten-3-ol	Crocoxanthin	→ Crocoxanthin			
3′,4′-Didehydro-β,ψ-caroten-3-ol	Anhydrosaproxanthin Celaxanthin	→ Anhydrosaproxanthin			
3′,4′-Didehydro-β,ψ-caroten-4-one	Anhydrodeoxyflexixanthin 4-Ketotorulene	→ Anhydrodeoxyflexixanthin			
3′,4′-Didehydrochlorobactene	3′,4′-Didehydrochlorobactene	—			
6,7-Didehydro-5,6-dihydro-β,β-carotene-3,3′-diol	Caloxanthin	→ Caloxanthin			
3′,4′-Didehydro-1′,2′-dihydro-β,ψ-carotene-1′,2′-diol	Plectaniaxanthin	→ Plectaniaxanthin			
3′,4′-Didehydro-1′,2′-dihydro-β,ψ-carotene-3,1′-diol	Saproxanthin	→ Saproxanthin			
(2′R)-1′,16′-Didehydro-1′,2′-dihydro-β,ψ-caroten-2′-ol	Aleuriaxanthin	→ Aleuriaxanthin			
3,4-Didehydro-1,2-dihydro-ψ,ψ-caroten-1-ol	3,4-Dehydrorhodopin "OH-P 481"	→ 3,4-Dehydrorhodopin			
7′,8′-Didehydro-5,6-dihydro-β,β-carotene-3,5,6,3′-tetrol	Heteroxanthin	→ Heteroxanthin			
6,7-Didehydro-5,6-dihydro-β,β-carotene-3,5,3′-triol	Deepoxyneoxanthin "Trollein"	→ "Trollein"*			

Semisystematic name	Trivial name		Formula	MW
7',8'-Didehydro-5,6-dihydro-β,β-carotene-3,5,3'-triol	Heteroxanthin	→ Heteroxanthin		
3',4'-Didehydro-1',2'-dihydro-β,ψ-carotene-2,1',2'-triol	2-Hydroxyplectaniaxanthin	→ 2-Hydroxyplectaniaxanthin		
3',4'-Didehydro-4-keto-γ-carotene	Anhydrodeoxyflexixanthin 4-Ketotorulene	→ Anhydrodeoxyflexixanthin		
3,4-Didehydrolycopene	3,4-Didehydrolycopene	—		
4',5'-Didehydro-4,5'-retro-β,β-carotene	"Dehydro-β-carotene" Dehydroretrocarotene Dehydroretro-β-carotene Retrodehydrocarotene Retrodehydro-β-carotene	→ "Dehydro-β-carotene"		
4',5'-Didehydro-4,5'-retro-β,ψ-carotene	Retrodehydro-γ-carotene	→ Retrodehydro-γ-carotene		
Didehydroretro-γ-carotene	Retrodehydro-γ-carotene	→ Retrodehydro-γ-carotene		
4',5'-Didehydro-4,5'-retro-β,β-carotene-3,3'-diol	Eschscholtzxanthin	→ Eschscholtzxanthin		
4',5'-Didehydro-4',5'-retro-β,β-carotene-3,3'-dione	Rhodoxanthin	→ Rhodoxanthin		
3,4-Didehydrorhodopin	3,4-Didehydrorhodopin	—	$C_{40}H_{56}O$	552.88
7,8-Didehydrozeaxanthin	Diatoxanthin	→ Diatoxanthin		
"Didemethylated spirilloxanthin"	Bacterioruberin Bacterioruberin α	→ Bacterioruberin α		
"Diepoxy-β-carotene"	β-Carotene-diepoxide β-Carotene-5,6,5',6'-diepoxide "Diepoxy-β-carotene" 5,6,5',6'-Diepoxy-β-carotene	→ β-Carotene-diepoxide		
5,6,5',6'-Diepoxy-β-carotene	β-Carotene-diepoxide β-Carotene-5,6,5',6'-diepoxide "Diepoxy-β-carotene"	→ β-Carotene-diepoxide		
5,6,5',6',6'-Diepoxy-5,6,5',6'-tetrahydro-β,β-carotene	β-Carotene-diepoxide β-Carotene-5,6,5',6'-diepoxide "Diepoxy-β-carotene" 5,6,5',6'-Diepoxy-β-carotene	→ β-Carotene-diepoxide		

Table I.3 (continued)
NAME LIST FOR CAROTENOIDS (TABLES I.5, I.6, I. PC, I. TLC, I. LC)

Pigment	Names mentioned in Tables I.5 through I. LC	All synonyms	Sum formula	Mol wt	Structure
5,6,5',8'-Diepoxy-5,6,5',8'-tetrahydro-β,β-carotene-3,3'-diol	Luteochrome	→ Luteochrome			
5,8,5',8'-Diepoxy-5,8,5',8'-tetrahydro-β,β-carotene-(3S,5R,6S,3'S,5'R,6'S)	Aurochrome	→ Aurochrome			
5,6,5',6'-Diepoxy-5,6,5',6'-tetrahydro-β,β-carotene-3,3'-diol	Violaxanthin Violeoxanthin 9-cis-Violaxanthin	→ Violaxanthin			
5,6,5',8'-Diepoxy-5,6,5',8'-tetrahydro-β,β-carotene-3,3'-diol	Luteoxanthin	→ Luteoxanthin			
5,8,5',8'-Diepoxy-5,8,5',8'-tetrahydro-β,β-carotene-3,3'-diol	Auroxanthin	→ Auroxanthin			
5,6,5',6'-Diepoxy-5,6,5',6'-tetrahydro-β,β-caroten-3-ol	Cryptoxanthindiepoxide β-Cryptoxanthin-5,6,5',6'-diepoxide	→ Cryptoxanthindiepoxide			
5,8,5',8'-Diepoxy-5,8,5',8'-tetrahydro-β,β-caroten-3-ol	Cryptochrome	→ Cryptochrome			
"5,8,5',8'-Diepoxyzeaxanthin"	Auroxanthin	→ Auroxanthin			
Di(β,D-glucosyl)-8,8'-diapo-carotene-8,8'-dioate	Crocetindi(β,D-glucosyl)-ester	→ Crocetindi(β,D-glucosyl)-ester			
7',8'-Dihydro-7'-apo-β-caroten-8'-one	Sintaxanthin	→ Sintaxanthin			
5',6'-Dihydro-5'-apo-18'-nor-β-caroten-6'-one	Citranaxanthin	→ Citranaxanthin			
β-Dihydrocarotene	7,7'-Dihydro-β-carotene	→ 7,7'-Dihydro-β-carotene			
"1,1'-Dihydro-β-carotene"	7,7'-Dihydro-β-carotene	→ 7,7'-Dihydro-β-carotene			

$C_{40}H_{58}$ 538.90

7,7'-Dihydro-β-carotene	7,7'-Dihydro-β-carotene	β-Dihydrocarotene "1,1'-Dihydro-β-carotene" 7,8-Dihydro-8,7'-retro-β,β-carotene
7',8'-Dihydro-β,ψ-carotene	β-Zeacarotene β₁-Zeacarotene	→ β-Zeacarotene
1',2'-Dihydro-β,ψ-caroten-1'-ol	1',2'-Dihydro-1'-hydroxy-γ-carotene	→ 1',2'-Dihydro-1'-hydroxy-γ-carotene
7',8'-Dihydro-γ-carotene	β-Zeacarotene β₁-Zeacarotene	→ β-Zeacarotene
7',8'-Dihydro-δ-carotene (6R)-7',8'-dihydro-ε,ψ-carotene	α-Zeacarotene α-Zeacarotene	→ α-Zeacarotene → α-Zeacarotene
1',2'-Dihydro-ψ,ψ-caroten-1'-ol	OH-Chlorobactene	→ OH-Chlorobactene
7,8-Dihydro-ψ,ψ-carotene	Flavorhodin Neurosporene Proneurosporene "Protetrahydrolycopene" "Tetrahydrolycopene" "5,6,5',6'-Tetrahydrolycopene"	→ Neurosporene
1,2-Dihydro-ψ,ψ-caroten-1-ol	OH-Lycopene Rhodopin	→ OH-Lycopene
1',2'-Dihydro-3',4'-dehydro-3,1'-dihydroxy-γ-carotene	Saproxanthin	→ Saproxanthin
1,2-Dihydro-3,4-dehydro-1-OH-lycopene	3,4-Dehydrorhodopin "OH-P-481"	→ 3,4-Dehydrorhodopin
7,8-Dihydro-4,4'-diapocarotene	4,4'-Diaponeurosporene	→ 4,4'-Diaponeurosporene
1',2'-Dihydro-3',4'-didehydro-3,1'-dihydroxy-γ-caroten-2'-yl-rhamnoside	Myxoxanthophyll	→ Myxoxanthophyll
1',2'-Dihydro-1'-glucosyl-3,4-dehydrotorulene	Myxobactin	→ Myxobactin
1',2'-Dihydro-1'-glucosyl-4-ketotorulene	Myxobactone	→ Myxobactone

Table I.3 (continued)
NAME LIST FOR CAROTENOIDS (TABLES I.5, I.6, I. PC, I. TLC, I. LC)

Pigment	Names mentioned in Tables I.5 through I. LC	All synonyms	Sum formula	Mol wt	Structure
1',2'-Dihydro-1'-hydroxy-γ-carotene	1',2'-Dihydro-1'-hydroxy-γ-carotene	1',2'-Dihydro-β,ψ-caroten-1'-ol Hydroxydihydro-γ-carotene 1'-Hydroxy-1',2'-dihydro-γ-carotene	$C_{40}H_{58}O$	554.90	
1',2'-Dihydro-1'-hydroxychlorobactene	OH-Chlorobactene	→ OH-Chlorobactene			
1',2'-Dihydro-1'-hydroxy-3,4-dehydrotorulene-glucoside	Myxobactin	→ Myxobactin			
1',2'-Dihydro-1'-hydroxy-4-keto-γ-carotene	1',2'-Dihydro-1'-hydroxy-4-keto-γ-carotene 4-Keto-1',2'-dihydro-1'-hydroxy-γ-carotene 1-Keto-1'-hydroxy-1',2'-dihydro-γ-carotene	→ 1-Keto-1'-hydroxy-1',2'-dihydro-γ-carotene			
1',2'-Dihydro-1'-hydroxy-4-ketotorulene	Deoxyflexixanthin 1',2'-Dihydro-1'-hydroxy-4-ketotorulene . 4-Keto-1',2'-dihydro-1'-hydroxytorulene	→ Deoxyflexixanthin			
1',2'-Dihydro-1'-hydroxy-4-ketotoruleneglucoside	Myxobactone	→ Myxobactone			
1',2'-Dihydro-1'-hydroxyspheroidenone	"OH-R" OH-Spheroidenone "Flavorhodin" .	→ OH-Spheroidenone			
7,8-Dihydrolycopene	Neurosporene Proneurosporene "Protetrahydrolycopene" "Tetrahydrolycopene" "5,6,5',6'-Tetrahydrolycopene"	→ Neurosporene			

"5,6-Dihydro-4-methoxy-lycopene-6-one	→ Spheroidenone		
1,2-Dihydro-1-OH-lycopene	→ OH-Lycopene		
Rhodopin			
		$C_{40}H_{54}O_4$	598.86
2'-Dihydrophillipsiaxanthin	1,1',2'-Trihydroxy-3,4,3',4'-tetradehydro-1,2,1',2'-tetrahydro-ψ,ψ-caroten-2-one		
2'-Dihydrophillipsiaxanthin			
Dihydrophytoene	→ Lycopersene		
7,8-Dihydro-8,7'-retro-β,β-carotene	→ 7,7'-Dihydro-β-carotene		
7',8'-Dihydrorhodovibrin	→ OH-Y		
OH-Spheroidene			
OH-Y			
"1',2'-Dihydrorubixanthin	→ Gazaniaxanthin		
11',12'-Dihydrospheroidene	→ P-412		
3,3'-Dihydroxycanthaxanthin	→ Astaxanthin		
3,3'-Dihydroxy-α-carotene	→ 3-Hydroxy-3'-hydroxy-α-carotene		
Lutein	→ Lutein		
"Xanthophyll"			
2,2'-Dihydroxy-β-carotene	—	$C_{40}H_{56}O_2$	568.88
3,4'-Dihydroxy-β-carotene	—	$C_{40}H_{56}O_2$	568.88

Table I.3 (continued)
NAME LIST FOR CAROTENOIDS (TABLES I.5, I.6, I. PC, I. TLC, I. LC)

Pigment	Names mentioned in Tables I.5 through I. LC	All synonyms	Sum formula	Mol wt	Structure
4,4′-Dihydroxy-β-carotene	4,4′-Dihydroxy-β-carotene Isozeaxanthin	"Aphanicol" β,β-Carotene-4,4′-diol Isozeaxanthin	$C_{40}H_{56}O_2$	568.88	
Dihydroxy-ζ-carotene (3S,3′S)-3,3′-Dihydroxy-β,β-carotene-4,4′-dione	Dihydroxy-ζ-carotene Astaxanthin	— → Astaxanthin			
(3S,5R,3′S,5′R)-3,3′-Dihydroxy-κ,κ-caroten-6,6′-dione	Capsorubin	→ Capsorubin			
(3R,3′S,5′R)-3,3′-Dihydroxy-β,κ-caroten-6′-one	Capsanthin	→ Capsanthin			
Dihydroxy-3,4-dehydro-α-carotene	Dihydroxy-3,4-dehydro-α-carotene	—			
"3,3′-Dihydroxydehydro-β-carotene"	Eschscholtzxanthin	→ Eschscholtzxanthin			
3,3′-Dihydroxy-7,8,7′,8′-dehydro-β-carotene	Alloxanthin Cynthiaxanthin	→ Alloxanthin			
3,3′-Dihydroxy-7,8-dehydro-β-caroten-5′,6′-epoxide	Diadinoxanthin	→ Diadinoxanthin			
3,3′-Dihydroxy-7,8-didehydro-β,β-carotene-4,4′-dione	Part of asterinic acid	→ Part of asterinic acid 7,8-Didehydroastaxanthin			
3,1′-Dihydroxy-3′,4′-didehydro-1′,2′-dihydro-β,ψ-caroten-4-one	Flexixanthin	→ Flexixanthin			
"3,3′-Dihydroxy-5,8,5′8′-diepoxy-β-carotene"	Auroxanthin	→ Auroxanthin			
(5R,6R)-5,6-Dihydroxy-5,6-dihydro-10′-apo-β-caroten-10′-oic acid	Azafrin	→ Azafrin			

3,3'-Dihydroxy-5,5'-dihydro-7,7'-didehydro-β-carotene	Nostoxanthin	→ Nostoxanthin
1',2'-Dihydroxy-1',2'-dihydrotorulene	Plectaniaxanthin	→ Plectaniaxanthin
3,1'-Dihydroxy-1',2'-dihydrotorulene	Saproxanthin	→ Saproxanthin
3,3'-Dihydroxy-4,4'-diketo-β-carotene	Astaxanthin	→ Astaxanthin
1,1'-Dihydroxy-2,2'-diketo-1,1'-2,2'-tetrahydro-3,3',4,4'-tetradehydrolycopene	2,2'-Diketobacterioruberin Phillipsiaxanthin	→ Phillipsiaxanthin
3,3'-Dihydroxy-2,2'-dinor-β,β-carotene-4,4'-dione-3,3'-diacylate	Actinioerythrin Actinioerythrol	→ Actinioerythrin
1,1'-Dihydroxy-2,2'-di-β-L-rhamnosyl-1,2,1',2'-tetrahydro-3,4,3',4'-tetradehydrolycopene	Oscillaxanthin	→ Oscillaxanthin
3,3'-Dihydroxy-5,6-epoxy-α-carotene	Isolutein Luteinepoxide Lutein-5,6-epoxide Taraxanthin Xanthophyllepoxide	→ Luteinepoxide
"3,3-Dihydroxy-5,8-epoxy-α-carotene"	Chrysanthemaxanthin Flavoxanthin	→ Chrysanthemaxanthin
3,3'-Dihydroxy-5,6-epoxy-β-carotene	Antheraxanthin	→ Antheraxanthin
3,3'-Dihydroxy-5,8-epoxy-β-carotene	Mutatoxanthin	→ Mutatoxanthin
3,3'-Dihydroxyluteochrome "Dihydroxylycopene"	Luteoxanthin	→ Luteoxanthin
1,2,1',2'-Tetrahydro-1,1'-dihydroxy-lycopene	1,2,1',2'-Tetrahydro-1,1'-dihydroxy-lycopene	→ 1,2,1',2'-Tetrahydro-1,1'-dihydroxy-lycopene
"3,3'-Dihydroxylycopene" "Dihydroxyneurosporene"	Lycophyll Chloroxanthin OH-Neurosporene	→ Lycophyll → Chloroxanthin

Table I.3 (continued)
NAME LIST FOR CAROTENOIDS (TABLES I.5, I.6, I. PC, I. TLC, I. LC)

Pigment	Names mentioned in Tables I.5 through I. LC	All synonyms	Sum formula	Mol wt	Structure
"3,3'-Dihydroxyretro-β-carotene"	Eschscholtzxanthin	→ Eschscholtzxanthin			
3,3'-Dihydroxy-7,8,7',8'-tetradehydro-β,β-carotene-4,4'-dione	Part of asterinic acid	→ Part of asterinic acid 4,4'-Diketocynthiaxanthin Diketotetradehydrozeaxanthin 3,3'-Dihydroxy-7,8,7',8'-tetradehydro-β,β-carotene-4,4'-dione			
1,1'-Dihydroxy-3,4,3',4'-tetradehydro-1,2,1',2'-tetrahydro-ψ,ψ-carotene-2,2'-dione	2,2'-Diketobacterioruberin Phillipsiaxanthin	→ Phillipsiaxanthin			
1,1'-Dihydroxy-1,2,1',2'-tetrahydro-ζ-carotene	1,1'-Dihydroxy-1,2,1',2'-tetrahydro-ζ-carotene	1,2,7,8,1',2',7',8'-Octahydro-ψ,ψ-carotene-1,1'-diol	$C_{40}H_{64}O_2$	576.94	
1,1'-Dihydroxy-1,2,1',2'-tetrahydrolycopene	"Dihydroxylycopene" 1,2,1',2'-Tetrahydro-1,1'-dihydroxy-lycopene	→ 1,2,1',2'-Tetrahydro-1,1'-dihydroxy-lycopene			
"Diketobacterioruberin"	"2,2'-Diketobacterioruberin" Phillipsiaxanthin	→ Phillipsiaxanthin			
"2,2'-Diketobacterioruberin"	"2,2'-Diketobacterioruberin	→ Phillipsiaxanthin			
3,4-Diketo-α-carotene	3,4-Diketo-α-carotene	—	$C_{40}H_{52}O_2$	564.85	
3,4-Diketo-β-carotene	Dehydrohydroxyechinenone "Euglenanone" Aphanicin	→ Dehydrohydroxyechinenone			
4,4'-Diketo-β-carotene	4,4'-Diketo-β-carotene Euglenanone	→ Canthaxanthin			

624.90

$C_{42}H_{56}O_4$

598.95

$C_{42}H_{62}O_2$

Name	Synonym	Derived / related
4,4'-Diketocynthiaxanthin	Part of asterinic acid	→ Part of asterinic acid 3,3'-Dihydroxy-7,8,7',8'-tetradehydro-β,β-carotene-4,4'-dione
"3,3'-Diketodehydro-β-carotene"	Rhodoxanthin	→ Rhodoxanthin
4,4'-Diketo-3-hydroxy-β-carotene "3,3'-Diketoretro-β-carotene"	3-Hydroxy-4,4'-diketo-β-carotene Rhodoxanthin	→ 3-Hydroxy-4,4'-diketo-β-carotene → Rhodoxanthin
2,2'-Diketospirilloxanthin P 518	2,2'-Diketospirilloxanthin P 518	1,1'-Dimethoxy-3,4,3',4'-tetradehydro-1,2,1',2'-tetrahydro-ψ,ψ-carotene-2,2'-dione "2-Ketospirilloxanthin" "P-512" P-518 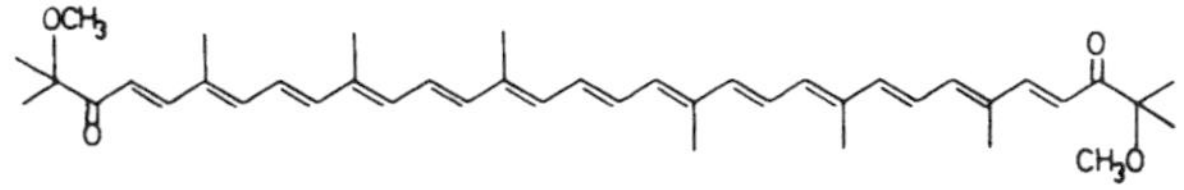
Diketotetradehydrozeaxanthin	Part of asterinic acid	→ Part of asterinic acid → 3,3'-Dihydroxy-7,8,7',8'-tetradehydro-β,β-carotene-4,4'-dione
"3,3'-Dimethoxy-γ-carotene"	Torulene	→ Torulene
"3,3'-Dimethoxy-3',4'-dehydro-γ-carotene"	Torulene	→ Torulene
1,1'-Dimethoxy-3',4'-didehydro-1,2,1',2'-tetrahydro-ψ,ψ-caroten-4-one	1,1'-Dimethoxy-3',4'-didehydro-1,2,1',2'-tetrahydro-ψ,ψ-caroten-4-one	1,1'-Dimethoxy-1,2,1',2'-tetrahydro-3',4'-didehydro-ψ,ψ-caroten-4-one 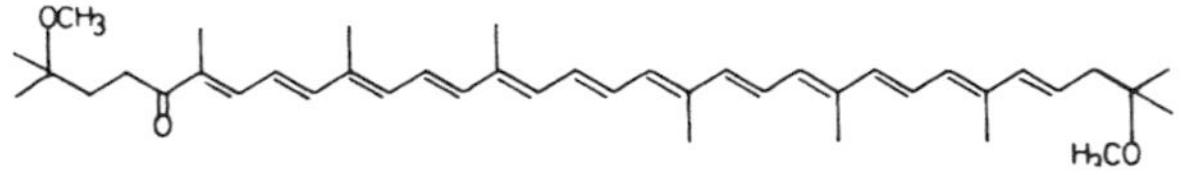
"3,3'-Dimethoxy"-structure	Rhodoviolascin Spirilloxanthin Rhodoviolascin	→ Rhodoviolascin
1,1'-Dimethoxy-3,4,3',4'-tetradehydro-1,2,1',2'-tetrahydro-ψ,ψ-carotene		→ Rhodoviolascin
1,1'-Dimethoxy-3,4,3',4'-tetradehydro-1,2,1',2'-tetrahydro-ψ,ψ-carotene-2,2'-dione	2,2'-Diketospirilloxanthin P-518	→ 2,2'-Diketospirilloxanthin

Table I.3 (continued)
NAME LIST FOR CAROTENOIDS (TABLES I.5, I.6, I. PC, I. TLC, I. LC)

Pigment	Names mentioned in Tables I.5 through I. LC	All synonyms	Sum formula	Mol wt	Structure
1,1'-Dimethoxy-1,2,1',2'-tetrahydro-ψ,ψ-carotene-4,4'-dione	1,1'-Dimethoxy-1,2,1',2'-tetrahydro-ψ,ψ-carotene-4,4'-dione	—	$C_{42}H_{60}O_4$	628.93	
1,1'-Dimethoxy-1,2,1',2'-tetrahydro-3',4'-didehydro-ψ,ψ-caroten-4-one	1,1'-Dimethoxy-3',4'-didehydro-ψ,ψ-caroten-4-one	→ 1,1'-Dimethoxy-3',4'-didehydro-1,2,1',2'-tetrahydro-ψ,ψ-caroten-4-one			
Dimethylcrocetin	Crocetindimethylester	→ Crocetindimethylester			
Dimethyl-6,6'-diapocarotene-6,6'-dioate	Methylbixin (*trans*)	→ Methylbixin (*trans*)			
Dimethyl-8,8'-diapocarotene-8,8'-dioate	Crocetindimethylester	→ Crocetindimethylester			
2,2'-Dinor-β,β-carotene-3,4,3',4'-tetrone	Violerythrin	→ Violerythrin			
Dinoxanthin	Dinoxanthin	(3S,5R,6R,3'S,5'R,6'S)-5',6'-Epoxy-6,7-didehydro-5,6,5',6'-tetrahydro-β,β-carotene-3,5,3'-triol-3-acetate Neoxanthin-3-acetate	$C_{42}H_{58}O_5$	642.92	
5,6-Dioxo-10'-apo-5,6-seco-β-caroten-10'-al	(β)-Carotenonaldehyde	→ (β)-Carotenonaldehyde			
5,6,5',6'-Diseco-β,β-carotene-5,6,5',6'-tetrone	β-Carotenone	→ β-Carotenone			
Dodecahydrolycopene	Phytofluene	→ Phytofluene			
β-Doradecin	Dehydroadonixanthin	→ Dehydroadonixanthin			
Echinenone	Aphanin Echinenone 4-Keto-β-carotene Myxoxanthin	Aphanin β,β-Caroten-4-one ''Echininone'' 4-Keto-β-carotene Myxoxanthin	$C_{40}H_{54}O$	550.87	

"Echininone"	Aphanin Echinenone 4-Keto-β-carotene Myxoxanthin	→ Echinenone
Eloxanthin	Isolutein Luteinepoxide Lutein-5,6-epoxide Taraxanthin Xanthophyllepoxide	→ Luteinepoxide
"5,8-Epoxy-α-carotene" 5,8-Epoxy-β-carotene	Flavochrome Flavacin Mutatochrome	→ Flavochrome → Flavacin
5,8-Epoxycryptoxanthin	Cryptoflavin Cryptoxanthin-5,8-epoxide	→ Cryptoflavin
5,6-Epoxy-7',8'-didehydro-5,6-dihydro-β,β-carotene-3,3'-diol	Diadinoxanthin	→ Diadinoxanthin
5',6'-Epoxy-6,7-didehydro-5,6,5',6'-tetrahydro-β,β-carotene-3,5,19',3'-tetrol	Vaucheriaxanthin	→ Vaucheriaxanthin
(3S,5R,6R,3'S,5'R,6'S)-5',6'-Epoxy-6,7-didehydro-5,6,5',6'-tetrahydro-β,β-carotene-3,5,3'-triol	Neoxanthin Trollixanthin	→ Neoxanthin
5',8'-Epoxy-6,7-didehydro-5,6,5',8'-tetrahydro-β,β-carotene-3,5,3'-triol	Neochrome Trollichrome	→ Neochrome
(3S,5R,6R,3'S,5'R,6'S,)-5',6',Epoxy-6,7-didehydro-5,6,5',6'-tetrahydro-β,β-carotene-3,5,3'-triol-3-acetate	Dinoxanthin	→ Dinoxanthin
5,6-Epoxy-5,6-dihydro-β-β-carotene	5,6-Monoepoxy-α-carotene	→ α-Carotene-epoxide
5,6-Epoxy-5,6-dihydro-β,β-carotene	β-Carotenemonoepoxide 5,6-Monoepoxy-β-carotene	→ β-Carotenemonoepoxide

Table I.3 (continued)
NAME LIST FOR CAROTENOIDS (TABLES I.5, I.6, I. PC, I. TLC, I. LC)

Pigment	Names mentioned in Tables I.5 through I. LC	All synonyms	Sum formula	Mol wt	Structure
5,6-Epoxy-5,6-dihydro-β,β-carotene-3,3'-diol	Antheraxanthin	→ Antheraxanthin			
5,8-Epoxy-5,8-dihydro-β,β-carotene-3,3'-diol	Mutatoxanthin	→ Mutatoxanthin			
5,6-Epoxy-5,6-dihydro-β,β-caroten-3-ol	Cryptoxanthinepoxide Cryptoxanthin-5,6-epoxide	→ Cryptoxanthinepoxide			
5,8-Epoxy-5,8-dihydro-β,β-carotene	Flavacin Mutatochrome	→ Flavacin			
5,8-Epoxy-5,8-dihydro-β,β-caroten-3-ol	Cryptoflavin Cryptoxanthin-5,8-epoxide	→ Cryptoflavin			
(3S,5R,6S,3'R,6'R)-5,6-Epoxy-5,6-dihydro-β,ε-caro-tene-3,3'-diol	Isolutein Luteinepoxide Lutein-5,6-epoxide Taraxanthin Xanthophyllepoxide	→ Luteinepoxide			
5,8-Epoxy-5,8-dihydro-β,ε-carotene	Flavochrome	→ Flavochrome			
5,8-Epoxy-5,8-dihydro-β,ε-carotene-3,3'-diol	Chrysanthemaxanthin Flavoxanthin	→ Chrysanthemaxanthin			
5,6-Epoxy-5,6-dihydro-β,ε-carotene-3,3',6'-triol	Trollixanthin	→ Trollixanthin			
5,8-Epoxy-5,8-dihydro-β,ε-carotene-3,3',6'-triol	Trollichrome	→ Trollichrome			
5,6-Epoxy-5,6-dihydro-β,ψ-caroten-3-ol	Rubixanthin-5,6-epoxide	→ Rubixanthin-5,6-epoxide			
5,8-Epoxy-5,8-dihydro-β,ψ-caroten-3-ol	Rubichrome	→ Rubichrome			
"5,8-Epoxy-3,3'-dihydroxy-α-carotene"	Chrysanthemaxanthin Flavoxanthin	→ Chrysanthemaxanthin			

$$C_{27}H_{36}O_1 \qquad 408.58$$

5',6'-Epoxy-3,3'-dihydroxy-7,8-didehydro-5',6'-dihydro-10,11,20-trinor-β,β-caroten-19',11'-olide	Pyrrhoxanthinol	→ Pyrrhoxanthinol
5',6'-Epoxy-3,3'-dihydroxy-7,8-didehydro-5',6'-dihydro-10,11,20-trinor-β,β-caroten-19',11'-olide-3-acetate	Pyrroxanthin	→ Pyrroxanthin
5,6-Epoxy-3,3'-dihydroxy-5,6-dihydro-β,κ-caroten-6'-one	Capsanthin-5,6-epoxide	→ Capsanthin-5,6-epoxide
5,8-Epoxy-3,3'-dihydroxy-5,8-dihydro-β,κ-caroten-6'-one	Capsochrome	→ Capsochrome
"5,8-Epoxy-3-hydroxy-γ-carotene"	Rubichrome	→ Rubichrome
5,6-Epoxy-3-hydroxy-5,6-dihydro-10'-apo-β-caroten-10'-al	5,6-Epoxy-3-hydroxy-5,6-dihydro-10'-apo-β-caroten-10'-al	"Apoviolaxanthinal" Apo-10'-violaxanthal	→ "Apoviolaxanthinal" Apo-10'-violaxanthal
5,6-Epoxy-3-hydroxy-5,6-dihydro-12'-apo-β-caroten-12'-al	Apo-12'-violaxanthal	→ Apo-12'-violaxanthal
5,6-Epoxy-lutein "5,8-Epoxy-lutein"	5,6-Monoepoxy-lutein Chrysanthemaxanthin Flavoxanthin	→ 5,6-Monoepoxy-lutein Chrysanthemaxanthin
"5,8-Epoxy-rubixanthin" (3S,5R,6S,3'S,5'R,6'R)-5,6-Epoxy-3,3',5'-trihydroxy-6',7'-didehydro-5,6,7,8,5',6'-hexahydro-β,β-caroten-8-one	Rubichrome Fucoxanthinol	→ Rubichrome → Fucoxanthinol
(3S,5R,6S,3'S,5'R,6'R)-5,6-Epoxy-3,3',5'-trihydroxy-6',7'-didehydro-5,6,7,8,5',6'-hexahydro-β,β-caroten-8-one-3'-acetate	Fucoxanthin "Fucoxanthol"	→ Fucoxanthin

Table I.3 (continued)
NAME LIST FOR CAROTENOIDS (TABLES I.5, I.6, I. PC, I. TLC, I. LC)

Pigment	Names mentioned in Tables I.5 through I. LC	All synonyms	Sum formula	Mol wt	Structure
5',6'-Epoxy-3,5,3'-trihydroxy-6,7-didehydro-5,6,5',6'-tetrahydro-10,11,20-trinor-β,β-caroten-19',11'-olide	Peridininol	→ Peridininol			
5',6'-Epoxy-3,5,3'-trihydroxy-6,7-didehydro-5,6,5',6'-Tetrahydro-10,11,20-trinor-β,β-caroten-19',11'-olide-3-acetate	Peridinin	→ Peridinin			
5,6-Epoxyzeaxanthin	Antheraxanthin	→ Antheraxanthin			
5,8-Epoxyzeaxanthin	Mutatoxanthin	→ Mutatoxanthin			
Eschscholtziaxanthin	Eschscholtzxanthin	→ Eschscholtzxanthin			
Eschscholtzxanthin	Eschscholtzxanthin	4',5'-Didehydro-4,5'-retro-β,β-carotene-3,3'-diol "3,3'-Dihydroxydehydro-β-carotene" "3,3'-Dihydroxyretro-β-carotene" Eschscholtziaxanthin "Retrodehydrozeaxanthin"	$C_{40}H_{54}O_2$	566.87	
4'-Ethoxy-β,β-caroten-4-one	4-Keto-4'-ethoxy-β-carotene	→ 4-Keto-4'-ethoxy-β-carotene			
4'-Ethoxy-4-keto-β-carotene	4-Keto-4'-ethoxy-β-carotene	→ 4-Keto-4'-ethoxy-β-carotene			
"Euglenanone"	Dehydrohydroxyechinenone	→ Dehydrohydroxyechinenone			
Euglenanone	Aphanicin	→ Canthaxanthin			
	Canthaxanthin				
	4,4'-Diketo-β-carotene				
"Euglenarhodon"	Dehydrohydroxyechinenone	→ Dehydrohydroxyechinenone			
	Astacene	→ Astacene			

Flavacin	Mutatochrome	β-Carotene-5,8-epoxide Carotene oxide Citroxanthin 5,8-Epoxy-β-carotene 5,8-Epoxy-5,8-dihydro-β,β-carotene Mutatochrome	$C_{40}H_{56}O$	552.88
Flavochrome	Flavochrome	"5,8-Epoxy-α-carotene" 5,8-Epoxy-5,8-dihydro-β,ε-carotene	$C_{40}H_{56}O$	552.88
Flavorhodin	Flavorhodin Neurosporene Proneurosporene "Protetrahydrolycopene" "Tetrahydrolycopene" "5,6,5',6'-Tetrahydrolycopene"	→ Neurosporene		
Flavoxanthin	Chrysanthemaxanthin Flavoxanthin	→ Chrysanthemaxanthin		
Flexixanthin	Flexixanthin	3,1'-Dihydroxy-3',4'-didehydro-1',2'-dihydro-β-ψ-caroten-4-one 4-Oxosaproxanthin	$C_{40}H_{54}O_3$	582.86
Foliachrome	Neochrome Trollichrome	→ Neochrome		
Foliaxanthin	Neoxanthin Trollixanthin	→ Neoxanthin		
Fucochrome	Neochrome Trollichrome	→ Neochrome		
Fucoxanthin	Fucoxanthin "Fucoxanthol"	(3S,5R,6S,3'S,5'R,6'R)-5,6-Epoxy-3,3',5'-trihydroxy-6',7'-didehydro-5,6,7,8,5',6'-hexahydro-β,β-caroten-8-one-3'-acetate "Fucoxanthol"	$C_{42}H_{58}O_6$	658.92

Table I.3 (continued)

NAME LIST FOR CAROTENOIDS (TABLES I.5, I.6, I. PC, I. TLC, I. LC)

Pigment	Names mentioned in Tables I.5 through I. LC	All synonyms	Sum formula	Mol wt	Structure
Fucoxanthinol	Fucoxanthinol	(3S,5R,6S,3'S,5'R,6'R)-5,6-Epoxy-3,3',5'-trihydroxy-6',7'-didehydro-5,6,7,8,5',6'-hexahydro-β,β-caroten-8-one	$C_{40}H_{56}O_5$	616.88	
"Fucoxanthol"	Fucoxanthin "Fucoxanthol"	→ Fucoxanthin			
Gazaniaxanthin	Gazaniaxanthin	(3R)-5'-*cis*-β,ψ-Caroten-3-ol "1',2'-Dihydrorubixanthin"	$C_{40}H_{56}O$	552.88	
β,D-Gentiobiosyl-β,D-glucosyl-8,8'-diapo-carotene-8,8'-dioate	Crocetin-(β,D-gentiobiosyl-β,D-glucosyl)-ester	→ Crocetin-(β,D-gentiobiosyl-β,D-glucosyl)-ester			
"4-(β-D-Glucopyranosyl-oxy)-4,4'-diaponeurosporene"	"4-(β-D-Glucopyranosyl-oxy)-4,4'-diaponeurosporene"	4-(β,D-Glucopyranosyl-oxy-7',8'-dihydro-4,4'-diapocarotene	$C_{36}H_{52}O_6$	580.80	
4-(β-D-Glucopyranosyloxy)-7',8'-dihydro-4,4'-diapocarotene	4-(β-D-Glucopyranosyloxy)-7',8'-dihydro-4,4'-diapocarotene	→ Phleixanthophyll			
1'-(β,D-Glucopyranosyloxy)-3',4'-didehydro-1',2'-dihydro-β,ψ-caroten-2'-ol	Phleixanthophyll	→ Phleixanthophyll			
4-(β,D-Glucopyranosyloxy)-4,4'-diapo-neurosporene"	"4-(β,D-Glucopyranosyloxy)-4,4'-diapo-neurosporene"	→ "4-(β,D-Glucopyranosyl-oxy)-4,4'-diapo-neurosporene"			
1'-(β,D-Glucopyranosyloxy)-2'-hydroxy-3',4'-didehydro-1',2'-dihydro-β,ψ-caroten-4-one	4-Keto-phleixanthophyll	→ 4-Keto-phleixanthophyll			
2-(4-(β,D-Glucopyranosyloxy)-3-methyl-2-butenyl)-2'-(4-hydroxy-3-methyl-2-butenyl)-ε,ε-carotene	Corynexanthin	→ Corynexanthin			
1'-Glucosyloxy-3',4'-didehydro-dro-β,ψ-carotene-4-one	Myxobactone	→ Myxobactone			

1'-Glucosyloxy-3,4,3',4'-tetradehydro-1',2'-dihydro-β,ψ-carotene	Myxobactin	→ Myxobactin	
Helenien Lutein dipalmitate	Helenien Lutein dipalmitate	(3R,3'R,6'R)-β,ε-Carotene-3,3'-diol dipalmitate Lutein dipalmitate Xanthophyll dipalmitate	$C_{72}H_{116}O_4$ 1045.71
Heteroxanthin*	Heteroxanthin	7',8'-Didehydro-5,6-dihydro-β,β-carotene-3,5,6,3'-tetrol 7',8'-Didehydro-5,6-dihydro-β,β-carotene-3,5,3'-triol? "3,3',5'-Trihydroxy-6'-hydro-7,8-dehydro-β-carotene" "*Vaucheria*-Heteroxanthin" 3,3',8-Trihydroxy-5,6-epoxy-β-carotene	$C_{40}H_{56}O_4$ 600.88
"Hexadecahydrolycopene" 2,3,2',3',4',5'-Hexadehydro-4,5'-retro-β,β-carotene	Phytoene Anhydroeschscholtzxanthin'	→ Phytoene → Anhydroeschscholtzxanthin	
15-cis-7,8,11,12,7',8'-Hexahydro-ψ,ψ-carotene	Phytofluene	→ Phytofluene	
7,8,11,12,7',8'-Hexahydro-4,4'-diapocarotene	4,4'-Diapophytofluene	→ 4,4'-Diapophytofluene	
7,8,11,12,7',8'-Hexahydrolycopene	Phytofluene	→ Phytofluene	
"1-Hexosyl-1,2-dihydro-3,4-didehydro-apo-8'-lycopenol"	"1-Hexosyl-1,2-dihydro-3,4-didehydro-apo-8'-lycopenol"	1-Mannosyloxy-3,4-didehydro-1,2-dihydro-8'-apo-ψ-caroten-8'-ol	$C_{36}H_{52}O_7$ 596.80
O-Hexosyl-4-keto-1'-hydroxy-1',2'-dihydro-γ-carotene	Myxobactone	→ Myxobactone	
"3-Hydroxy-β-apo-2'-carotenal"	β-Citraurin	→ β-Citraurin	
3-Hydroxy-β-apo-10'-carotenal	3-Hydroxy-10'-apo-β-caroten-10'-al	→ 3-hydroxy-10'-apo-β-caroten-10'-al	

Table I.3 (continued)
NAME LIST FOR CAROTENOIDS (TABLES I.5, I.6, I. PC, I. TLC, I. LC)

Pigment	Names mentioned in Tables I.5 through I. LC	All synonyms	Sum formula	Mol wt	Structure
(3R)-3-Hydroxy-8'-apo-β-caroten-8'-al	β-Citraurin	→ β-Citraurin			
3-Hydroxy-10'-apo-β-caroten-10'-al	3-Hydroxy-10'-apo-β-caroten-10'-al	"Apo-3-zeaxanthinal"; 3-Hydroxy-β-apo-10'-carotenal; → "Apo-4-zeaxanthinal"	$C_{27}H_{36}O_2$	392.58	
3-Hydroxy-12'-apo-β-caroten-12'-al	3-Hydroxy-12'-apo-β-caroten-12'-al	→ α-Citraurin			
3-Hydroxy-8'-apo-ε-caroten-8'-al	α-Citraurin	→ α-Citraurin			
3-Hydroxycanthaxanthin	4,4'-Diketo-3-hydroxy-β-carotene; 3-Hydroxy-4,4'-diketo-β-carotene	3-Hydroxy-4,4'-diketo-β-carotene			
2-Hydroxy-α-carotene	2-Hydroxy-α-carotene; α-Cryptoxanthin; Zeinoxanthin	—			
3'-Hydroxy-α-carotene	α-Cryptoxanthin	→ α-Cryptoxanthin; → α-Cryptoxanthin			
4-Hydroxy-α-carotene	4-Hydroxy-α-carotene	β,ε-Caroten-4-ol			
2-Hydroxy-β-carotene	2-Hydroxy-β-carotene; Cryptoxanthin; β-Cryptoxanthin	—; → β-Cryptoxanthin	$C_{40}H_{56}O$	552.88	
4-Hydroxy-β-carotene	4-Hydroxy-β-carotene; Isocryptoxanthin; "Myxoxanthol"	"Aphanol"; β,β-Caroten-4-ol; Isocryptoxanthin; "Myxoxanthol"	$C_{40}H_{56}O$	552.88	
3-Hydroxy-γ-carotene	Rubixanthin	→ Rubixanthin			
3-Hydroxy-β,β-carotene-4,4'-dione	4,4'-Diketo-3-hydroxy-β-carotene; 3-Hydroxy-4,4'-diketo-β-carotene	→ 3-Hydroxy-4,4'-diketo-β-carotene			

$C_{30}H_{42}O$

418.66

Semisystematic name	Trivial name	
3-Hydroxy-β,β-caroten-4-one	Hydroxyechinenone 4'-hydroxyechinenone	→ Hydroxyechinenone
3'-Hydroxy-β,β-caroten-4-one	3'-Hydroxyechinenone	→ 3'-Hydroxyechinenone
4'-Hydroxy-β,β-caroten-4-one	4-Keto-3'-hydroxy-β-carotene	→ 4-Hydroxy-4'-keto-β-carotene
3-Hydroxy-β,ε-caroten-3'-one	4'-Hydroxy-4'-keto-β-carotene	→ 3-Hydroxy-3'-keto-α-carotene
(3'S,5'R)-3'-Hydroxy-β,κ-caroten-6'-one	3'-Hydroxy-3'-keto-α-carotene Philosamiaxanthin	→ Cryptocapsin
3-Hydroxycitranaxanthin	Cryptocapsin	→ Reticulataxanthin
3-Hydroxy-7,8-dehydro-α-carotene	Reticulataxanthin	→ Crocoxanthin
3-Hydroxy-3',4'-dehydro-γ-carotene	Crocoxanthin	→ Anhydrosaproxanthin
"4-Hydroxy-4,4'-diapaneurosporene"	Anhydrosaproxanthin Celaxanthin	→ 4-Hydroxy-7'-8'-dihydro-4,4'-diapo-carotene 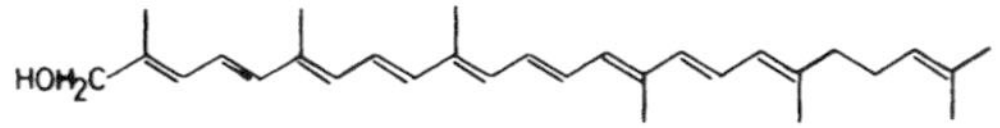
1'-Hydroxy-3',4'-didehydro-1',2'-dihydro-β-ψ-caroten-2'-one	"4-Hydroxy-4,4'-diaponeurosporene"	→ 2'-Dehydroplectaniaxanthin
1'-Hydroxy-3',4'-didehydro-1',2'-dihydro-β,ψ-caroten-4-one	2'-Dehydroplectaniaxanthin	→ Deoxyflexixanthin
"3-Hydroxy-5,8,5',8'-di-epoxy-β-carotene"	Deoxyflexixanthin 1',2'-Dihydro-1'-hydroxy-4-keto-torulene	→ Cryptochrome
(3R)-3-Hydroxy-5',6'-dihydro-5'-apo-18'-nor-β-caroten-6'-one	4-Keto-1',2'-dihydro-1'-hydroxy-torulene Cryptochrome	→ Reticulataxanthin
Hydroxydihydro-γ-carotene	Reticulataxanthin 1',2'-Dihydro-1'-hydroxy-γ-carotene	→ 1',2'-Dihydro-1'-hydroxy-γ-carotene

Table I.3 (continued)
NAME LIST FOR CAROTENOIDS (TABLES I.5, I.6, I. PC, I. TLC, I. LC)

Pigment	Names mentioned in Tables I.5 through I. LC	All synonyms	Sum formula	Mol wt	Structure
1'-Hydroxy-1',2'-dihydro-γ-carotene	1'-Hydroxy-1',2'-dihydro-γ-carotene	→ 1',2'-Dihydro-1'-hydroxy-γ-carotene			
1'-Hydroxy-1',2'-dihydro-γ-carotene	1',2'-Dihydro-1'-hydroxy-γ-carotene	→ 1',2'-Dihydro-1'-hydroxy-γ-carotene			
1'-Hydroxy-1',2'-dihydro-β,ψ-caroten-4-one	1',2'-Dihydro-1'-hydroxy-4-keto-γ-carotene	→ 4-Keto-1'-hydroxy-1',2'-dihydro-γ-carotene			
	4-Keto-1',2'-dihydro-1'-hydroxy-γ-carotene				
	4-Keto-1'-hydroxy-1',2'-dihydro-γ-carotene				
4-Hydroxy-7',8'-dihydro-4,4'-diapocarotene	4-Hydroxy-4,4'-diaponeurosporene	→ 4-Hydroxy-4,4'-diaponeurosporene			
1-Hydroxy-1,2-dihydrolycopene	OH-Lycopene	→ OH-Lycopene			
"1-Hydroxy-1,2-dihydroneurosporene"	Rhodopin				
1-Hydroxy-1',2'-dihydroneurosporene	Chloroxanthin	→ Chloroxanthin			
1-Hydroxy-1,2-dihydrophytofluene	OH-Neurosporene				
1'-Hydroxy-1',2'-dihydrophytofluene	Chloroxanthin	→ Chloroxanthin			
1'-Hydroxy-1'-2'-dihydrospheroidene	OH-Neurosporene				
1'-Hydroxy-1'-2'-dihydrospheroidenone	OH-Phytofluene	→ OH-Phytofluene			
3'-Hydroxy-3,4-diketo-β-carotene	Phytofluenol				
	OH-Phytofluene	→ OH-Phytofluene			
	Phytofluenol				
	7',8'-Dihydrorhodovibrin	→ OH-Y			
	OH-Spheroidene				
	OH-Y				
	"OH-R"	→ PH-Spheroidenone			
	OH-Spheroidenone				
	Dehydroadonixanthin	→ Dehydroadonixanthin			

3'-Hydroxy-4,4'-diketo-β-carotene	4,4'-Diketo-3-hydroxy-β-carotene 3-Hydroxy-4,4'-diketo-β-carotene	Adonirubin 4,4'-Diketo-3-hydroxy-β-carotene 3-Hydroxycanthaxanthin 3-Hydroxy-β,β-carotene-4,4'-dione "Metridene" Phoenicoxanthin	$C_{40}H_{52}O_3$	580.85
Hydroxyechinenone	Hydroxyechinenone	3-Hydroxy-β,β-caroten-4-one 3-Hydroxy-4-keto-β-carotene	$C_{40}H_{54}O_2$	566.87
3'-Hydroxyechinenone	3'-Hydroxyechinenone 4-Keto-3'-hydroxy-β-carotene	Asteroidenone 3'-Hydroxy-β,β-caroten-4-one 3'-Hydroxy-4-oxo-β-carotene 4-Keto-3'-hydroxy-β-carotene	$C_{40}H_{54}O_2$	566.87
4'-Hydroxyechinenone	4'-Hydroxyechinenone 4-Hydroxy-4'-keto-β-carotene	→ 4-Hydroxy-4'-keto-β-carotene		
3-Hydroxy-5,8-epoxy-β-carotene	Cryptoflavin Cryptoxanthin-5,8-epoxide	→ Cryptoflavin		
"3'-Hydroxyeuglenanone" 3-Hydroxy-3'-hydroxy-α-carotene	Dehydroadonixanthin Lutein "Xanthophyll"	→ Dehydroadonixanthin → Lutein		
3-Hydroxy-3'-keto-α-carotene	3-Hydroxy-3'-keto-α-carotene Philosamiaxanthin	3'-Dehydrolutein 3-Hydroxy-β,ε-caroten-3'-one 3-Hydroxy-3'-keto-α-carotene Philosamiaxanthin	$C_{40}H_{54}O_2$	566.87
3-Hydroxy-4-keto-β-carotene	Hydroxyechinenone	→ Hydroxyechinenone		

Table I.3 (continued)
NAME LIST FOR CAROTENOIDS (TABLES I.5, I.6, I. PC, I. TLC, I. LC)

Pigment	Names mentioned in Tables I.5 through I. LC	All synonyms	Sum formula	Mol wt	Structure
4-Hydroxy-4'-keto-β-carotene	4'-Hydroxyechinenone	4'-Hydroxy-β,β-caroten-4-one	$C_{40}H_{54}O_2$	566.87	
1'-Hydroxy-2'-keto-1',2'-dihydrotorulene	4-Hydroxy-4'-keto-β-carotene	4'-Hydroxyechinenone			
19-Hydroxylutein	2'-Dehydroplectaniaxanthin	→ 2'-Dehydroplectaniaxanthin			
16-Hydroxylycopene	Pyrenoxanthin "Trollein"	→ "Trollein"*			
3-Hydroxy-3'-methoxy-α-carotene	Lycoxanthin	→ Lycoxanthin			
1'-Hydroxy-1-methoxy-3,4-didehydro-1,2,1',2',7',8'-hexahydro-ψ,ψ-caroten-2-one	3-Hydroxy-3'-methoxy-α-carotene	—	$C_{41}H_{58}O_2$	582.91	
1'-Hydroxy-1-methoxy-3,4,3',4'-tetradehydro-1,2,1',2'-tetrahydro-ψ,ψ-caroten-2-one	"OH-R" OH-Spheroidenone	→ OH-Spheroidenone			
1'-Hydroxy-1-methoxy-1,2,1',2'-tetrahydro-ψ,ψ-caroten-4-one	2-Ketorhodovibrin	→ 2-Ketorhodovibrin			
2'-(4-Hydroxy-3-methyl-2-butenyl)-2-(3-methyl-2-butenyl)-ε,ε-caroten-18-ol	1-methoxy-1'-hydroxy-1,2,1',2'-tetrahydro-ψ,ψ-caroten-4-one	→ 1-methoxy-1'-hydroxy-1,2,1',2'-tetrahydro-ψ,ψ-caroten-4-one			
4-Hydroxymyxoxanthophyll	1'-Hydroxy-1-methoxy-1,2,1',2'-tetrahydro-ψ,ψ-caroten-4-one				
Hydroxyneurosporene	Thiothece-OH-484				
3'-Hydroxy-4-oxo-β-carotene	Sarcinaxanthin	→ Sarcinaxanthin			
	Aphanizophyll	→ Aphanizophyll			
	Chloroxanthin	→ Chloroxanthin			
	OH-Neurosporene				
	3'-Hydroxyechinenone	→ 3'-Hydroxyechinenone			
	4-Keto-3'-hydroxy-β-carotene				

584.88

Entry	Synonyms	Cross-references
3-Hydroxy-4-oxo-2,3-dehydro-β-carotene	Dehydrohydroxyechinenone "Euglenanone"	→ Dehydrohydroxyechinenone
Hydroxyphytofluene	OH-Phytofluene Phytofluenol	→ OH-Phytofluene
2-Hydroxyplectaniaxanthin	2-Hydroxyplectaniaxanthin $C_{40}H_{56}O_3$ 3',4'-Didehydro-1',2'-dihydro-β,ψ-carotene-2,1',2'-triol Rhodoauranxanthin	
"Hydroxy-R"	"OH-R"	
	OH-Spheroidenone	→ OH-Spheroidenone
	Triphasiaxanthin	→ Triphasiaxanthin
3'-Hydroxy-5,6-seco-β,β-carotene-5,6-dione		
3-Hydroxysemi-β-carotenone	Triphasiaxanthin	→ Triphasiaxanthin
Hydroxyspheroidene	7',8'-Dihydrorhodovibrin OH-Spheroidene OH-Y	→ OH-Y
"Hydroxyspheroidenone"	OH-R OH-Spheroidenone	→ OH-Spheroidenone
Hydroxyspirilloxanthin	Hydroxyspirilloxanthin OH-Spheroidenone	→ OH-Spirilloxanthin
1-Hydroxy-1,2,7',8'-tetrahydrolycopene	Chloroxanthin	→ Chloroxanthin
3-Hydroxytorulene	OH-Neurosporene Anhydrosaproxanthin Celaxanthin	→ Anhydrosaproxanthin
Isocarotene	"Dehydro-β-carotene" Dehydro-retro-β-carotene Retrodehydro-carotene Retrodehydro-β-carotene	→ "Dehydro-β-carotene"
Iso-ζ-carotene	7,8,11,12-Tetrahydro-ψ,ψ-carotene 7,8,11,12-Tetrahydrolycopene	→ Asym. ζ-carotene
Isocrocetin	Crocetin	→ Crocetin
Isocryptoxanthin	4-Hydroxy-β-carotene Isocryptoxanthin "Myxoxanthol"	→ 4-Hydroxy-β-carotene

Table I.3 (continued)
NAME LIST FOR CAROTENOIDS (TABLES I.5, I.6, I. PC, I. TLC, I. LC)

Pigment	Names mentioned in Tables I.5 through I. LC	All synonyms	Sum formula	Mol wt	Structure
Isofucoxanthin	Isofucoxanthin	3,5,3',5'-Tetrahydroxy-6',7'-didehydro-5,8,5',6'-tetrahydro-β,β-caroten-8-one-3'-acetate	$C_{42}H_{58}O_6$	658.92	
Isofucoxanthinol	Isofucozanthinol	Pentaxanthin ? 3,5,3',5'-Tetrahydroxy-6',7'-didehydro-5,8,5',6'-tetrahydro-β,β-caroten-8-one	$C_{40}H_{56}O_5$	616.88	
Isolutein	Isolutein Luteinepoxide Lutein-5,6-epoxide Taraxanthin Xanthophyllepoxide	→ Luteinepoxide			
Isomethylbixin	Methylbixin (*trans*)	→ Methylbixin (trans)			
Isorenieratene	Isorenieratene Leprotene	φ,φ-Carotene Leprotene	$C_{40}H_{48}$	528.82	
β-Isorenieratene	β-Isorenieratene	β,φ-Carotene	$C_{40}H_{52}$	532.85	
Isozeaxanthin	4,4'-Dihydroxy-β-carotene Isozeaxanthin	→ 4,4'-Dihydroxy-β-carotene			
4-Keto-α-carotene	4-Keto-α-carotene	β,ψ-Caroten-4-one	$C_{40}H_{54}O$	550.87	
4-Keto-β-carotene	Aphanin Echinenone 4-Keto-β-carotene Myxoxanthin	→ Echinenone			

4-Keto-γ-carotene 4-Keto-1′,2′-dihydro-1′-hydroxy-γ-carotene	4-Keto-γ-carotene 1′,2′-Dihydro-1′-hydroxy-4-keto-γ-carotene 4-Keto-1′,2′-dihydro-1′-hydroxy-γ-carotene 4-Keto-1′-hydroxy-1′,2′-dihydro-γ-carotene	β-ψ-Carotene-4-one → 4-Keto-1′-hydroxy-1′,2′-dihydro-γ-carotene	$C_{40}H_{54}O$	550.87	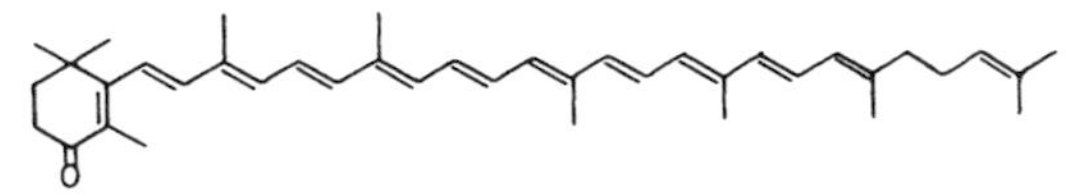
4-Keto-1′,2′-dihydro-1′-hydroxy-torulene	Deoxyflexixanthin 4-Keto-1′,2′-dihydro-1′-hydroxy-torulene	→ Deoxyflexixanthin			
"2-Keto-7′,8′-dihydrorhodovibrin"	"OH-R" OH-Spheroidenone	→ OH-Spheroidenone			
4-Keto-4′-ethoxy-β-carotene	4-Keto-4′-ethoxy-β-carotene	4′-Ethoxy-β,β-caroten-4-one 4′-Ethoxy-4-keto-β-carotene	$C_{42}H_{58}O_2$	594.92	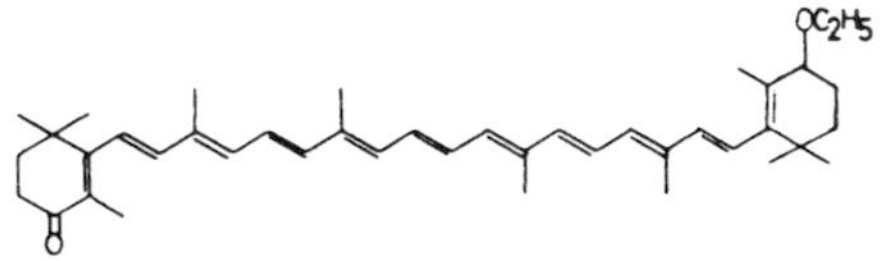
4-Keto-3′-hydroxy-β-carotene	3′-Hydroxyechinenone 4-Keto-3′-hydroxy-β-carotene	→ 3′-Hydroxyechinenone			
4-Keto-4′-hydroxy-β-carotene	4-Keto-4′-hydroxy-β-carotene	—	$C_{40}H_{54}O_2$	566.87	
4-Keto-1′-hydroxy-1′,2′-dihydro-γ-carotene	1′,2′-Dihydro-1′-hydroxy-4-keto-γ-carotene 4-Keto-1′,2′-dihydro-1′-hydroxy-γ-carotene 4-Keto-1′-hydroxy-1′,2′-dihydro-γ-carotene	1′,2′-Dihydro-1′-hydroxy-4-keto-γ-carotene 1′-Hydroxy-1′,2′-dihydro-β,ψ-caroten-4-one 4-Keto-1′,2′-dihydro-1′-hydroxy-γ-carotene 4-Keto-1′-hydroxy-1′,2′-dihydro-γ-carotene	$C_{40}H_{56}O_2$	568.88	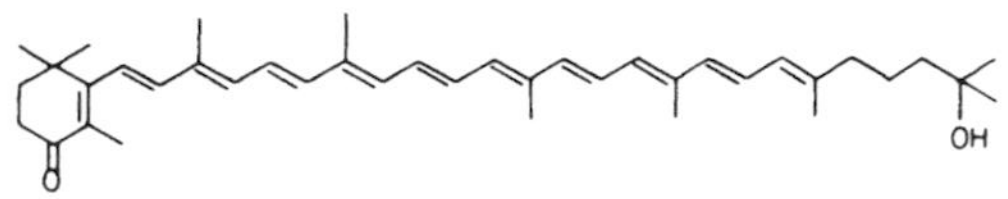
2-Keto-OH-spirilloxanthin	2-Ketorhodovibrin	→ 2-Ketorhodovibrin			
4-Ketophleixanthophyll	4-Ketophleixanthophyll	1′-(β,D-Glucopyranosyloxy)-2′-hydroxy-3′,4′-didehydro-1′,2′-dihydro-β,ψ,caroten-4-one	$C_{46}H_{64}O_8$	745.01	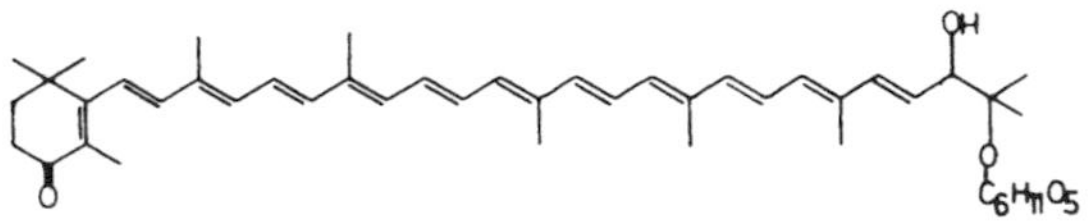

Table I.3 (continued)
NAME LIST FOR CAROTENOIDS (TABLES I.5, I.6, I. PC, I. TLC, I. LC)

Pigment	Names mentioned in Tables I.5 through I. LC	All synonyms	Sum formula	Mol wt	Structure
2-Ketorhodovibrin	2-Ketorhodovibrin	1'-Hydroxy-1-methoxy-3,4,3',4'-tetradehydro-1,2,1',2'-tetrahydro-ψ,ψ-caro-ten-2-one 2-Keto-OH-spirilloxanthin OH-P-511	$C_{41}H_{56}O_3$	596.89	
"2-Ketospirilloxanthin"	2,2'-Diketospirilloxanthin P-518	→ 2,2'-Diketospirilloxanthin			
4-Ketotorulene	Anhydro-deoxy-flexixanthin 4-Ketotorulene	→ Anhydro-deoxy-flexixanthin			
Leprotene	Isorenieratene Leprotene	→ Isorenieratene			
Loroxanthin	"Trollein" Pyrenoxanthin Torularhodin	→ "Trollein"*			
"Lusomycin"	Torularhodin	→ Torularhodin			
Lutein	3-Hydroxy-3'-hydroxy-α-carotene Lutein "Xanthophyll"	$(3R,3'R,6'R)$-β,ε-Carotene-3,3'-diol "Cucurbitaxanthin" 3,3'-Dihydroxy-α-carotene 3-Hydroxy-3'-hydroxy-α-carotene "Luteol" "Xanthophyll" → Helenien	$C_{40}H_{56}O_2$	568.88	
Lutein dipalmitate	Helenien Lutein dipalmitate				

Luteinepoxide	Isolutein Luteinepoxide Lutein-5,6-epoxide Taraxanthin Xanthophyllepoxide	3,3′-Dihydroxy-5,6-epoxy-α-carotene Eloxanthin (3S,5R,6S,3′R,6′R)-5,6-Epoxy-5,6-dihydro-β,ε-carotene-3,3′-diol Lutein-5,6-epoxide Taraxanthin Tareoxanthin	$C_{40}H_{56}O_3$	584.88
Lutein-5,6-epoxide	Isolutein Luteinepoxide Lutein-5,6-epoxide Taraxanthin Xanthophyllepoxide	→ Luteinepoxide		
Luteochrome	Luteochrome	5,6,5′,8′-Diepoxy-5,6,5′,8′-tetrahydro-β,β-carotene → Lutein	$C_{40}H_{56}O_2$	568.88
"Luteol"	3-Hydroxy-3′-hydroxy-α-carotene Lutein "Xanthophyll"			
Luteoxanthin	Luteoxanthin	5,6,5′,8′-Diepoxy-5,6,5′,8′-tetrahydro-β,β-carotene-3,3′-diol 3,3′-Dihydroxyluteochrome → Apo-6′-lycopenal	$C_{40}H_{56}O_4$	600.88
"Lycopenal"	Apo-2-lycopenal Apo-6′-lycopenal			
Lycopene	Lycopene Poly-*cis*-lycopene Prolycopene Rhodopurpurin Lycophyll Lycoxanthin	ψ,ψ-Carotene Poly-*cis*-lycopene Prolycopene Rhodopurpurin → Lycophyll → Lycoxanthin	$C_{40}H_{56}$	536.88
Lycopene-16,16′-diol Lycopene-16-ol				

Table I.3 (continued)
NAME LIST FOR CAROTENOIDS (TABLES I.5, I.6, I. PC, I. TLC, I. LC)

Pigment	Names mentioned in Tables I.5 through I. LC	All synonyms	Sum formula	Mol wt	Structure
Lycopersene	Lycopersene	7,8,11,12,15,7',8',11', 12',15'-Decahydro-ψψ-carotene; Decahydrolycopene; Dihydrophytoene	$C_{40}H_{66}$	546.96	
Lycophyll	Lycophyll	ψ,ψ-Carotene-16,16'-diol; "3,3'-Dihydroxylycopene"; Lycopene-16,16'-diol	$C_{40}H_{56}O_2$	568.88	
Lycoxanthin	Lycoxanthin	ψ,ψ-Caroten-16-ol; 16-Hydroxy-lycopene; Lycopen-16-ol	$C_{40}H_{56}O$	552.88	
1-Mannosyloxy-3,4-didehydro-1,2-dihydro-8'-apo-ψ-caroten-8'-ol	"1-Hexosyl-1,2-dihydro-3,4-didehydro-8'-apo-ψ-lycopenol"	→ "1-Hexosyl-1,2-dihydro-3,4-didehydro-apo-8'-lycopenol"			
1-Methoxy-3,4-didehydro-1,2-dihydro-ψ,ψ-carotene	Anhydrorhodovibrin	→ Anhydrorhodovibrin			
1-Methoxy-3,4-didehydro-1,2,7',8',11',12'-hexahydro-ψ,ψ-carotene	"P-481"; "P-412"	→ "P-412"			
1'-Methoxy-3',4'-didehydro-1,2,7,8,1',2'-hexahydro-ψ,ψ-caroten-1-ol	7',8'-Dihydrorhodovibrin; OH-Spheroidene; OH-Y	→ OH-Y			
1-Methoxy-3,4-didehydro-1,2,7',8'-tetrahydro-ψ,ψ-carotene	"P-450"; Spheroidene	→ "P-450"			
1'-Methoxy-3',4'-didehydro-1,2,1',2'-tetrahydro-ψ,ψ-caroten-1-ol	Rhodovibrin	→ Rhodovibrin			

582.91 $C_{41}H_{58}O_2$

586.94 $C_{41}H_{62}O_2$

600.92 $C_{41}H_{60}O_3$

1-Methoxy-3,4-didehydro-1,2,7',8'-tetrahydro-ψ,ψ-caroten-2-one	Pigment R Spheroidenone	→ Spheroidenone
1'-Methoxy-1',2'-dihydro-β,ψ-caroten-4'-one	Thiothece 474	→ Thiothece 474
1'-Methoxy-1',2'-dihydro-χ,ψ-caroten-4'-one	Okenone	→ Okenone
1'-Methoxy-1',2'-dihydro-φ,ψ-caroten-4'-one (old)	Thiothece-474	→ Thiothece-474
1-Methoxy-1,2-dihydro-ψ,ψ-caroten-4-one	1-Methoxy-1,2-dihydro-ψ,ψ-caroten-4-one	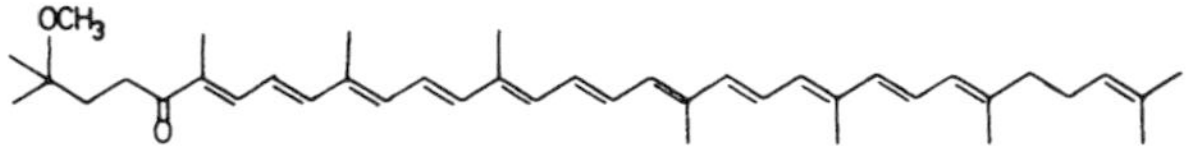 —
1-Methoxy-1,2-dihydro-3,4-dehydrolycopene	Anhydrorhodovibrin "P-481" "P-450" Spheroidene	→ Anhydrorhodovibrin
"4-Methoxy-5,6-dihydrolycopene"		→ "P-450"
1-Methoxy-1,2,7',8',11',12'-hexahydro-ψ,ψ-caroten-4-one	1-Methoxy-1,2,7',8',11',12'-hexahydro-ψ,ψ-caroten-4-one	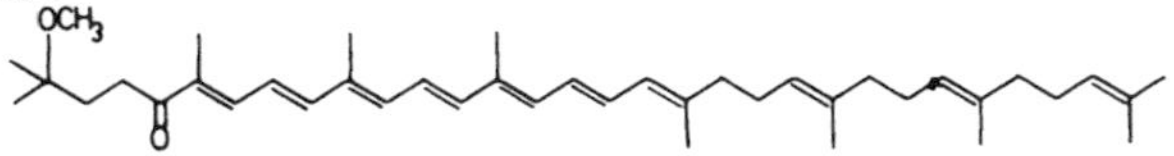 —
"1-Methoxy-1'-hydroxy-1,2,1',2'-tetrahydro-ψ,ψ-caroten-4-one"	"1-Methoxy-1'-hydroxy-1,2,1',2'-tetrahydro-ψ,ψ-caroten-4-one" Thiothece-OH-484	1'-Hydroxy-1-methoxy-1,2,1',2'-tetrahydro-ψ,ψ-caroten-4-one "OH-Okenone" Thiothece-OH-484 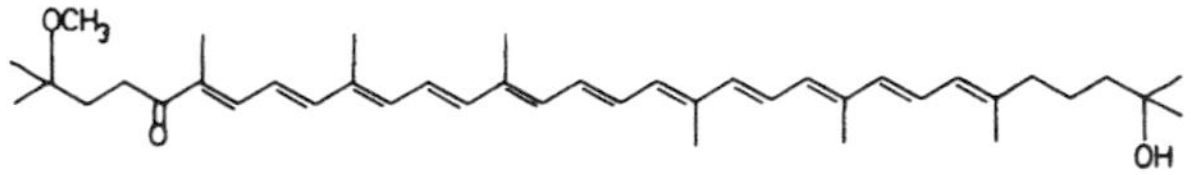
1-Methoxy-2-keto-7',8'-dihydro-3,4-dehydrolycopene	Pigment R Spheroidenone	→ Spheroidenone
1-Methoxy-4-oxo-1,2-dihydro-8'-apo-ψ-caroten-8'-al	Thiothece-460	→ Thiothece-460

Table I.3 (continued)
NAME LIST FOR CAROTENOIDS (TABLES I.5, I.6, I. PC, I. TLC, I. LC)

Pigment	Names mentioned in Tables I.5 through I. LC	All synonyms	Sum formula	Mol wt	Structure
1-Methoxy-1,2,7′,8′-tetrahydro-ψ,ψ-caroten-4-one	1-Methoxy-1,2,7′,8′-tetrahydro-ψ,ψ-caroten-4-one	—	$C_{41}H_{60}O_2$	584.92	
1-Methoxy-1,2,7′,8′-tetrahydro-3,4-dehydrolycopene	"P-450"	→ "P-450"			
1′-Methoxy-3,4,3′,4′-tetradehydro-1,2,1′,2′-tetrahydro-ψ,ψ-caroten-1-ol	Spheroidene				
	Hydroxyspirilloxanthin	→ OH-Spirilloxanthin			
	OH-Spirilloxanthin				
Methyl-4′-apo-β-caroten-4′-oate	Neurosporaxanthin-methylester	→ Neurosporaxanthin-methylester			
Methyl-6′-apo-ψ-caroten-6′-oate	Methyl-apo-6′-lycopenoate	→ Methyl-apo-6′-lycopenoate			
Methyl-apo-6′-lycopenoate	Methyl-6′-apo-ψ-caroten-6′-oate	→ Methyl-6′-apo-ψ-caroten-6′-oate	$C_{33}H_{44}O_2$	472.71	
Methylbixin (*trans*)	Methylbixin (*trans*)	Dimethyl-6,6′-diapocarotene-6,6′-dioate Isomethylbixin	$C_{26}H_{33}O_4$	409.54	
Methyl-3′,4′-didehydro-β,ψ-caroten-16′-oate	Torularhodin-methylester	→ Torularhodin-methylester			
"Methyl-1-hexosyl-1,2-dihydro-3,4-didehydroapo-8′-lycopenoate"	"Methyl-1-hexosyl-1,2-dihydro-3,4-didehydroapo-8′-lycopenoate"	Methyl-1-mannosyloxy-3,4-didehydro-1,2-dihydro-8′-apo-ψ-caroten-8′-oate	$C_{37}H_{52}O_8$	624.81	
Methyl-1-mannosyloxy-3,4-didehydro-1,2-dihydro-8′-apo-ψ-caroten-8′-oate	"Methyl-1-hexosyl-1,2-dihydro-3,4-didehydroapo-8′-lycopenoate"	→ "Methyl-1-hexosyl-1,2-dihydro-3,4-didehydroapo-8′-lycopenoate"			
Methyl-1′-methoxy-4′-oxo-1′,2′-dihydro-χ,ψ-caroten-16(or 17 or 18)-oate	Thiothece-484	→ Thiothece-484			

Name	Synonyms	Formula	MW
2'-(O-Methyl-5-C-methylpentosyloxy)-3',4'-didehydro-1',2'-dihydro-β,ψ-carotene-3,1'-diol "Metridene"	Myxol-2'-O-methyl-methylpentoside → Myxol-2'-O-methyl-methyl-pentoside		
	3-Hydroxy-4,4'-diketo-β-carotene → 3-Hydroxy-4,4'-diketo-β-carotene		
	4,4'-Diketo-3-hydroxy-β-carotene		
Micronone	Micronone —		
Microxanthin	Microxanthin —		
Monadoxanthin	Monadoxanthin 7,8-Didehydro-β,ε-carotene-3,3'-diol	$C_{40}H_{54}O_2$	566.87
3,4-Monodehydro-β-carotene	3,4-Dehydro-β-carotene "Dehydrocarotene II"	$C_{40}H_{54}$	534.87
	3,4-Dehydro-β-carotene		
	3,4-Dehydro-β,β-carotene		
	3,4-Monodehydro-β-carotene		
Monodehydrolycopene	3,4-Dehydrolycopene → 3,4-Dehydrolycopene		
Monodemethylated spirilloxanthin	Hydroxyspirilloxanthin		
	OH-Spirilloxanthin → OH-Spirilloxanthin		
5,6-Monoepoxy-α-carotene	5,6-Monoepoxy-α-carotene → α-Caroten-epoxide		
5,6-Monoepoxy-β-carotene	β-Carotene-monoepoxide → β-Carotene-monoepoxide		
	5,6-Monoepoxy-β-carotene		
5,6-Monoepoxylutein	5,6-Epoxy-lutein → Luteinepoxide		
Flavacin	Flavacin → Flavacin		
Mutatochrome	Mutatochrome		
Mutatoxanthin	Mutatoxanthin 3,3'-Dihydroxy-5,8-epoxy-β-carotene	$C_{40}H_{56}O_3$	584.88
	5,8-Epoxy-5,8-dihydro-β,β-carotene-3,3'-diol		
	5,8-Epoxyzeaxanthin		
	Zeaxanthin-5,8-epoxide		
	Zeaxanthinfuranoide		

Table I.3 (continued)
NAME LIST FOR CAROTENOIDS (TABLES I.5, I.6, I. PC, I. TLC, I. LC)

Pigment	Names mentioned in Tables I.5 through I. LC	All synonyms	Sum formula	Mol wt	Structure
Myxobactin	Myxobactin	1′,2′-Dihydro-1′-glucosyl-3,4-dehydrotorulene 1′,2′-Dihydro-1′-hydroxy-3,4-dehydrotorulene-glucoside 1′-Glucosyloxy-3,4,3′,4′-tetradehydro-1′,2′-dihydro-β,ψ-carotene	$C_{46}H_{64}O_6$	713.01	
Myxobactone	Myxobactone	1′,2′-Dihydro-1′-glucosyl-4-ketotorulene 1′,2′-Dehydro-1′-hydroxy-4-ketotorulene-glucoside 1′-Glucosyloxy-3′,4′-didehydro-1′,2′-dihydro-β,ψ-caroten-4-one O-Hexosyl-4-keto-1′-hydroxy-1′,2′-dihydro-γ-carotene	$C_{46}H_{64}O_7$	729.01	
Myxol-2′-O-methyl-methylpentoside	Myxol-2′-O-methylpentoside P-476	2′-(O-Methyl-5-C-methylpentosyloxy-3′,4′-didehydro-1′,2′-dihydro-β,ψ-carotene-3,1′-diol P-476	$C_{47}H_{68}O_7$	745.05	
Myxol-2′-rhamnoside Myxoxanthin	Myxoxanthophyll Aphanin Echinenone 4-Keto-β-carotene Myxoxanthin	→ Myxoxanthophyll →Echinenone			
"Myxoxanthol"	4-Hydroxy-β-carotene Isocryptoxanthin "Myxoxanthol"	→ 4-Hydroxy-β-carotene			

Name	Semisystematic name / synonyms	Molecular formula	M.W.	Structure
Myxoxanthophyll	1',2'-Dihydro-3',4'-didehydro-3,1'-dihydroxy-γ-caroten-2'-yl-rhamnoside Myxol-2'-rhamnoside 2'-(β,L-Rhamnopyranosyloxy)-3',4'-didehydro-1',2'-dihydro-β,ψ-carotene-3,1'-diol 2'-O-Rhamnosylmyxol	$C_{46}H_{66}O_7$	731.02	
Neo-β-carotene	→ β-Carotene β,β-Carotene			
Neochrome Trollichrome	5',8'-Epoxy-6,7-didehydro-5,6,5',8'-tetrahydro-β,β-carotene-3,5,3'-triol Foliachrome Fucochrome Trollichrome* Trolliflavin	$C_{40}H_{56}O_4$	600.88	
Neofucoxanthin A,B Neoluteoxanthin U	— —			
Neoxanthin	(3S,5R,6R,3'S,5'R,6'S)-5',6'-Epoxy-6,7-didehydro-5,6,5',6'-tetrahydro-β,β-carotene-3,5,3'-triol Foliaxanthin "3,3',5'-Trihydroxy-5',6'-dihydro-5',6'-epoxy-β-carotene" Trolliflor	$C_{40}H_{56}O_4$	600.88	
Neoxanthin-3-acetate	→ Dinoxanthin			
Neurosporaxanthin	β-Apo-4'-carotenoic acid 4'-Apo-β-caroten-4-oic acid	$C_{35}H_{46}O_2$	498.75	
Neurosporaxanthin-methylester	Methyl-4'-apo-β-caroten-4'-oate	$C_{36}H_{48}O_2$	512.77	

Table I.3 (continued)
NAME LIST FOR CAROTENOIDS (TABLES I.5, I.6, I. PC, I. TLC, I. LC)

Pigment	Names mentioned in Tables I.5 through I. LC	All synonyms	Sum formula	Mol wt	Structure
Neurosporene	Flavorhodin Neurosporene Proneurosporene "Protetrahydrolycopene" "Tetrahydrolycopene" "5,6,5',6'-Tetrahydrolycopene"	φ-Carotene "ψ-Carotene" 7,8-Dihydro-ψ,ψ-carotene 7,8-Dihydrolycopene Flavorhodin "Poly-cis-ψ-carotene" Proneurosporene "Protetrahydrolycopene" "Tetrahydrolycopene" "5,6,5',6'-Tetrahydrolycopene"	$C_{40}H_{58}$	538.90	
Nostoxanthin	Nostoxanthin	3,3'-Dihydroxy-5,5'-dihydro-7,7'-didehydro-β-carotene 6,7,6',7'-Tetradehydro-5,6,5',6'-tetrahydro-β,β-caro-tene-3,3'-diol	$C_{40}H_{56}O_2$	568.88	
15-cis-7,8,11,12,7',8',11',12'-Octahydro-ψ,ψ-carotene	Phytoene	→ Phytoene			
1,2,7,8,1',2',7',8'-Octa-hydro-ψ,ψ-carotene-1,1'-diol	1,1'-Dihydroxy-1,2,1',2'-tetrahydro-ζ-carotene	→ 1,1'-Dihydroxy-1,2,1',2'-te-trahydro-ζ-carotene			
1,2,7,8,7',8',11',12'-Octa-hydro-ψ,ψ-caroten-1-ol	OH-Phytofluene Phytofluenol	→ OH-Phytofluene			
1,2,7,8,11,12,7',8'-Octa-hydro-ψ,ψ-caroten-1-ol	OH-Phytofluene Phytofluenol	→ OH-Phytofluene			
7,8,11,12,7',8',11',12'-Oc-tahydro-4,4'-diapocarotene	4,4'-Diapophytoene	→ 4,4'-Diapophytoene			
"Octahydrolycopene"	Phytofluene ζ-Carotene	→ Phytofluene → ζ-Carotene			
5,6,7,8,5',6',7',8'-Octahydrolycopene	7,8,7',8'-Tetrahydro-ψ,ψ-carotene	→ 7,8,7',8'-Tetrahydro-ψ,ψ-carotene			
7,8,11,12,7',8',11',12'-Octahydrolycopene	Phytoene	→ Phytoene			

			M.W.	Formula
OH-ζ-Carotene	OH-ζ-Carotene	—		
OH-Chlorobactene	OH-Chlorobactene	1′,2′-Dihydro-φ,ψ-caroten-1′-ol 1′,2′-Dihydro-1′-hydroxychlorobactene	550.87	$C_{40}H_{54}O$
OH-Lycopene	OH-Lycopene Rhodopin	"Bacterioerythrin β" 1,2-Dihydro-ψ,ψ-caroten-1-ol 1,2-Dihydro-1-OH-lycopene 1-Hydroxy-1,2-dihydrolycopene Rhodopin	554.90	$C_{40}H_{58}O$
OH-Neurosporene	Chloroxanthin OH-Neurosporene	→ Chloroxanthin		
"OH-Okenone"	"1-Methoxy-1′-hydroxy-1,2,1′,2′-tetrahydro-ψ,ψ-caroten-4-one" Thiothece-OH-484	"1-Methoxy-1′-hydroxy-1,2,1′,2′-tetrahydro-ψ,ψ-caroten-4-one"		
"OH P-481"*	3,4-Dehydrorhodopin Rhodovibrin OHP-481	→ 3,4-Dehydrorhodopin → Rhodovibrin		
"OH P-482" OH P-511	3,4-Dehydrorhodopin 2-Ketorhodovibrin	→ 3,4-Dehydrorhodopin → 2-Ketorhodovibrin		
OH-Phytofluene	1-Hydroxy-1,2-dihydrophytofluene Phytofluenol	1-Hydroxy-1,2-dihydrophytofluene 1′-Hydroxy-1′,2′-dihydrophytofluene Hydroxyphytofluene	560.95	$C_{40}H_{64}O$
OH-Pigment Y	7′,8′-Dihydrorhodovibrin OH-Spheroidene OH-Y	1,2,7,8,7′,8′,11′,12′-Octahydro-ψ,ψ-caroten-1-ol → OH-Y		
"OH-R"	"OH-R" OH-Spheroidenone	→ OH-Spheroidenone		
OH-Rhodopin	"Dihydroxylycopene" 1,2,1′,2′-Tetrahydro-1,1′-dihydroxy-lycopene	"1,2,1′,2′-Tetrahydro-1,1′-dihydroxy-lycopene"		
OH-Spheroidene	7′,8′-Dihydrorhodovibrin OH-Spheroidene OH-Y	→ OH-Y		

Table I.3 (continued)
NAME LIST FOR CAROTENOIDS (TABLES I.5, I.6, I. PC, I. TLC, I. LC)

Pigment	Names mentioned in Tables I.5 through I. LC	All synonyms	Sum formula	Mol wt	Structure
OH-Spheroidenone	"Hydroxyspheroidenone" OH-Spheroidenone "OH-R"	1',2'-Dihydro-1'-hydroxyspheroidenone 1'-Hydroxy-1',2'-dihydrospheroidenone 1'-Hydroxy-1-methoxy-3,4-didehydro-1,2,1',2',7',8'-hexahydro-ψ,ψ-caroten-2-one "Hydroxy-R" "Hydroxyspheroidenone" "2-Keto-7',8'-dihydrorhodovibrin" "OH-R"	$C_{41}H_{60}O_3$	600.92	
OH-Spirilloxanthin	Hydroxyspirilloxanthin OH-Spirilloxanthin	"α-Bacterioruberin-monomethyl-ether" Dehydrorhodovibrin Hydroxyspirilloxanthin 1'-Methoxy-3,4,3',4'-tetradehydro-1,2,1',2'-tetrahydro-ψ,ψ-caroten-1-ol Monodemethylated spirilloxanthin	$C_{41}H_{58}O_2$	582.91	
OH-Y	7',8'-Dihydrorhodovibrin OH-Spheroidene OH-Y	7',8'-Dihydrorhodovibrin 1'-Hydroxy-1',2'-dihydrospheroidene Hydroxyspheroidene 1'-Methoxy-3',4'-didehydro-1,2,7,8,1',2'-hexahydro-ψ,ψ-caroten-1-ol OH-Pigment Y OH-Spheroidene	$C_{41}H_{62}O_2$	586.94	

Okenone	Okenone	1′-Methoxy-1′,2′-dihydro-χ,ψ-caroten-4′-one	$C_{41}H_{54}O_2$	578.88 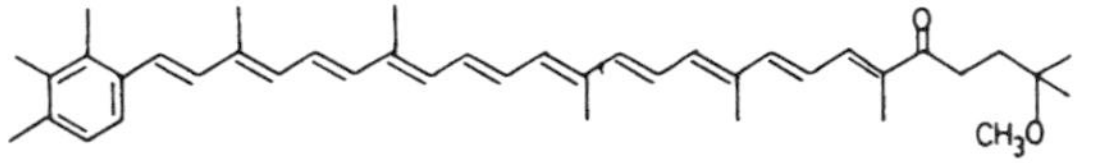
2′-(*O*-Methyl-5-*C*-methylpentosyloxy)-3′,4′-didehydro-1′,2′-dihydro-β,ψ-carotene-3,1′-diol	Myxol-2′-*O*-methyl-methylpentoside P-476	→ Myxol-2′-*O*-methyl-methylpentoside		
Oscillaxanthin	Oscillaxanthin	2,2′-Bis-(β,L-rhamnopyranosyloxy)-3,4,3′,4′-tetradehydro-1,2,1′,2′-tetrahydro-ψ,ψ-carotene-1,1′-diol "1,1′-Dihydroxy-2,2′-di-β-L-rhamnosyl-1,2,1′,2′-tetrahydro-3,4,3′,4′-tetradehydrolycopene" Oscillol-2,2′-dirhamnoside	$C_{52}H_{76}O_{12}$	893.17 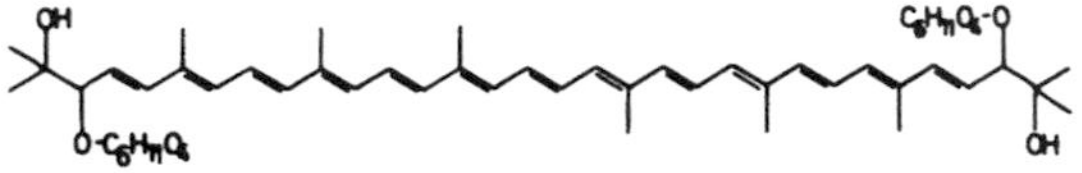
Oscillol-2,2′-di-(*O*-methyl-methylpentoside)	Oscillol-2,2′-di-(*O*-methyl-methylpentoside)	2,2′-Bis-(*O*-methyl-5-C-methyl-pentosyloxy)-3,4,3′,4′-tetradehydro-1,2,1′,2′-tetrahydro-ψ,ψ-carotene-1,1′-diol P-496	$C_{54}H_{80}O_{12}$	921.22 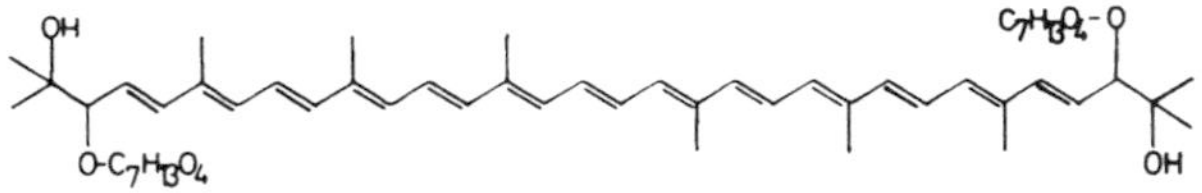
Oscillol-2,2′-dirhamnoside	Oscillaxanthin	→ Oscillaxanthin		
"Ovoester"	Astaxanthin	→ Astaxanthin		
3-Oxocanthaxanthin	Dehydroadonirubin	→ Dehydroadonirubin		
3-Oxoechinenone	Dehydrohydroxyechinenone Euglenanone	→ Dehydrohydroxyechinenone		
4-Oxosaproxanthin	Flexixanthin	→ Flexixanthin		
16′-Oxotorulene	Torularhodin-aldehyde	— Torularhodin-aldehyde		
"P-412"	"P-412"	11′,12′-Dihydrospheroidene 1-Methoxy-3,4-didehydro-1,2,7′,8′,11′,12′-hexahydro-ψ,ψ-carotene	$C_{41}H_{62}O$	570.94 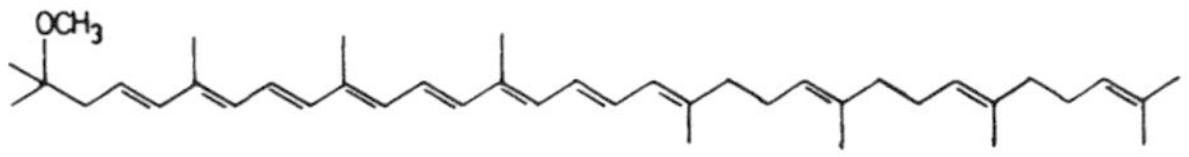

Table I.3 (continued)
NAME LIST FOR CAROTENOIDS (TABLES I.5, I.6, I. PC, I. TLC, I. LC)

Pigment	Names mentioned in Tables I.5 through I. LC	All synonyms	Sum formula	Mol wt	Structure
"P-450"	"P-450" Spheroidene	1-Methoxy-3,4-didehydro-1,2,7',8'-tetrahydro-ψ,ψ-carotene "4-Methoxy-5,6-dihydrolycopene" 1-Methoxy-1,2,7',8'-tetrahydro-3,4-dehydrolycopene "Pigment Y" Spheroidene	$C_{41}H_{60}O$	568.92	
P-476	Myxol-2'-O-methyl-methylpentoside P-476	→ Myxol-2'-O-methyl-methylpentoside			
"P-481"	Anhydrorhodovibrin "P-481"	→ Anhydrorhodovibrin			
P-496	P-496	→ Oscillol-2,2'-di-(O-methyl-methylpentoside)			
"P-512"	2,2'-Diketospirilloxanthin P-518	→ 2,2'-Diketospirilloxanthin			
P-518	2,2'-Diketospirilloxanthin P-518	→ 2,2'-Diketospirilloxanthin			
Pectenoxanthin	Alloxanthin Cynthiaxanthin Isofucoxanthinol	→ Alloxanthin			
Pentaxanthin		→ Isofucoxanthinol			
Peridinin	Peridinin	5',6'-Epoxy-3,5,3'-trihydroxy-6,7-didehydro-5,6,5',6'-tetrahydro-10,11,20-trinor-β,β-caroten-19',11'-olide-3-acetate Sulcatoxanthin	$C_{39}H_{50}O_7$	630.82	

Name	Synonyms/Systematic name	Formula	M.W.
Peridininol	5',6'-Epoxy-3,5,3'-trihydroxy-6,7-didehydro-5,6,5',6'-tetrahydro-10,11,20-trinor-β,β-caroten-19',11'-olide	$C_{37}H_{48}O_6$	588.78
Phillipsiaxanthin	"2,2'-Diketobacterioruberin" 1,1'-Dihydroxy-2,2'-diketo-1,2,1',2'-tetrahydro-3,4,3',4'-tetradehydrolycopene 1,1'-Dihydroxy-3,4,3',4'-tetradehydro-1,2,1',2'-tetrahydro-ψ,ψ-carotene-2,2'-dione "2,2'-Diketobacterioruberin"	$C_{40}H_{52}O_4$	596.85
Philosamiaxanthin	3-Hydroxy-3'-keto-α-carotene → 3-Hydroxy-3'-keto-α-carotene		
Phleixanthophyll	1'-(β,D-Glucopyranosyloxy)-3',4'-didehydro-1',2'-dihydro-β,ψ-caroten-2'-ol	$C_{46}H_{66}O_7$	731.02
Phoeniconone Phoenicopterone Phoenicoxanthin*	→ Dehydroadonirubin → 4-Keto-α-carotene → Dehydroadonirubin 4,4'-Diketo-3-hydroxy-β-carotene 3-Hydroxy-4,4'-diketo-β-carotene		
Physalien	(3R,3'R)-β,β-Carotene-3,3'-diol dipalmitate Zeaxanthin dipalmitate	$C_{72}H_{116}O_4$	1045.71
Physoxanthin*	→ α-Cryptoxanthin* → β-Cryptoxanthin		

Table I.3 (continued)
NAME LIST FOR CAROTENOIDS (TABLES I.5, I.6, I. PC, I. TLC, I. LC)

Pigment	Names mentioned in Tables I.5 through I. LC	All synonyms	Sum formula	Mol wt	Structure
Phytoene	Phytoene	Dehydrogenans-phytoene 15,15'-Dehydrolycopersene "Hexadecahydrolycopene" 15-cis-7,8,11,12,7',8',11',12'-Octahydro-ψ,ψ-carotene 7,8,11,12,7',8',11',12'-Octahydrolycopene	$C_{40}H_{64}$	544.95	
Phytofluene	Phytofluene	"Decahydro-β-carotene" Dehydrogenans-Phytofluene 11,12-Dehydrophytoene Dodecahydrolycopene 15-cis-7,8,11,12,7',8'-hexa-hydro-ψ,ψ-carotene 7,8,11,12,7',8'-hexahydrolycopene "Octahydrolycopene"	$C_{40}H_{62}$	542.93	
Phytofluenol	OH-Phytofluene Phytofluenol	→ OH-Phytofluene			
Pigment R	Pigment R Spheroidenone	→ Spheroidenone			
"Pigment X"	β-Zeacarotene β,-Zeacarotene	→ β-Zeacarotene			
"Pigment Y"	"P-450" Spheroidene	→ "P-450"			
Plectaniaxanthin	Plectaniaxanthin	3',4'-didehydro-1',2'-dihydro-β,ψ-carotene-1',2'-dihydroxy-1',2'-dihydrotorulene	$C_{40}H_{56}O_2$	568.88	
Poly-cis-γ-carotene	β,ψ-Carotene γ-Carotene Pro-γ-carotene	→ γ-Carotene			

"Poly-cis-ψ-carotene"	"Flavorhodin"	→ Neurosporene
	Neurosporene	
	Proneurosporene	
	"Protetrahydrolycopene"	
	"Tetrahydrolycopene"	
	"5,6,5',6'-tetrahydrolycopene"	
Polycislycopene	Lycopene	→ Lycopene
	Poly-cislycopene	
	Prolycopene	
	Rhodopurpurin	
"Pro-γ-carotene"	β,ψ-Carotene	→ γ-Carotene
	γ-Carotene	
	Pro-γ-carotene	
Prolycopene	Lycopene	→ Lycopene
	Poly-cis-lycopene	
	Prolycopene	
	Rhodopurpurin	
Proneurosporene	"Flavorhodin"	→ Neurosporene
	Neurosporene	
	Proneurosporene	
	"Protetrahydrolycopene"	
	"Tetrahydrolycopene"	
	"5,6,5',6'-Tetrahydrolycopene"	
"Protetrahydrolycopene"	"Flavorhodin"	→ Neurosporene
	Neurosporene	
	Proneurosporene	
	"Protetrahydrolycopene"	
	"Tetrahydrolycopene"	
	"5,6,5',6'-Tetrahydrolycopene"	
Pseudo-α-carotene	β-Carotene	→ β-Carotene
	β,β-Carotene	
Pyrenoxanthin*	"Trollein"	→ "Trollein"*
		β,ε-Carotene-3,20,3'-triol

$C_{40}H_{56}O_3$

584.88

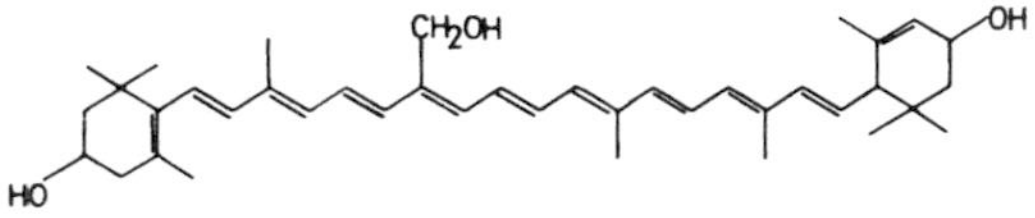

Table I.3 (continued)
NAME LIST FOR CAROTENOIDS (TABLES I.5, I.6, I. PC, I. TLC, I. LC)

Pigment	Names mentioned in Tables I.5 through I. LC	All synonyms	Sum formula	Mol wt	Structure
Pyrrhoxanthin	Pyrrhoxanthin	5',6'-Epoxy-3,3'-dihydroxy-7,8-didehydro-5',6'-dihydro-10,11,20-trinor-β,β-caroten-19',11'-olide-3-acetate	$C_{39}H_{48}O_6$	612.80	
Pyrrhoxanthinol	Pyrrhoxanthinol	5',6'-Epoxy-3,3'-dihydroxy-7,8-didehydro-5',6'-dihydro-10,11,20-trinor-β,β-caroten-19',11'-olide	$C_{37}H_{46}O_5$	570.77	
Renierapurpurin	Renierapurpurin	χ,χ-Carotene	$C_{40}H_{48}$	528.82	
Renieratene	Renieratene	φ,χ-Carotene	$C_{40}H_{48}$	528.82	
Reticulataxanthin	Reticulataxanthin	3-Hydroxycitranaxanthin (3R)-3-Hydroxy-5',6'-dihydro-5'-apo-18'-nor-β-caroten-6'-one	$C_{33}H_{44}O_2$	472.71	
Retinylidenetiglic acid	β-Apo-12'-carotenoic acid 3,4,3',4'-Bisdehydro-β-carotene	→ β-Apo-12'-carotenoic acid → 3,4,3',4'-Bisdehydro-β-carotene			
Retrobisdehydro-β-carotene	''Dehydro-β-carotene'' Dehydro-retro-β-carotene Retrodehydrocarotene Retrodehydro-β-carotene				
Retrodehydrocarotene	''Dehydro-β-carotene'' Dehydro-retro-β-carotene Retrodehydrocarotene Retrodehydro-β-carotene	→ ''Dehydro-β-carotene''			
Retrodehydro-β-carotene		→ ''Dehydro-β-carotene''			

Retrodehydro-γ-carotene	Retrodehydro-γ-carotene	Didehydroretro-γ-carotene 4',5'-Didehydro-4,5'-retro-β,ψ-carotene	$C_{40}H_{54}$	534.87
"Retrodehydrozeaxanthin" 2'-(β,L-Rhamnopyranosyloxy-)3',4'-didehydro-1',2'-dihydro-β,ψ-carotene-3,1'-diol	Eschscholtzxanthin	→ Eschscholtzxanthin		
	Myxoxanthophyll	→ Myxoxanthophyll		
2'-(β,L-Rhamnopyranosyloxy-)3',4'-didehydro-1',2'-dihydro-β,ψ-carotene-3,4,1'-triol	Aphanizophyll	→ Aphanizophyll		
2'-O-Rhamnosylmyxol	Myxoxanthophyll	→ Myxoxanthophyll		
Rhodoauranxanthin	2-Hydroxyplectaniaxanthin	→ 2-Hydroxyplectaniaxanthin		
Rhodopin	OH-Lycopene	→ OH-Lycopene		
	Rhodopin			
Rhodopurpurin	Lycopene	Lycopene		
	Poly-cis-lycopene			
	Prolycopene			
	Rhodopurpurin			
"Rhodotorulene"	Torularhodin	→ Torularhodin		
Rhodovibrin	Rhodovibrin	1'-Methoxy-3',4'-didehydro-1,2,1',2'-tetrahydro-ψ,ψ-caroten-1-ol "OH-P-481"	$C_{41}H_{60}O_2$	584.92
Rhodoviolascin	Rhodoviolascin	Bacterioerythrin α Bacterioprupurin α "3,3'-Dimethoxy"-structure 1,1'-Dimethoxy-3,4,3',4'-tetradehydro-1,2,1',2'-tetrahydro-ψ,ψ-carotene "Spirillotoxin" Spirilloxanthin	$C_{42}H_{60}O_2$	596.93
	Spirilloxanthin			

Table I.3 (continued)
NAME LIST FOR CAROTENOIDS (TABLES I.5, I.6, I. PC, I. TLC, I. LC)

Pigment	Names mentioned in Tables I.5 through I. LC	All synonyms	Sum formula	Mol wt	Structure
Rhodoxanthin	Rhodoxanthin	4′,5′-Didehydro-4′,5′-retro-β,β-carotene-3,3′-dione "3,3′-Diketodehydro-β-carotene" "3,3′-Diketoretro-β-carotene" 3,3′-Diketoretrodehydro-β-carotene	$C_{40}H_{50}O_2$	562.83	
Rubichrome	Rubichrome	5,8-Epoxy-5,8-dihydro-β,ψ-caroten-3-ol "5,8-Epoxy-3-hydroxy-γ-carotene" "5,8-Epoxyrubixanthin"	$C_{40}H_{56}O_2$	568.88	
Rubixanthin	Rubixanthin	(3R-)β,ψ-Caroten-3-ol 3-Hydroxy-γ-carotene	$C_{40}H_{56}O$	552.88	
Rubixanthinepoxide	Rubixanthinepoxide	—	$C_{40}H_{56}O_2$	568.88	
Rubixanthin-5,6-epoxide	Rubixanthin-5,6-epoxide	5,6-Epoxy-5,6-dihydro-β,ψ-caroten-3-ol	$C_{40}H_{56}O_2$	568.88	
"Salmon acid" Saprospira grandis 434 Saprospira grandis 460 Saprospira grandis 500	Astacene S.g. 434 S.g. 460 S.g. 500	→ Astacene			
Saproxanthin	Saproxanthin	3′,4′-Didehydro-1′,2′-dihydro-β,ψ-carotene-3,1′-diol 1′,2′-Dihydro-3′,4′-dehydro-3,1′-dihydroxy-γ-carotene 3,1′-Dihydroxy-1′,2′-dihydrotorulene	$C_{40}H_{56}O_2$	568.88	

Sarcinaxanthin*	Sarcinaxanthin	(2R,6S,2'R,6'S)-2,2'-Bis-(4-hydroxy-3-methyl-2-butenyl)-γ,γ-carotene	C₅₀H₇₂O₂	705.12
		2'-(4-Hydroxy-3-methyl-2-butenyl)2-(3-methyl-2-butenyl)-ε,ε-caroten-18-ol		
		(2R,6R,2'R,6'R)-2,2'-Bis-(4-hydroxy-3-methyl-2-butenyl)-ε,ε-carotene		
		Decaprenoxanthin		
		Dehydrogenans-P-439		
5,6-Seco-β,β-carotene-5,6-dione	Semi-β-carotenone	→ Semi-β-carotenone		
(6'R)-5,6-Seco-β,ε-carotene-5,6-dione	Semi-α-carotenone	→ Semi-α-carotenone		
Semi-α-carotenone	Semi-α-carotenone	(6'R)-5,6-Seco-β,ε-carotene-5,6-dione	C₄₀H₅₆O₂	568.88
Semi-β-carotenone	Semi-β-carotenone	5,6-Seco-β,β-carotene-5,6-dione	C₄₀H₅₆O₂	568.88
S.g. 434	S.g. 434	→ Saprospira grandis 434		
S.g. 460	S.g. 460	→ S. grandis 460		
S.g. 500	S.g. 500	→ S. grandis 500		
Sintaxanthin	Sintaxanthin	7',8'-Dihydro-7'-apo-β-caroten-8'-one	C₃₁H₄₂O	430.67

Table I.3 (continued)
NAME LIST FOR CAROTENOIDS (TABLES I.5, I.6, I. PC, I. TLC, I. LC)

Pigment	Names mentioned in Tables I.5 through I. LC	All synonyms	Sum formula	Mol wt	Structure
Siphonaxanthin	Siphonaxanthin "Xanthophyll K₁S"	3,19,3'-Trihydroxy-7,8-dihydro-β,ε-caroten-8-one 3,3',19-Trihydroxy-7,8-dihydro-8-oxo-α-carotene "Xanthophyll K₁S" → Siphonein	$C_{40}H_{56}O_4$	600.88	
Siphonaxanthin-monolaurate	Siphonaxanthin-monolaurate Siphonein "Xanthophyll K₁"				
Siphonein	Siphonaxanthin-monolaurate "Xanthophyll K₁" Siphonein	Siphonaxanthin-monolaurate 3,19,3'-Trihydroxy-7,8-dihydro-β,ε-caroten-8-one-19-laurate "Xanthophyll K₁" → γ-Carotene	$C_{52}H_{78}O_5$	783.19	
"Sphaerobolin"	β,ψ-Carotene γ-Carotene Pro-γ-carotene				
Spheroidene	"P-450" Spheroidene	→ "P-450"			
Spheroidenone	Pigment R Spheroidenone	"5,6-Dihydro-4-methoxy-lycopen-6-one" 1-Methoxy-3,4-didehydro-1,2,7',8'-tetrahydro-ψ,ψ-caroten-2-one "1-Methoxy-2-keto-7',8'-dihydro-3,4-dehydrolycopene" Pigment R	$C_{41}H_{58}O_2$	582.91	
"Spirillotoxin"	Rhodoviolascin Spirilloxanthin	→ Rhodoviolascin			

$$C_{40}H_{48}O_4 \qquad 592.88$$

Spirilloxanthin	Rhodoviolascin Spirilloxanthin	→ Rhodoviolascin
Sulcatoxanthin	Peridinin	→ Peridinin
Taraxanthin	Isolutein Luteinepoxide Lutein-5,6-epoxide Taraxanthin Xanthophyllepoxide	→ Luteinepoxide
Tareoxanthin	Isolutein Luteinepoxide Lutein-5,6-epoxide Taraxanthin Xanthophyllepoxide	→ Luteinepoxide
7,8,7',8'-Tetradehydroastaxanthin	Part of asterinic acid	→ Asterinic acid 3,3'-Dihydroxy-7,8,7',8'-tetradehydro-β,β-carotene-4,4'-dione 4,4'-Diketotetradehydroxanthin Diketotetradehydrozeaxanthin
3,4,3',4'-Tetradehydro-β,β-carotene	3,4,3',4'-Bisdehydro-β-carotene	→ 3,4,3',4'-Bisdehydro-β-carotene
3,4,3',4'-Tetradehydro-ψ,ψ-carotene	"Dehydrolycopene"	→ "Dehydrolycopene"
(3R,3'R-)7,8,7',8'-Tetradehydro-β,β-carotene-3,3'-diol	Alloxanthin Cynthiaxanthin "Dehydrolycopene"	→ Alloxanthin → "Dehydrolycopene"
3,4,3',4'-Tetradehydrolycopene		
6,7,6',7'-Tetradehydro-5,6,5',6'-tetrahydro-β,β-carotene-3,3'-diol	Nostoxanthin	→ Nostoxanthin
7,8,7',8'-Tetradehydrozeaxanthin	Alloxanthin Cynthiaxanthin	→ Alloxanthin
7',8',11',12'-Tetrahydro-β,ψ-carotene	7',8',11',12'-Tetrahydro-γ-carotene	→ 7',8',11',12'-Tetrahydro-γ-carotene

Table I.3 (continued)
NAME LIST FOR CAROTENOIDS (TABLES I.5, I.6, I. PC, I. TLC, I. LC)

Pigment	Names mentioned in Tables I.5 through I. LC	All synonyms	Sum formula	Mol wt	Structure
7',8',11',12'-Tetrahydro-γ-carotene	7',8',11',12'-Tetrahydro-γ-carotene	7',8',11',12'-Tetrahydro-β,ψ-carotene	$C_{40}H_{60}$	540.91	
7,8,11,12-Tetrahydro-ψ,ψ-carotene	Asym. ζ-carotene 7,8,11,12-Tetrahydro-ψ,ψ-carotene 7,8,11,12-Tetrahydrolycopene	→ Asym. ζ-carotene			
7,8,7',8'-Tetrahydro-ψ,ψ-carotene	ζ-Carotene 7,8,7',8'-Tetrahydro-ψ,ψ-carotene	→ ζ-Carotene			
1,2,1',2'-Tetrahydro-ψ,ψ-carotene-1,1'-diol	"Dihydroxylycopene" 1,2,1',2'-Tetrahydro-1,1'-dihydroxylycopene	→ 1,2,1',2'-Tetrahydro-1,1'-dihydroxylycopene			
1,2,7',8'-Tetrahydro-4,4-caroten-1-ol	Chloroxanthin OH-Neurosporene	→ Chloroxanthin			
7,8,7',8'-Tetrahydro-4,4'-diapocarotene	4,4'-Diapo-ζ-carotene	→ 4,4'-Diapo-ζ-carotene			
7,8,11,12-Tetrahydro-4,4'-diapocarotene	4,4'-Diapo-7,8,11,12-tetrahydrolycopene	→ 4,4'-Diapo-7,8,11,12-tetrahydrolycopene			
1,2,1',2'-Tetrahydro-1,1'-dihydroxy-lycopene	"Dihydroxylycopene" 1,2,1',2'-Tetrahydro-1,1'-dihydroxy-lycopene	"Dihydroxylycopene" 1,1'-Dihydroxy-1,2,1',2'-tetrahydrolycopene OH-Rhodopin 1,2,1',2'-Tetrahydro-ψ,ψ-carotene-1,1'-diol 1,2,1',2'-Tetrahydrolycopene-1,1'-diol	$C_{40}H_{60}O_2$	572.91	

"Tetrahydrolycopene"	"Flavorhodin" Neurosporene Proneurosporene "Protetrahydrolycopene" "Tetrahydrolycopene" "5,6,5',6'-Tetrahydrolycopene"	→ Neurosporene
"5,6,5',6'-Tetra-hydrolycopene"	"Flavorhodin" Neurosporene Proneurosporene "Protetrahydrolycopene" "Tetrahydrolycopene" "5,6,5',6'-Tetrahydrolycopene"	→ Neurosporene
7,8,7',8'-Tetrahydrolycopene	ζ-Carotene 7,8,7',8'-Tetrahydro-ψ,ψ-carotene	→ ζ-Carotene
7,8,11,12-Tetra-hydrolycopene	Asym. ζ-carotene 7,8,11,12-Tetrahydro-ψ,ψ-carotene 7,8,11,12-Tetrahydrolycopene	→ Asym. ζ-carotene
"7',8',11',12'-Tetrahydrolycopene"	Asym. ζ-carotene 7,8,11,12-Tetrahydro-ψ,ψ-carotene 7,8,11,12-Tetrahydrolycopene	→ Asym. ζ-carotene
1,2,1',2'-Tetrahydrolyco-pene-1,1'-diol	"Dihydroxylycopene" 1,2,1',2'-Tetrahydro-1,1'-dihydroxylycopene	→ 1,2,1',2'-Tetrahydro-1,1'-dihydroxylycopene
3,5,3',5'-Tetrahydroxy-6',7'-didehydro-5,8,5',6'-tetrahy-dro-β,β-caroten-8-one	Isofucoxanthinol	→ Isofucoxanthinol
3,5,3',5'-Tetrahydroxy-6',7'-didehydro-5,8,5',6'-tetrahy-dro-β,β-caroten-8-one-3-acetate	Isofucoxanthin	→ Isofucoxanthin
3,4,3',4'-Tetraketo-β-carotene	Astacene	→ Astacene
"4,5,4',5'-Tetraketo-β-carotene"	Astacene	→ Astacene

Table I.3 (continued)
NAME LIST FOR CAROTENOIDS (TABLES I.5, I.6, I. PC, I. TLC, I. LC)

Pigment	Names mentioned in Tables I.5 through I. LC	All synonyms	Sum formula	Mol wt	Structure
Thiothece-460	*Thiothece-460*	1-Methoxy-4-oxo-1,2-dihydro-8'-apo-ψ-caroten-8'-al	$C_{31}H_{42}O_3$	462.67	
Thiothece-474	*Thiothece-474*	1'-Methoxy-1',2'-dihydro-β,ψ-caroten-4'-one 1'-Methoxy-1',2'-dihydro-φ,ψ-caroten-4'-one "Thiothece-478" → *Thiothece-474*	$C_{41}H_{58}O_2$	582.91	
"*Thiothece-478*"	*Thiothece-474*				
Thiothece-484	*Thiothece-484*	Methyl-1'-methoxy-4'-oxo-1',2'-dihydro-χ,ψ-caroten-16 (or 17 or 18)-oate	$C_{42}H_{54}O_4$	622.89	R,R',R''= 2CH₃, 1COOCH₃
Thiothece-OH-484	1-Methoxy-1'-hydroxy-1,2,1',2'-tetrahydro-ψ,ψ-caroten-4-one Thiothece-OH-484	→ 1-Methoxy-1'-hydroxy-1,2,1',2'-tetrahydro-ψ,ψ-caroten-4-one			
Torularhodin	Torularhodin	"16'-Carboxyl-3',4'-dehydro-γ-carotene" 3',4'-Didehydro-β,ψ-caroten-16'-oic acid "Lusomycin''? "Torulene-carboxylic-(16'-)acid"	$C_{40}H_{52}O_2$	564.85	
Torularhodin-aldehyde	Torularhodin-aldehyde	"3',4'-Dehydro-17'-oxo-γ-carotene" "3,4-Dehydro-18'-oxo-γ-carotene"	$C_{40}H_{52}O$	548.85	

	3',4'-Didehydro-β,ψ-caroten-16'-al		
	16'-Oxotorulene		
	Torulenal		
Torularhodin-methylester	Methyl-3',4'-didehydro-β,ψ-caroten-16'-oate	$C_{41}H_{54}O_2$	578.88
Torulene	3',4'-Dehydro-γ-carotene	$C_{40}H_{54}$	534.87
	3',4'-Didehydro-β,ψ-carotene		
	"3,3'-Dimethoxy-γ-carotene"		
	"3,3'-Dimethoxy-3',4'-dehydro-γ-carotene"		
"Torulene-carboxylic-(16'-)acid	→ Torularhodin		
Trihydroxy-α-carotene	Trihydroxy-α-carotene		
3,19,3'-Trihydroxy-7,8-dihydro-β,ε-caroten-8-one	Siphonaxanthin	—	
	"Xanthophyll K,S"		
3,19,3'-Trihydroxy-7,8-dihydro-β,ε-caroten-8-one-19-laurate	Siphonaxanthin-monolaurate	→ Siphonaxanthin	
	Siphonein	→ Siphonein	
	"Xanthophyll K₁"		
3,3',5'-Trihydroxy-5',6'-dihydro-5',6'-epoxy-β-carotene	Neoxanthin	→ Neoxanthin	
	Trollixanthin		
3,19,3'-Trihydroxy-7,8-dihydro-8-oxo-α-carotene	Siphonaxanthin	→ Siphonaxanthin	
	"Xanthophyll K,S"		
3,3',6'-Trihydroxy-5,6-epoxy-α-carotene	Trollixanthin	→ Trollixanthin	
3,3',6'-Trihydroxy-5,8-epoxy-α-carotene	Trollichrome	→ Trollichrome	
3,8,3'-Trihydroxy-5,6-epoxy-β-carotene	Heteroxanthin	→ Heteroxanthin	
"3,3',5'-Trihydroxy-6'-hydro-7,8-dehydro-β-carotene"	Heteroxanthin	→ Heteroxanthin	
1,1',2'-Trihydroxy-3,4,3',4'-tetradehydro-1,2,1',2'-tetrahydro-ψ,ψ-caroten-2-one	2'-Dihydrophillipsiaxanthin	→ 2'-Dihydrophillipsiaxanthin	
3,4,4'-Triketo-β-carotene	Dehydroadonirubin	→ Dehydroadonirubin	

Table I.3 (continued)

NAME LIST FOR CAROTENOIDS (TABLES I.5, I.6, I. PC, I. TLC, I. LC)

Pigment	Names mentioned in Tables I.5 through I. LC	All synonyms	Sum formula	Mol wt	Structure
Triphasiaxanthin	Triphasiaxanthin	3'-Hydroxy-5,6-seco-β,β-caro- tene-5,6-dione 3-Hydroxy-semi-β-carotenone	$C_{40}H_{56}O_3$	584.88	
"Trollein"*	Deepoxyneoxanthin	β,ε-Carotene-3,19,3'-triol Deepoxyneoxanthin 6,7-Didehydro-5,6-dihydro- β,β-carotene-3,5,3'-triol 19-Hydroxy-lutein Loroxanthin Pyrenoxanthin*	$C_{40}H_{56}O_3$	584.88	
Trollichrome*	Trollichrome	6'-Oxychrysanthemaxanthin 3,3',6'-Trihydroxy-5,8-epoxy- α-carotene 5,8-Epoxy-5,8-dihydro-β,ε-car- otene-3,3',6'-triol	$C_{40}H_{56}O_4$	600.88	
Trolliflavin	Neochrome Trollichrome Neoxanthin Trollixanthin	→ Neochrome → Neochrome → Neoxanthin			
Trolliflor					
Trollixanthin	Trollixanthin	3,3',6'-Trihydroxy-5,6-epoxy- α-carotene 5,6-Epoxy-5,6-dihydro-β,ε-car- otene-3,3',6'-triol	$C_{40}H_{56}O_4$	600.88	
"Unidentified II"	β-Zeacarotene β₁-Zeacarotene	→ β-Zeacarotene			
"Vaucheria-Heteroxanthin"	Heteroxanthin	→ Heteroxanthin			

Name	Synonyms	Systematic name	Formula	M.W.
Vaucheriaxanthin	Vaucheriaxanthin	5',6'-Epoxy-6,7-didehydro-5,6,5',6'-tetrahydro-β,β-carotene-3,5,19',3'-tetrol	$C_{40}H_{56}O_5$	616.88
Violaxanthin	Violaxanthin 9-cis-Violaxanthin Violeoxanthin	(3S,5R,6S,3'S,5'R,6'S)-5,6,5',6'-Diepoxy-5,6,5',6'-tetrahydro-β,β-carotene-3,3'-diol 9-cis-Violaxanthin Violeoxanthin → Violaxanthin	$C_{40}H_{56}O_4$	600.88
9-cis-Violaxanthin	Violaxanthin Violeoxanthin 9-cis-Violaxanthin	→ Violaxanthin		
13-cis-Violaxanthin	13-cis-Violaxanthin Violaxanthin 9-cis-Violaxanthin Violeoxanthin	→ Violaxanthin		
Violerythrin	Violerythrin	2,2'-Dinor-β,β-carotene-3,4,3',4'-tetrone	$C_{38}H_{44}O_4$	564.76
"Xanthophyll"	3-Hydroxy-3'-hydroxy-α-carotene Lutein	→ Lutein		
"Xanthophyll K₁"	"Xanthophyll" Siphonaxanthinmonolaurate Siphonein	→ Siphonein		
"Xanthophyl K,S"	Siphonaxanthin Siphonein	→ Siphonaxanthin		
Xanthophyll dipalmitate	Helenien Lutein dipalmitate	→ Helenien		
Xanthophyllepoxide	Isolutein Luteinepoxide Lutein-5,6-epoxide Taraxanthin Xanthophyllepoxide			

Table I.3 (continued)
NAME LIST FOR CAROTENOIDS (TABLES I.5, I.6, I. PC, I. TLC, I. LC)

Pigment	Names mentioned in Tables I.5 through I. LC	All synonyms	Sum formula	Mol wt	Structure
α-Zeacarotene	α-Zeacarotene	7',8'-Dihydro-δ-carotene (6R)-7',8'-Dihydro-ε,ψ-carotene	$C_{40}H_{58}$	538.90	
β-Zeacarotene	β-Zeacarotene β₁-Zeacarotene	"Carotene X'''? "Pigment X'''? 7',8'-Dihydro-β,ψ-carotene 7',8'-Dihydro-γ-carotene β₁-Zeacarotene → β-Zeacarotene	$C_{40}H_{58}$	538.90	
β₁-Zeacarotene	β-Zeacarotene β₁-Zeacarotene	β-Zeacarotene β₁-Zeacarotene			
"Zeaxanthene" Zeaxanthin	Zeaxanthin Zeaxanthin	→ Zeaxanthin Anchovyxanthin (3R,3'R)-β,β-Carotene-3,3'-diol "Zeaxanthene" "Zeaxanthol"	$C_{40}H_{56}O_2$	568.88	
Zeaxanthin dipalmitate Zeaxanthin-5,6-epoxide Zeaxanthin-5,8-epoxide Zeaxanthinfuranoide "Zeaxanthol" Zeinoxanthin	Physalien Antheraxanthin Mutatoxanthin Mutatoxanthin Zeaxanthin α-Cryptoxanthin Zeinoxanthin	→ Physalien → Antheraxanthin → Mutatoxanthin → Mutatoxanthin → Zeaxanthin → α-Cryptoxanthin			

Table I.4
NAME LIST FOR CAROTENOIDS (HPLC AND GC TABLES)

Antheraxanthin
Auroxanthin epimer 1
Auroxanthin epimer 2
Bacterioruberin
Bacterioruberin neo A
Bacterioruberin neo U
Bacterioruberin neo V
Bacterioruberin neo W
6-But-2-enylidene-1,5,5-trimethyl-cyclo-hex-1-ene
Canthaxanthin
α-Carotene
β-Carotene
β-Carotene, *cis* isomer 1
β-Carotene, *cis* isomer 2
(Z-)β-Carotene
β,β-Carotene
β,ε-Carotene
β,ψ-Carotene
(Z)-β,ψ-Carotene
γ-Carotene
ε,ψ-Carotene
(Z-)ε,ψ-Carotene
ζ-Carotene
(Z-)ζ-Carotene
β-Citraurin
Cryptoxanthin
all-(E-)β-Cryptoxanthin
(Z-)β-Cryptoxanthin
Cryptoxanthinester
3,4-Dehydrorhodopin
Diadinoxanthin
Diatoxanthin
Dinoxanthin
2,2'-Diol
Echinenone
Fucoxanthin
β-Ionone
Lutein
Lutein neo A
Lutein neo B
Lutein neo U
Lutein neo V
Lutein epoxide
Lutein-5,6-epoxide
Lutein-3'-ether epimer 1
Lutein-3'-ether epimer 2
Lycopene

all-(E-)Lycopene
Lycopersene
(8R-)Mutatoxanthin
(8S-)Mutatoxanthin
Neochrome
Neochrome epimer 1
Neochrome epimer 2
Neofucoxanthin A
Neofucoxanthin B
Neolutein A
Neolutein B
Neoperidinin
Neoxanthin
Neoxanthin neo A
Neoxanthin X
Neurosporene
all-(E-)Neurosporene
(5Z-)Neurosporene
Okenone
Peridinin
Phytoene
Phytofluene
all-(E-)Phytofluene
Phytol
Rhodopin
all-(E-)rubixanthin
(5'Z-)Rubixanthin (Gazaniaxanthin)
(9'Z-)Rubixanthin
(13Z-)Rubixanthin
(13'Z-)Rubixanthin
(5'Z,13Z-) or (5'Z,13'Z-)Rubixanthin
Spheroidenone
Spirilloxanthin
Squalene
Tetrahydrospirilloxanthin
3,5,6,3'-Tetrol
Tetrofuranoxyd
Torulene
Triacontane
Violaxanthin
Violacanthin-9-*cis*
Violaxanthin-13-*cis*
Violeoxanthin
Xanthophyllester
Zeaxanthin
(9Z-)Zeaxanthin
(13Z-)Zeaxanthin

Table I.4 (continued)
NAME LIST FOR CAROTENOIDS (HPLC AND GC TABLES)

Hydrogenated Carotenoids

H_2-β-apo-4'-carotenal
H_2-β-apo-10'-carotenal
H_2-β-apo-8'-carotenoic acid
H_2-β-apo-8'-carotenoic acid ethyl ester
H_2-β-apo-8'-carotenoic acid methyl ester
H_2-β-apo-8'-carotenal
H_2-astacene
H_2-azafrin
H_2-bixin
H_2-canthaxanthin
H_2-capsanthin
H_2-α-carotene
H_2-β-carotene
H_2-γ-carotene
H_2-ζ-carotene
H_2-β-carotenone
H_2-carotinin
H_2-crocetin
H_2-cryptoxanthin
H_2-cryptoxanthin, Ac
H_2-cryptoxanthin, TMS
H_2-decapreno-β-carotene
H_2-3,4-dehydro-β-apo-8'-carotenal
H_2-dehydro-β-carotene
H_2-4,4'-diapo-ζ-carotene
H_2-4,4'-diaponeurosporene
H_2-4,4'-diaponeurosporen-4-oate methylester
H_2-4,4'-diaponeurosporen-4-oic acid
H_2-4,4'-diapophytoene
H_2-4,4'-diapophytofluene
H_2-diethylcrocetin
H_2-dihydrosqualene
H_2-dimethoxyisozeaxanthin

H_2-dimethoxyzeaxanthin
H_2-dimethylcrocetin
H_2-echinenone
H_2-fucoxanthin
H_2-4-hydroxy-4,4'-diaponeurosporene
H_2-isocryptoxanthin
H_2-isocryptoxanthin, Ac
H_2-isocryptoxanthin, TMS
H_2-isozeaxanthin
H_2-isozeaxanthin, diAc
H_2-isozeaxanthin, diTMS
H_2-lycopene
H_2-lycopersene
H_2-methylazafrin
H_2-methylbixin
H_2-neo-α-carotene
H_2-neo-β-carotene
H_2-neurosporene
H_2-physalien; C_{40}-fragment
H_2-phytoene
H_2-phytofluene
H_2-phytol
H_2-retinaldehyde
H_2-retinol
H_2-rubixanthin
H_2-squalene
H_2-tetrahydrosqualene
H_2-torularhodin
H_2-β-zeacarotene
H_2-zeaxanthin
H_2-zeaxanthin, diAc
H_2-zeaxanthin, diTMS

Table I.5
QUALITATIVE SPECTROSCOPIC DATA: ABSORPTION MAXIMA IN DIFFERENT SOLVENTS

NOTES

In Table I.5, the pigment names have been listed alphabetically (first vertical row). Since a number of carotenoids have been named differently by different authors, the same pigment may appear under two or even more names. If the desired pigment is not found, refer to row 1 (vertical) of the name list (Table I.3).

Second and following rows: literature references (superscripted) refer to absorption maxima (usually three) given in each row. Absorption maxima in parentheses indicate shoulders.

"Benzine" (solvent 3) is a mixture of hydrocarbons similar to petroleum ether (solvent 10), but in the literature (Karrer and Jucker[1a] and Frye[14]) is nonspecified.

Row 10, "petroleum ether", combines solvents with a boiling range of approximately 40 to 80°C (that is, it combines values recorded in light petroleum, bp 60—70 and 70—80°C).

Table I.5
QUALITATIVE SPECTROSCOPIC DATA: ABSORPTION MAXIMA IN DIFFERENT SOLVENTS

Pigment	Solvent 1 Acetone	Solvent 2 Benzene	Solvent 3 "Benzine"	Solvent 4 Carbon disulfide	Solvent 5 Chloroform	Solvent 6 Cyclohexane	Solvent 7 Ethanol	Solvent 8 Ether	Solvent 9 Hexane	Solvent 10 Petroleum ether	Solvent 11 Methanol	Solvent 12 Pyridine
Actinioerythrin	(480), 508, (538)	—	—	495, 533, 574[1]	518, (550)[1]	—	—	—	—	(470), 496, 529[1]	—	—
Actinioerythrol	—	—	—	—	(490), 518, (550)[1]	—	—	—	—	—	—	—
Alloxanthin	—	434, 464, 496[1a]	—	454, 488, 518[1a]	(436), 460, 489[2]	—	(427), 450, 478[2]	—	—	458, 488[1a]	—	—
Alloxanthin (trans)	—	—	—	—	—	—	451, 480[3]	—	451, (467), 480[4]	—	—	—
Anhydrodeoxyflexixanthin (trans)	493, 522.5[5]	—	—	—	—	—	—	—	—	489.5, 519.5[5]	490, (516)[5]	—
Anhydroeschscholtzxanthin	(480), 500, (431)[6]	—	—	503, 539, 578[7]	484, 516, 549[7]	474, 503, 538[7]	—	—	—	474, 496, 529[6]	—	—
Anhydrorhodovibrin	460, 485, 520[8]	444, 482, 516[9]	—	—	—	358, 374, 455, 483, 517[10]	—	—	—	—	375, 455, 483, 517[11]	—
Anhydrosaproxanthin	460, 488, 520[12]	—	—	—	—	—	—	—	—	—	—	—
Antheraxanthin	—	433, 457, 488[13]	—	445, 478, 510[14]	430, 456, 484[15]	—	422, 444, 472[15]	—	424, 442, 470[19]	—	—	—
Antheraxanthin (cis)	—	457, 487[16]	—	376, 506[16]	—	—	419, 443, 469[17]	—	—	—	—	—
Antheraxanthin (trans)	—	—	—	—	—	—	421, 443, 473[18]	—	—	—	—	—
Aphanicin	—	—	—	494, 533[16]	474, 504[16]	—	—	—	462, 494[16]	—	—	—
Aphanizophyll	—	—	—	454, 484, 518[16]	453, 482, 510[20]	—	445, 472, 502[20]	—	—	—	—	—
Aphanizophyll (trans)	365, 450, 476, 507[21]	—	—	—	—	—	—	—	—	350, 362, 446, 472, 502[21]	—	—

Compound										
β-Apo-4-carotenal	—	—	—	—	—	—	—	ca. 442[1a]	—	—
β-Apo-2'-carotenal	—	—	—	—	—	—	498[22]	—	—	—
β-Apo-8'-carotenal	—	—	—	—	—	447—498[1a]	457[22]	454, 484[1a]	—	—
β-Apo-10'-carotenal	—	—	—	—	—	—	435[22]	435[23]	—	—
β-Apo-12'-carotenal	—	—	—	—	—	—	—	410[24]	—	—
β-Apo-4'-carotenoic acid	473, (496)[25]	—	—	—	—	—	—	477, 505[25]	470, 495[25]	—
β-Apo-8'-carotenoic acid	—	—	—	—	—	—	—	450[24]	—	—
β-Apo-10'-carotenoic acid	—	—	—	—	—	—	—	425[24]	—	—
β-Apo-2'-carotenol	—	—	—	—	—	—	473[22]	—	—	—
β-Apo-8'-carotenol	—	—	—	—	—	—	426[22]	—	—	—
β-Apo-10'-carotenol	—	—	—	—	—	—	403[22]	—	—	—
Apo-2-lycopenal	—	—	455.5, 490.5, 525.5[1a]	—	280, 449, 475, 505[10]	—	—	—	—	—
Apo-3-lycopenal	488, 518[1a]	—	474, 502[1a]	—	—	—	—	—	—	—
Apo-6'-lycopenal	(448), 473[26]	—	—	—	—	—	—	(462), 479, (515)[a,26]	477[26]	—
Apo-8'-lycopenal	470[26]	488[22]	—	—	—	—	—	468[26]	470[26]	500 (broad)[1a]
Astacene	—	—	510 (ca.)[1a]	—	—	478[27]	467[28]	—	—	493[29]
Astaxanthin	—	485[19]	503[30]	472[30]	—	—	—	—	472[30]	—
Astaxanthin-diacetate	—	478[31]	—	—	—	474[31]	—	—	—	—
Astaxanthin-diester (cis)	—	—	—	—	—	471[32]	—	—	—	—
Astaxanthin-diester (trans)	—	—	—	—	—	478[32]	—	—	—	—
Astaxanthin-monoester (cis)	—	—	—	—	—	471[32]	—	—	—	—
Astaxanthin-monoester (trans)	—	—	—	—	—	478[32]	—	—	—	—
Asterinic acid	—	—	(480), 518, 541[29]	428[14]	—	—	—	—	—	—
Aurochrome	440[14]	—	428, 457[14]	—	—	—	—	437[14]	—	—
Auroxanthin	—	—	401, 432, 451[33]	335, 385, 410[33]	—	382, 403, 428[14]	379.5, 401, 425.5[a,34]	379, 399, 423[33]	—	(478), 495, 522[29]
Auroxanthin (cis)	—	—	—	—	—	375, 400, 425[18]	—	—	—	—

Table I.5 (continued)
QUALITATIVE SPECTROSCOPIC DATA: ABSORPTION MAXIMA IN DIFFERENT SOLVENTS

Pigment	Solvent 1 Acetone	Solvent 2 Benzene	Solvent 3 "Benzine"	Solvent 4 Carbon disulfide	Solvent 5 Chloroform	Solvent 6 Cyclohexane	Solvent 7 Ethanol	Solvent 8 Ether	Solvent 9 Hexane	Solvent 10 Petroleum ether	Solvent 11 Methanol	Solvent 12 Pyridine
Auroxanthin (*trans*)	—	—	—	—	—	—	380, 402, 428[18]	—	—	—	—	—
Azafrin	—	—	—	—	428, 458[1a]	—	—	—	—	—	—	428, 458[1a]
Bacterioruberin α	374, 389, 466, 499, 533.5[35]	378, 398, 481, 511, 549[35]	—	418, 500.5, 533.5, 572[35]	380, 397, 475, 506, 544[35]	—	—	—	—	369, 385, 461, 494, 528[35]	—	—
3,4,3',4'-Bisdehydro-β-carotene	—	—	—	—	—	—	—	—	462, 490, 522[36]	471[23]	—	—
Caloxanthin	—	—	—	—	(432), 458, 484[20]	—	(426), 449, 475[20]	—	—	—	—	—
Canthaxanthin	—	—	—	500[37]	483[20]	—	477[32]	456[21]	462[37]	465—467[23]	—	—
Canthaxanthin (*cis*)	—	—	—	—	—	467[38]	472[40]	—	—	356, 465[39]	—	—
Capsanthin	—	486, 520[1a]	475, 505[1a]	503, 542[1a]	—	—	472[40]	—	(450), 474, 504[40]	—	—	—
Capsanthin-diester	—	485, 518[40]	—	—	—	—	474[40]	—	(450), 474, 503[40]	—	—	—
Capsanthin-5,6-epoxide	—	483, 514[1a]	—	499, 534[1a]	481, 511[1a]	—	—	—	—	—	—	—
Capsanthin-5,6-epoxide-diester	—	—	—	—	—	—	473[40]	—	—	473, 502[40]	—	—
Capsanthin-monoester	—	—	—	—	—	—	474[40]	—	(450), 474, 504[40]	—	—	—
Capsochrome	—	464, 496[1a]	—	482, 515[1a]	462, 492[1a]	—	—	—	—	—	—	—
Capsorubin	—	455, 486, 520[1a]	444, 474, 506[1a]	468, 503, 541.5[1a]	—	—	—	—	444, 474, 506[16]	—	—	—
Capsorubin-diester	—	457, 488, 522[40]	—	—	—	—	485[40]	—	452, 477, 510[40]	—	—	—
α-Carotene	424, 448, 476[41]	—	447.5, 478[14]	450, 475, 505[37]	432, 457, 485[37]	—	423, 444, 473[2]	(420), 443, 472[42]	422, 445, 475[37]	421, 445, 474[43]	—	—

	1	2	3	4	5	6	7	8	9
β-Carotene	(428), 452, 477[44]	463, 492[45]	426, 452, 483.5[14]	450, 485, 520[14]	463, 493[33]	457, 485[23]	426, 451, 478[46]	425, 450, 478[33]	426, 451, 477[33]
β,β-Carotene	(432), 453, 479[48]	—	—	—	—	—	—	421, 449, 477[a,47]	(427), 448, 474[49]
β,ε-Carotene	—	—	—	—	—	—	—	—	421, 443, 473[49]
γ-Carotene	—	447, 477, 519[1a]	431, 462, 495[1a]	463, 496, 533.5[14]	446, 475, 508.5[14]	—	440, 460, 489[51]	431, 462, 494[14]	(410), 435, 460, 489[52]
γ-Carotene (trans)	—	440, 465, 495[53]	—	—	—	—	—	—	—
δ-Carotene	—	—	—	457, 490, 526[14]	440, 470, 503[14]	—	—	428, 458, 490[14]	281, 431, 456, 489[54]
δ-Carotene (trans)	—	—	—	—	—	—	—	—	431, 456, 487[54]
ε-Carotene	—	(414), 434, 458, 488.5[54]	—	—	—	—	417, 440, 469[2]	—	267, 417, 441, 471[54]
ε₁-Carotene	—	—	—	470, 501[14]	452, 483[14]	419, 440, 475[23]	—	—	439, 470[14]
ζ-Carotene	—	—	—	—	—	—	—	378, 400, 425[52]	380, 387, 400, 412, 426[55]
ζ-Carotene (equilibrium mixture)	—	—	—	—	—	—	—	—	285, 296, 359, 377, 398, 423[52]
ζ-Carotene (trans)	—	—	—	—	—	—	—	—	296, 361, 379.5, 400.5, 424.5[56]
ð-Carotene	—	—	—	—	—	—	—	—	377, 382, 398, 410, 422[55]
β-Carotene-diepoxide	—	456, 485[14]	—	472, 502[14]	456, 484[14]	—	—	—	417, 439, 470[2]
α-Carotene-5,6-epoxide	—	455, 484[1a]	—	471, 503[1a]	454, 483[1a]	—	—	—	442, 471[1a]
β-Carotene-monoepoxide	—	—	—	479, 511[14]	459, 492[14]	—	—	—	(420), 442, 470[2]
β,β-Carotene-2,3,3'-triol	—	—	—	—	—	—	450, 478[50]	—	—
Carotenonaldehyde	—	420, 446, 476[1a]	406, 432, 461[1a]	430, 459, 491[1a]	423, 450, 482[1a]	—	(442), (473)[1a]	405, 431, 458[1a]	404, 430, 457[1a]
β-Carotenone	—	453, 486, 522[1a]	440, 468, 502[1a]	466, 499, 538[1a]	454, 489, 527[1a]	—	—	436, 466, 500[1a]	—

Table I.5 (continued)
QUALITATIVE SPECTROSCOPIC DATA: ABSORPTION MAXIMA IN DIFFERENT SOLVENTS

Pigment	Solvent 1 Acetone	Solvent 2 Benzene	Solvent 3 "Benzine"	Solvent 4 Carbon disulfide	Solvent 5 Chloroform	Solvent 6 Cyclohexane	Solvent 7 Ethanol	Solvent 8 Ether	Solvent 9 Hexane	Solvent 10 Petroleum ether	Solvent 11 Methanol	Solvent 12 Pyridine
Celaxanthin	—	—	—	487, 521, 562[14]	—	—	455, 488, 520.5[1a]	—	456, 486.5, 520[16]	456, 486.5, 520[1a]	—	—
Chlorobactene	—	—	—	—	—	—	—	435, 461, 491[57]	—	435, 461, 491[53]	—	—
Chlorobactene (*trans*)	440, 465, 495[53]	—	—	—	—	—	—	—	—	—	—	—
Chloroxanthin	—	—	—	—	—	—	—	—	—	—	—	—
Chloroxanthin (*trans*)	—	425, 452, 482[9]	—	—	—	—	—	—	—	417, 440, 470[11]	—	—
Chrysanthemaxanthin	—	—	—	451, 480.5[16]	430, 459[1a]	—	421, 448[1a]	—	421, 450[16]	421, 450[1a]	—	—
Citranaxanthin	—	—	—	449, 480, 514[1a]	—	—	475, (489)[58]	—	—	463, 495[58]	—	—
α-Citraurin	—	—	—	—	—	—	—	—	—	438, 477[1a]	—	—
β-Citraurin	—	467, 497[1a]	459, 488[1a]	457, 490, 525[1a]	—	—	Broad[1a]	—	458, 487[1a]	—	—	—
Corynexanthin	—	—	—	435, 466, 495[16]	423, 447, 478[16]	—	415, 437, 467[16]	—	—	—	—	—
Crocetin (stable *trans*)	—	—	424.5, 450.5[1a]	426, 453, 482[1a]	434.5, 463[1a]	—	—	—	—	—	—	—
Crocetindialdehyde	—	—	—	—	—	—	—	—	—	408, 430, 458[59]	—	411, 436, 464[1a]
Crocetin-di-(β-D-gluco-syl)-ester	—	—	—	—	—	—	—	—	—	—	—	421, 443, 471[60]
Crocetindimethylester	—	—	424.5, 450.5[1a]	—	434.5, 463[1a]	—	—	—	—	400, 422, 450[23]	—	—
Crocetin-(β-D-gentio-biosyl)-(β-D-glucosyl) ester	—	—	—	—	—	—	—	—	—	—	—	421, 443, 471[60]

Compound							
Crocoxanthin		422, (427), 445, (462), 475[61]	(421), 443, 472[2]	(430), 454, 482[2]			
Cryptocapsin		(445), 470, 497[40]					486, 519[62]
Cryptocapsone							491[62]
Cryptochrome				424, 456[14]	388, 409, 439[15]	424, 456[14]	
Cryptoflavin		439, 470[14]		459, 490[14]	438, 468[14]	439, 470[14]	
Cryptoxanthin	425, 447, 481[19]	423, 451, 484[1a]	424, 452, 486[1a]	452, 483, 519[14]	433, 463, 497[14]		424, 452, 485.5[14]
α-Cryptoxanthin	420, 446, 477[64]	227, 267, 299, 333, 347, 421, 427, 446, 462, 475[64], 446, 471[1a,34]					
β-Cryptoxanthin	420, 452, 485.5[65]	423, 451, 484[65], 425, 451, 476[1a,34]	(428), 449, 473[2]	453, 483, 518[65]	433, 463, 497[65]		
Cryptoxanthin-diepoxide		455, 486[1a]	442, 473[1a]	473, 503[1a]	453, 482[1a]		
β-Cryptoxanthin-5,6-5',6'-diepoxide			423, 442, 472[15]		432, 452, 480[15]		
Cryptoxanthin-5,6-epoxide		422, 443, 472[62]					
β-Cryptoxanthin-5',6'-monoepoxide			424, 445, 477[15]		(434), 456, 483[15]		
Cryptoxanthin-5',6'-monoepoxide		461, 494[2]	424, 445, 477[2]	479, 512[2]	456, 488[2]		
Cryptoxanthin-5,8-epoxide		428, 454[66]					
β-Cryptoxanthin-monoester		424, 445, 477[40]					
Cynthiaxanthin							463, 492[31]
Dehydroadonirubin		466[28]	477[27]				
Dehydroadonixanthin		455[28]	466[27]				
Dehydro-β-carotene		447, 475, 504[1a]		472, 504, 543[1a]	455, 485, 518[1a]		
Dehydro-β-carotene (trans)		445, 471, 502[67]					

Table I.5 (continued)
QUALITATIVE SPECTROSCOPIC DATA: ABSORPTION MAXIMA IN DIFFERENT SOLVENTS

Pigment	Solvent 1 Acetone	Solvent 2 Benzene	Solvent 3 "Benzine"	Solvent 4 Carbon disulfide	Solvent 5 Chloroform	Solvent 6 Cyclohexane	Solvent 7 Ethanol	Solvent 8 Ether	Solvent 9 Hexane	Solvent 10 Petroleum ether	Solvent 11 Methanol	Solvent 12 Pyridine
3,4-Dehydro-β-carotene	—	—	—	—	—	—	—	—	425, 462, 491[36]	—	—	—
Dehydro-hydroxy-echinenone	—	—	—	—	—	—	466[27]	—	—	—	—	—
Dehydrolycopene	—	493, 531, 570[1a]	—	520, 557, 601[14]	493, 528, 567[1a]	—	—	—	476, 504, 542[1a]	—	—	498, 534, 574[1a]
3,4-Dehydrolycopene	—	—	—	—	—	—	—	—	—	(430), 452, 486, 520[68]	—	—
2'-Dehydro-plectaniaxanthin	495, (522)[69]	—	—	—	—	—	—	—	—	—	—	—
Dehydro-retrocarotene	—	—	—	—	—	—	—	—	—	446, 472, 502[70]	—	—
3,4-Dehydrorhodopin	—	—	—	—	—	—	—	—	—	(358), (374), 455, 483, 517[71]	—	—
Deoxyflexixanthin	480.5, 508[5]	—	—	—	—	—	—	—	—	476.5, 503[5]	477.5, (500)[5]	—
Diadinoxanthin	(428), 449, 479[49]	—	—	~449, 474, 506[72]	432, 455, 482[15]	—	424, 446, 476[72]	—	421, 445, 475[72]	—	—	—
Diadinoxanthin (*trans*)	340, 426, 447.5, 478[44]	—	—	—	—	—	425, 446, 477[3]	—	—	—	338, 444.5, 474[44]	—
4,4'-Diapo-ζ-carotene	—	—	—	—	—	—	—	—	—	(358), 378, 400, 425[73]	—	—
4,4'-Diapolycopen-4-al	—	—	—	—	488[74]	—	477[74]	—	—	454, 476, 506[74]	—	—
4,4'-Diapophytoene	—	—	—	—	—	—	—	—	—	275, 285.5, 297.5[73]	—	—
4,4'-Diapophytofluene	—	—	—	—	—	—	—	—	—	(315.5), 330, 346.5, 366[73]	—	—

Compound	1	2	3	4	5	6	7
4,4'-Diapo-7,8,11,12-tetrahydrolycopene	—	—	—	—	—	—	(354), 374, 395, 419[73]
Diatoxanthin	(434), 454, 482[49]	—	(433), 458, 486[15]	—	(425), 449, 475[15]	—	430, 452, 483[75]
Diatoxanthin (trans)	—	—	—	—	428, 452, 479[3]	450, (469), 479[4]	—
3',4'-Didehydrochlorobactene (trans)	(465), 491, 524[76]	—	—	—	—	—	—
3,4-Didehydrolycopene	—	—	—	—	—	—	458, 491, 524[77]
7,7'-Dihydro-β-carotene	—	—	—	—	—	—	382, 405, 429[23]
1'2,'-Dihydro-1'-hydroxy-γ-carotene	—	—	—	—	(440), 460, 489[51]	—	—
1',2'-Dihydro-1'-hydroxy-γ-carotene	—	—	—	—	—	—	350, (438), 459, 488[78]
1',2'-Dihydro-1'-hydroxy-γ-carotene (trans)	471, (491)[78]	—	—	—	—	—	—
1',2'-Dihydro-1'-hydroxy-4-keto-γ-carotene	—	—	—	—	—	—	465, (490)[78]
1',2'-Dihydro-1'-hydroxy-4-keto-γ-carotene (trans)	469, (489)[78]	—	—	—	—	—	—
1',2'-Dihydro-1'-hydroxy-4-keto-torulene	—	457, 486, 522[11]	439, 465.5, 498[11]	—	460, 480, (504)[51]	—	—
7',8'-Dihydrorhodovibrin	439, 467, 501[11]	—	—	—	—	—	429, 454, 486[11]
3,4'-Dihydroxy-β-carotene (cis)	—	—	—	—	—	449, 478[66]	—
3,4'-Dihydroxy-β-carotene (trans)	—	—	—	—	—	451, 480[66]	—
4,4'-Dihydroxy-β-carotene	—	—	—	(427), 450, 478[38]	—	—	—
Dihydroxy-ζ-carotene	—	—	—	—	—	—	378, 400, 425[79]
2,2'-Diketo-bacterioruberin	—	500[81]	361, 543[80]	—	—	—	—
3,4-Diketo-α-carotene	—	—	476[81]	—	—	462[81]	—
4,4'-Diketo-β-carotene	480[82]	—	478[82]	—	—	466[82]	—

Table I.5 (continued)
QUALITATIVE SPECTROSCOPIC DATA: ABSORPTION MAXIMA IN DIFFERENT SOLVENTS

Pigment	Solvent 1 Acetone	Solvent 2 Benzene	Solvent 3 "Benzine"	Solvent 4 Carbon disulfide	Solvent 5 Chloroform	Solvent 6 Cyclohexane	Solvent 7 Ethanol	Solvent 8 Ether	Solvent 9 Hexane	Solvent 10 Petroleum ether	Solvent 11 Methanol	Solvent 12 Pyridine
2,2'-Diketo-spirilloxanthin	—	—	—	530, 561, 603[83]	—	—	—	—	—	—	—	—
1,1'-Dimethoxy-1,2-1',2'-tetrahydro-ψ,ψ-carotene-4,4'-dione	—	—	—	—	—	—	—	—	(370), 387, 461, 489, 522.5[84]	—	—	—
1,1'-Dimethoxy-1,2-1',2'-tetrahydro-3',4'-didehydro-ψ,ψ-caroten-4-one	—	—	—	—	—	—	—	—	(368), 387, (470), 494.5, 527[84]	—	—	—
Dinoxanthin	418, 442, 470[44]	—	—	441, 467, 498[72]	—	—	419, 441, 470[72]	—	416, 439, 469[72]	—	416, 438, 467[72]	—
Echinenone	—	472[23]	—	488—494[16]	473[16]	—	458, 459[85]	455[21]	453[16]	452—456[85]	—	—
5,6-Epoxy-3-hydroxy-5,6-dihydro-10'-apo-β-caroten-10'-al (*cis*)	—	439, (468)[86], 437[86]	—	—	—	—	—	—	—	—	—	—
5,6-Epoxy-3-hydroxy-5,6-dihydro-10'-apo-β-caroten-10'-al (*trans*)	—	442, (470)[86], 419, (442)[86]	—	—	—	—	—	—	—	—	—	—
5,6-Epoxy-3-hydroxy-5,6-dihydro-12'-apo-β-caroten-12'-al (*trans*)	—	414, (440)[86], 392, (412)[86]	—	—	—	—	—	—	—	—	—	—
Eschscholtzxanthin	—	459, 486, 520[7]	—	474, 507, 542[7]	456, 488, 520[7]	451, 478, 510[7]	448, 476, 505[7]	—	438, 464, 495[23]	442, 472, 502[7]	—	464, 489, 521[6]
Flavacin	—	—	—	—	—	—	—	415, 428, 452[21]	—	—	—	—
Flavochrome	—	434, 462[1a]	—	451, 482[14]	433, 461[14]	—	—	—	—	—	—	—
Flavorhodin	—	—	—	472, 502[87]	—	—	—	—	422, 450[16]	422, 450[1a]	—	—
Flavoxanthin	—	432, 481[16]	421, 450[16]	449, 479[88]	430, 459[1a]	—	400, 423, 448[89]	—	—	421, 450[1a]	—	—

Flexixanthin	483, 510[5]	—	—	—	—	—	—
Fucoxanthin	(427), 449, 471[48]	445, 477, 510[1a]	457, 492[1a]	(330), (426), 449, (465)[2], 398, 423, 448[2]	—	—	(425), 447.5, 475[90]
Fucoxanthol	—	—	—	—	—	—	—
Gazaniaxanthin	447.5, 476, 509[1a]	461, 494.5, 531[1a]	—	434.5, 462, 494.5[1a]	—	434.5, 462.2, 494.5[16]	434.5, 462.5, 494.5[1a]
Gazaniaxanthin (*trans*)	435, 462, 492[91]	—	—	—	—	—	—
4-(β-D-Glucopyranosyl)-oxy-4,4'-diaponeurosporene	—	—	400, 424, 450, 479[73]	391, 416, 439, 468[73]	390, 414, 437, 466[73]	—	—
Helenien	278, 306, 342, 356, 433, 439, 457, 474, 487[64]	—	—	—	—	226, 267, 298, 331, 348, 420, 427, 445, 462, 475[64]	—
Heteroxanthin	—	—	434, 456, 485[15]	426, 448, 477[15]	—	—	—
1-Hexosyl-1,2-dihydro-3,4-didehydro-apo-8'-lycopenol	—	—	—	—	427, 452, 482[92]	—	—
3-Hydroxy-10'-apo-β-caroten-10'-al (*cis*)	—	—	—	—	—	427, (452)[86]	—
3-Hydroxy-10'-apo-β-caroten-10'-al (*trans*)	—	—	—	—	—	433, (458)[86]	—
3-Hydroxy-12'-apo-β-caroten-12'-al (*cis*)	—	—	—	—	—	404, (425)[86]	—
3-Hydroxy-12'-apo-β-caroten-12'-al (*trans*)	—	—	(427), 451, 479[38]	—	—	408, (428)[86]	—
4-Hydroxy-β-carotene	—	—	—	—	—	—	(386), 413, 435, 465[74]
4-Hydroxy-4,4'-diaponeurosporene	494[81]	—	423, 449, 477[74]	415, 438, 467[74]	—	—	—
1-Hydroxy-1,2-dihydro-γ-carotene	—	—	—	—	—	—	444, 462, 494[93]
3-Hydroxy-4,4'-diketo-β-carotene	494[81]	—	483[81]	—	—	467[81]	—
Hydroxyechinenone	—	—	474–481[85]	462[85]	—	—	460[85]
3'-Hydroxyechinenone	—	—	471[20]	(462)[20]	—	—	—

Table I.5 (continued)
QUALITATIVE SPECTROSCOPIC DATA: ABSORPTION MAXIMA IN DIFFERENT SOLVENTS

Pigment	Solvent 1 Acetone	Solvent 2 Benzene	Solvent 3 "Benzine"	Solvent 4 Carbon disulfide	Solvent 5 Chloroform	Solvent 6 Cyclohexane	Solvent 7 Ethanol	Solvent 8 Ether	Solvent 9 Hexane	Solvent 10 Petroleum ether	Solvent 11 Methanol	Solvent 12 Pyridine
3-Hydroxy-3'-keto-α-carotene	(425), 449, 477[94]	—	—	—	—	—	—	—	—	—	—	—
4-Hydroxy-4'-keto-β-carotene	—	—	—	—	470[20]	—	(460)[20]	—	—	454, (473)[39]	—	—
4-Hydroxy-4'-keto-β-carotene (trans)	460, (480)[95]	—	—	—	—	—	—	—	—	454, (474)[95]	—	—
Isocryptoxanthin	—	—	—	—	—	—	—	—	451, 479[96]	(424), 449, 478[39]	—	—
Isofucoxanthin	—	—	—	—	—	—	—	—	—	(430), 453, 482[97]	—	—
Isofucoxanthinol	—	424, 456, 487[1a]	—	444, 474, 506[1a]	—	—	—	—	—	—	(445), 475, (505)[1a]	—
Isorenieratene	—	430, 463, 492[98]	—	452, 484, 520[98]	—	—	—	—	—	(425), 452, 480[99]	—	—
β-Isorenieratene	—	—	—	456, 487, 508[100]	—	—	—	—	—	—	—	—
β-Isorenieratene (trans)	(430), 453, 480[99]	—	—	—	—	—	—	—	—	—	—	—
Isozeaxanthin	—	—	—	—	—	(428), 451, 478.5[38]	451, 478[28]	—	—	(428), 451, 479[95]	—	—
4-Keto-α-carotene	—	—	—	—	—	(425), 451, (474)[38]	—	—	453.5, 470[101]	—	—	—
4-Keto-β-carotene	—	—	—	—	—	432, 458.5, 483[38]	—	—	458[82]	—	—	—
4-Keto-γ-carotene	—	—	—	—	—	—	467, (483)[51]	—	—	465, (490)[78]	—	—
4-Keto-γ-carotene (cis)	—	—	—	—	—	—	—	—	—	350, 461[102]	—	—
4-Keto-γ-carotene (trans)	471, (491)[78]	—	—	—	—	—	—	—	—	462[102]	—	—

4-Keto-1',2'-dihydro-1'-hydroxy-γ-carotene	—	—	—	—	—	—	—	—	—
4-Keto-1',2'-dihydro-1'-hydroxytorulene	—	—	—	—	—	—	—	—	—
4-Keto-4'-ethoxy-β-carotene	—	458[82]	—	—	—	—	—	—	—
4-Keto-3'-hydroxy-β-carotene	—	452[103]	—	—	—	—	—	—	—
4-Keto-1'-hydroxy-1',2'-dihydro-γ-carotene (cis)	—	—	—	—	—	—	—	—	365, 474[102]
4-Keto-1'-hydroxy-1',2'-dihydro-γ-carotene (trans)	—	462[102]	—	—	—	—	—	—	—
4-Keto-4'-hydroxy-β-carotene	—	—	458[82]	(431), 457.5, 481[38]	—	—	—	—	—
4-Ketophleixantophyll	—	—	502[22]	—	—	—	—	—	480, (507)[6]
2-Ketorhodovibrin	—	485, 511, 547[104]	—	520[104]	—	—	—	—	485, 512, (540)[104]
4-Ketotorulene	—	456, 483, 518[6]	—	470, 490, (516)[51]	—	—	—	—	465, 490, 523[6]
Leprotene	—	—	425, 452, 484[16]	—	428, 460, 495[16]	477, 499, 517[16]	—	—	—
Lutein	—	420, 444, 474[33]	420, (425), 445, (460), 474[4]; 421, 446, 474[a,47]	422, 445, 474[105]	421, 445, 473[46]	428, 454, 483[33]	446, 475, 505[33]	447, 475.5, 506[14]	429, 456, 485[62]
Lutein (trans)	—	—	421, 445, 474[a,34]	—	—	—	—	—	—
Luteinepoxide	—	—	—	420, 442, 469[89]	440, 470, 501[89]	—	—	—	—
Lutein-5,6-epoxide	—	—	442, 471[16]	(417), 440, 469[46]	472, 502[16]	—	—	—	453, 482[16]
Luteochrome	—	—	395, 421, 448[a,34]	396, 420, 446[2]	451, 482[14]	—	—	—	407, 432, 460[62]
Luteoxanthin	—	—	448[a,34]	—	—	—	—	—	—
Lycopene	461, 490, 526[107]	445, 471, 502[107]	448, 472, 503[36]	443, 472, 503[14]	453, 480, 517[14]	477, 507.5, 548[14]	—	448, 474, 505.5[107]	455, 487, 522[14]

Table I.5 (continued)
QUALITATIVE SPECTROSCOPIC DATA: ABSORPTION MAXIMA IN DIFFERENT SOLVENTS

Pigment	Solvent 1 Acetone	Solvent 2 Benzene	Solvent 3 "Benzine"	Solvent 4 Carbon disulfide	Solvent 5 Chloroform	Solvent 6 Cyclohexane	Solvent 7 Ethanol	Solvent 8 Ether	Solvent 9 Hexane	Solvent 10 Petroleum ether	Solvent 11 Methanol	Solvent 12 Pyridine
Lycopene (*trans*)	446, 472, 504[26]	—	—	—	—	—	—	—	286, 295, 425, 448, 476, 507[54]	445, 474, 505[8]	—	—
Lycophyll	—	456, 487, 521[14]	444, 473, 504[14]	472, 506, 546[14]	—	—	444, 474, 505[14]	—	447, 473, 504[16]	441, 468, 501[108]	—	—
Lycoxanthin	448, 474, 505[109]	456, 487, 521[14]	444, 473, 504[14]	473, 507, 547[14]	—	—	444, 474, 505[14]	—	—	443, 469, 500[109]	—	—
1-Methoxy-1,2-dihydro-ψ,ψ-caroten-4-one	—	—	—	—	—	—	—	—	(360), 375, (462), 488, 520[84]	—	—	—
1-Methoxy-1,2-7',8',11',12'-hexahydro-ψ,ψ-caroten-4-one	—	—	—	—	—	—	—	—	398.5, 420, 446[84]	—	—	—
1-Methoxy-1'-hydroxy-1,2,1',2'-tetrahydro-ψ,ψ-caroten-4-one	—	—	—	—	—	—	—	—	(360), 377, (463), 487, 519[84]	—	—	—
1-Methoxy-1,2-7',8'-tetrahydro-ψ,ψ-caroten-4-one	—	—	—	—	—	—	—	—	(330), 345, (436.5), 460.5, 490[84]	—	—	—
Methyl-apo-6'-lycopenoate	445, 469, (495)[26]	—	—	—	—	—	—	—	—	(448), 471, 503[26]	—	—
Methyl-apo-6'-lycopenoate (*trans*)	—	—	—	—	—	—	—	—	—	446, 472, 504[26]	—	—
Methylbixin (*trans*)	—	—	—	—	—	—	—	—	—	432, 456, 490[23]	—	—
Methyl-1-hexosyl-1,2-dihydro-3,4-didehydro-apo-8'-lycopenoate	469, 497[92]	—	—	—	—	—	—	—	—	—	—	—
Monadoxanthin	—	—	—	—	—	—	—	—	422, (427), 445, (461), 475[61]	—	—	—

Compound										
3,4-Mono-dehydro-β-carotene	—	—	—	—	—	—	—	461[23]	—	—
5,6-Mono-epoxy-β-carotene	—	460, 492[1a]	—	—	—	—	416, 439, 467[62]	—	—	—
5,6-Mono-epoxy-β-carotene (*trans*)	—	—	479, 511[1a]	—	—	459, 492[1a]	—	420, 445, 472[18]	—	—
5,6-Mono-epoxylutein (*trans*)	—	—	—	—	417, 442, 471[18]	—	—	—	—	—
Mutatochrome	—	440, 470[16]	459, 489.5[16]	435, 469[16]	—	(418), 428, 436[42]	397, 422, 450[62]	409, 428, 452[42]	—	—
Mutatochrome (equilibrium mixture)	—	—	—	—	—	—	—	310, (402), 423, 449[42]	—	—
Mutatochrome (*trans*)	—	—	—	—	—	(407), 428, 454[42]	—	—	—	—
Mutatoxanthin	—	439, 468[1a]	431, 459, 488[88]	437, 468[1a]	427, 457[88]	—	426, 456[19]	426, 456[14]	—	443, 473[1a]
Myxobactin	—	—	—	—	454, 477, 506[51]	—	—	—	—	—
Myxobacton	—	—	—	—	460, 480, (504)[51]	—	—	—	—	—
Myxoxanthin	—	—	488[110]	473[110]	470[110]	—	—	465[110]	—	—
Myxoxanthol	—	—	464, 494, 529[110]	441, 474, 508[110]	—	—	—	431, 465, 495[110]	—	—
Myxoxanthophyll	—	462, 488, 522[76]	—	(460), 488, 522[76]	448, 475, 505[111]	—	—	—	—	(465) 493, 528[76]
Myxoxanthophyll (*trans*)	—	—	—	—	—	—	—	—	—	—
Neochrome	(350), (363), (450), 474, 505[76]	(450), 478, 510[76]	—	—	(375), 398, 421, 449[68]	—	—	—	—	—
Neofucoxanthin A	—	—	—	—	447[105]	—	445[16]	—	—	—
Neofucoxanthin A and B	—	—	—	—	—	—	—	—	—	—
Neoxanthin	413, 438, 466[49]	420, 441, 477[62]	466, 497[33]	422, 449, 466[33]	417, 438, 467[33]	414, 437, 466[105]	416, 437, 466[33]; 414, 436, 465a[34]; 472[113]	415, 439, 467[33]	440, 470[112]	—
Neurosporaxanthin	471.5, (496)[25]	—	—	—	—	—	477.5, 505[25]	471, 505[25]	464[113]	470[113]

Table I.5 (continued)
QUALITATIVE SPECTROSCOPIC DATA: ABSORPTION MAXIMA IN DIFFERENT SOLVENTS

Pigment	Solvent 1 Acetone	Solvent 2 Benzene	Solvent 3 "Benzine"	Solvent 4 Carbon disulfide	Solvent 5 Chloroform	Solvent 6 Cyclohexane	Solvent 7 Ethanol	Solvent 8 Ether	Solvent 9 Hexane	Solvent 10 Petroleum ether	Solvent 11 Methanol	Solvent 12 Pyridine
Neurosporaxanthin (equilibrium mixture)	—	486[113]	—	—	—	—	—	—	—	—	—	—
Neurosporaxanthin-methylester	—	—	—	—	—	—	—	—	—	474, 505[25]	472, (495)[25]	—
Neurosporaxanthin-methylester (*trans*)	—	—	—	—	—	—	—	—	—	(360), 473.5, 504[25]	473[25]	—
Neurosporene	—	—	—	—	—	—	—	—	416, 440, 470[23]	(392), 413, 438, 469[52]	—	—
Nostoxanthin	—	—	—	—	(432), 457, 485[20]	—	(426), 448, 475[20]	—	—	—	—	—
OH-ζ-Carotene	—	—	—	—	—	—	—	—	—	376, 396, 415[53]	—	—
OH-Chlorobactene	—	—	—	—	—	—	—	—	—	435, 461, 491[53]	—	—
OH-Chlorobactene (*trans*)	(350), (440), 464, 494[53]	—	—	—	—	—	—	—	—	—	—	—
OH-Lycopene	—	—	—	—	—	—	—	—	—	445, 474, 506[114]	—	—
OH-Neurosporene	—	—	—	—	—	—	—	—	—	414, 437, 466[114]	—	—
OH-P-481	—	—	—	—	—	—	—	—	—	455, 582, 515[114]	—	—
OH-Phytofluene	—	—	—	—	—	—	—	—	—	332, 347, 367[53]	—	—
OH-R	—	—	—	—	—	—	—	—	—	(460), 482, 515[115]	—	—
OH-Spheroidene	—	440.5, 467.5, 501.5[83]	—	—	—	—	—	—	—	—	—	—

OH-Spheroidenone	484, (505)[104]	501, 530[83]	—	—	—	487[104]	—	460, 483, 516[104]	—
OH-Spirilloxanthin	—	529, 565[114]	—	—	—	—	—	489, 523[114]	—
OH-Spirilloxanthin (*trans*)	373, 389, 467[116]	—	—	—	—	—	—	369, 385, 462, 494, 528[116]	—
Okenone	—	—	—	—	—	—	484[22]	—	—
Okenone (*trans*)	(460), 484, 512[22]	—	—	—	—	—	—	—	—
Oscillaxanthin	466, 490, 522[103]	—	—	494, 528, 568[16]	476, 501, 534[20]	468, 492, 526[20]	—	—	—
Oscillaxanthin (*cis*)	—	—	—	—	—	—	—	—	—
Oscillaxanthin (*trans*)	390, 470, 499, 532[103]	—	—	—	—	—	—	—	—
P-412	—	—	—	—	—	—	—	412, 438[114]	—
P-450	—	—	—	—	—	—	—	423, 450, 480[114]	—
P-481	—	—	—	—	—	—	—	455, 482, 514[114]	—
P-518	495, 528, 559[115]	510, 539, 575[115]	—	528, 562, 601[115]	360, 543[80]	522[115]	—	487.5, 518, 555[115]	—
Peridinin	471[48]	465, 502[117]	—	454, 480, 512[72]	—	472[72]	431, 454, 484[72]	455, 485[122]	467[72]
Peridininol	446[44]	—	—	—	—	—	—	—	—
Phillipsiaxanthin	(495), 528, 559[118]	—	—	—	—	—	—	488, 518, 554[118]	—
Phleixanthophyll	454, 478, 509[6]	—	—	—	(465), 489, 522[6]	—	—	—	(465), 493, 526[6]
Phleixanthophyll (*trans*)	(368), (428), 456, 478, 509[119]	—	—	—	—	—	—	—	—
Physalien	—	—	—	—	—	—	—	452, 480[23]	—
Phytoene	—	—	—	—	—	276, 286, 296[51]	275, 286, 298[120]	(264), 275, 286, 296[52]	—
Phytoene (*trans*)	—	—	—	—	—	—	286, 298[56]	286, 296[52]	—
Phytofluene	338, 355, 374[16]	—	—	—	—	332, 348, 366[51]	331, 347, 367[62]	(317), 331, 346, 368[52]	—
Phytofluene (*cis*)	—	—	—	—	—	—	—	249, 257, 304, 318, 331.5, 347.5, 367[56]	—

Table I.5 (continued)
QUALITATIVE SPECTROSCOPIC DATA: ABSORPTION MAXIMA IN DIFFERENT SOLVENTS

Pigment	Solvent 1 Acetone	Solvent 2 Benzene	Solvent 3 "Benzine"	Solvent 4 Carbon disulfide	Solvent 5 Chloroform	Solvent 6 Cyclohexane	Solvent 7 Ethanol	Solvent 8 Ether	Solvent 9 Hexane	Solvent 10 Petroleum ether	Solvent 11 Methanol	Solvent 12 Pyridine
Phytofluene (trans)	—	—	—	—	—	—	—	—	—	332, 348, 368[56]	—	—
Phytofluenol	—	—	—	—	—	—	—	—	332, 348, 368[16]	—	—	—
Pigment R	—	475, 499, 530[121]	—	495, 519.5, 553[121]	499[121]	—	488[121]	—	—	460, 482, 513[121]	—	—
Plectaniaxanthin	454, 478, 508[69]	—	—	—	—	—	—	—	—	—	—	—
Pro-γ-carotene	—	447.5, 477[1a]	—	460.5, 493.5[1a]	(444), 473[1a]	—	(437), (465)[1a]	—	—	(435), 464[1a]	—	—
Prolycopene	—	455.5, 485[1a]	—	469.5, 500.5[1a]	453.5, 484[1a]	—	445, (471)[1a]	—	—	(414), 436, (463)[123]	—	—
Prolycopene (cis)	—	—	—	—	—	—	—	—	—	440, 468[18]	—	—
Proneurosporene	—	—	—	—	—	—	—	—	—	(404), 436, 463[123]	—	—
Protetrahydro-lycopene	—	—	—	—	—	—	—	—	407, 430[16]	—	—	—
Pyrrhoxanthin	457.5[44]	—	—	512, 586[72]	—	—	471[72]	—	459, 487[72]	—	—	—
Pyrrhoxanthinol	457.5[44]	—	—	—	—	—	—	—	—	—	—	—
Renierapurpurin	—	464, 487, 519[100]	—	477, 504, 544[100]	—	—	—	—	—	—	—	—
Renieratene	—	457, 476, 507[98]	—	463, 496, 532[98]	—	—	—	—	—	—	—	—
Reticulataxanthin	—	—	—	—	—	—	—	—	—	463, 490[124]	—	—
Retro-dehydro-γ-carotene (trans)	(380), (462), 487, 518[119]	—	—	—	—	—	—	—	—	—	—	—
Rhodopin	445, 472, 504[8]	—	—	478, 508, 547[1a]	453, 486, 521[1a]	296, 447, 474, 507	(445), 474, 505[1a]	—	440, 470, 501[16]	447, 470, 501[1a]	—	—
Rhodopin (trans)	—	—	—	511, 550[87]	—	—	—	—	—	—	—	—
Rhodopurpurin	—	—	—	(408), 491.	370, 385.	—	—	—	—	—	—	—
Rhodovibrin	363, 378, 460. 488, 522[125]	372, 388. 473, 503. 535[125]	—	522, 559[125]	469, 498. 532[125]	—	—	—	483[22]	358, 374. 455, 483. 516[125]	—	—

Compound	C1	C2	C3	C4	C5	C6	C7	C8	C9	C10
Rhodoviolascin	—	482, 511, 548[9]	496, 534, 573.5[1a]	476, 507, 544[1a]	(465), 491, 526[1a]	—	—	—	—	—
Rhodoxanthin	—	474, 503.5, 542[1a]	491, 525, 564[1a]	482, 510, 546[1a]	438, 495[1a]	—	522, 489, 458[36]	456, 487, 521[1a]—	—	—
Rubichrome	—	—	476, 506[14]	—	—	—	448, 480[14]	—	—	—
Rubixanthin	—	—	461, 494, 533[14]	439, 474, 509[14]	433, 463, 496[14]	432, 462, 494[65]	—	432, 463, 495.5[65]	—	—
Rubixanthin (cis)	—	—	—	—	—	—	—	430, 455, 485[18]	—	—
Rubixanthin (trans)	(350), (440), 465, 495[53]	—	—	—	—	—	—	—	—	—
Rubixanthinepoxide	—	—	461, 491, 526[14]	474, 504[14]	—	—	—	—	—	—
Saproxanthin	451, 478.5, 509[107]	—	—	(370), 460, 486, 518[107]	—	448, 472.5, 503[107]	—	(360.5), 445, 470, 500[107]	—	(372), 490, 552[107]
Saproxanthin (trans)	(350), (363), 450, 473, 503[76]	—	—	—	—	—	—	—	—	—
Sarcinaxanthin	—	424, 451, 481[1a]	436, 466.5, 499[1a]	423, 451, 480[1a]	(415), 441, 469.5[1a]	—	415, 440, 469[16]	(415), 440, 469[1a]	—	—
Semi-α-carotenone	—	—	499, 533[1a]	—	—	—	—	—	—	—
Semi-β-carotenone	—	458, 486, 518[1a]	499, 538[1a]	(487), (519)[1a]	—	—	443, 469, 500[1a]	446, 470, 501[1a]	—	—
Sintaxanthin	—	422[126]	—	—	—	—	(425), 448, 475[126]	—	—	—
Siphonaxanthin	—	—	—	466[90]	—	—	445—451[90]	(427), 450, 478[90]	—	—
Siphonaxanthin-monolaurate	—	—	—	—	—	—	455[127]	450, (473)[127]	—	—
Siphonein	—	—	—	466[90]	—	—	452—464[90]	(427), 450, 478[90]	—	—
Spheroidene	—	440, 468, 500[9]	—	—	—	—	—	405, 427, 453, 484[74]	—	—
Spheroidene (trans)	—	438, 467, 499[9]	—	—	—	—	—	—	—	—
Spheroidenone	484, (505)[128]	(475), 502, 530[128]	(490), 520, 553[128]	—	—	—	483[129]	—	488[130]	—
Spheroidenone (cis)	—	378, 492.5[9]	—	—	—	—	—	—	—	—
Spheroidenone (trans)	(455), 484, (505)[115]	(475), 501, 530[115]	(490), 520, 553[115]	488[115]	—	—	—	460, 482.5, 515[115]	—	—

Table 1.5 (continued)
QUALITATIVE SPECTROSCOPIC DATA: ABSORPTION MAXIMA IN DIFFERENT SOLVENTS

Pigment	Solvent 1 Acetone	Solvent 2 Benzene	Solvent 3 "Benzine"	Solvent 4 Carbon disulfide	Solvent 5 Chloroform	Solvent 6 Cyclohexane	Solvent 7 Ethanol	Solvent 8 Ether	Solvent 9 Hexane	Solvent 10 Petroleum ether	Solvent 11 Methanol	Solvent 12 Pyridine
Spirilloxanthin	—	378, 395, 479, 510, 548.5[35]	—	(418), 495, 532, 571.5[35]	(475), 505, 543[35]	—	—	463, 493, 528[57]	493[15]	368, 384, 461, 493, 528[35]	—	—
Spirilloxanthin (trans)	373, 389, 468, 498, 533[116]	—	—	496.5, 534, 573.5[16]	476, 507, 544[16]	—	—	—	—	369, 385, 462, 494, 528[116]	—	—
Taraxanthin	—	428.5, 455, 485[131]	443, 472[1a]	441, 469, 501[1a]	—	—	—	—	443, 472[16]	—	—	—
Taraxanthin (trans)	—	—	—	—	—	—	420, 441, 469[106]	—	—	—	—	—
7,8,7',8'-Tetrahydro-ψ,ψ-carotene	—	—	—	—	—	—	—	—	375.5, 397, 421[52]	—	—	—
7,8,11,12-Tetrahydro-ψ,ψ-carotene	—	—	—	—	—	—	—	—	—	374.5, 395, 419[84]	—	—
1,2,1',2'-Tetrahydro-1,1'-dihydroxylyc-opene	—	—	—	—	—	—	445, 469, 500[51]	—	—	—	—	—
1,2,1',2'-Tetrahydro-1,1'-dihydroxylyco-pene (trans)	365, 448, 474, 506.5[8]	458, 485, 520[8]	—	479, 506, 543[8]	458, 485, 518[8]	—	—	—	—	363, 446, 473, 504[8]	—	—
Tetrahydrolycopene	—	—	—	—	—	—	—	—	410, 433[16]	—	—	—
7,8,11,12-Tetra-hydrolycopene	—	—	—	—	—	—	—	—	—	354, 374, 394.5, 418.5[52]	—	—
7,8,11,12-Tetrahydroly-copene (equilibrium mixture)	—	—	—	—	—	—	—	—	—	285, 296, 350, 372, 391, 415[52]	—	—
5,6,5',6'-Tetra-hydrolycopene	—	—	—	—	—	—	—	—	—	413.5, 437.5, 467.5[23]	—	—
Thiothece-460 (trans)	(445), 471, 506[132]	—	—	—	—	—	—	—	—	(412), 435, 460.5, 494.5[132]	(445), 473.5, (500)[132]	—

Thiothece-474 (trans)	(450), 477, (504)[132]	—	—	—	—	—	—	474, 488, 500[123]	477[132]	—
Thiothece-484 (trans)	(465), 478.5, 518[132]	—	—	—	—	—	—	458, 484, 513.5[132]	(462), 486, (511)[132]	—
Thiothece-OH-484 (trans)	460, 484.5, 515[132]	—	—	—	—	—	—	459, 484, 515[132]	(460), 483, (510)[132]	—
Torularhodin	485, 519, 557[16]	—	—	500, 541, 582[16]	—	—	—	468, 502, 539[36]	515[133]	—
Torulene	(380), (463)[5]	470, 503, 541[1a]	491, 525, 565[1a]	469, 501, 539[1a]	456, 486, 520[1a]	—	—	(378), 457.5, 483, 516[5]	483, 516[5]	475, 508, 545[1a]
Triphasiaxanthin	—	—	480, 510[134]	—	—	—	—	440, 467, 495[134]	—	—
Trollichrome	—	432, 459[1a]	450, 479[1a]	430, 458[1a]	—	—	—	—	—	—
Trollixanthin	—	457, 483[1a]	473, 501[1a]	447, 474[1a]	—	—	—	—	—	—
Vaucheriaxanthin	—	—	—	419, 443, 471[135]	—	—	—	—	—	—
Violaxanthin	426, 453, 482[62]	417.5, 443, 472[14]	—	424, 451.5, 482[14]	430, 453, 483[45]	417.5, 442.5, 471[14]	421, 441, 471[105]	417, 440, 470[17] 417, 441, 469[a,47]	420, 443, 472[33]	415, 440, 469[14]
Violaxanthin (cis)	—	—	—	—	—	327, 415, 438, 468[17]	(328), 417, 440, 471[106]	309, 408, 430, 457[17]	—	—
Violaxanthin (trans)	—	—	—	—	—	—	—	417.5, 440, 470[a,34]	—	—
Violaxanthin (9-cis)	—	—	—	—	423, 448, 479[45]	—	—	—	—	—
Violaxanthin (13-cis)	—	—	—	—	422, 448, 478[45]	—	—	—	—	—
Violeoxanthin	—	—	472, 501.5[1a]	—	414, 436, 465[46]	—	—	—	—	—
Violerythrin	556[16]	—	—	—	580[16]	—	—	—	—	—
Xanthophyll	420, 447.5, 477.5[1a]	—	447, 474, 505[37]	—	420, 446.5, 476[1a]	—	—	420, 440, 467[37]	—	418, 444, 473.5[1a]
Xanthophyll K1	—	—	467[90]	454, 466[90]	—	—	—	—	(428), 450, 477[90]	—
Xanthophyll K1S	—	—	458, 468[90]	444, 452[90]	—	—	—	—	(428), 450, 478[90]	—
Xanthophyllepoxide	430, 456, 482.5[131]	—	—	424, 446, 473[33]	—	—	—	—	442, 471[1a]	—
α-Zeacarotene	—	—	—	—	—	—	—	399, 421, 449[137]	—	—

Table I.5 (continued)
QUALITATIVE SPECTROSCOPIC DATA: ABSORPTION MAXIMA IN DIFFERENT SOLVENTS

Pigment	Solvent 1 Acetone	Solvent 2 Benzene	Solvent 3 "Benzine"	Solvent 4 Carbon disulfide	Solvent 5 Chloroform	Solvent 6 Cyclohexane	Solvent 7 Ethanol	Solvent 8 Ether	Solvent 9 Hexane	Solvent 10 Petroleum ether	Solvent 11 Methanol	Solvent 12 Pyridine
β₁-Zeacarotene										427[13]		
Zeaxanthin	(430), 453, 479[103]	432, 459, 488[62]	423, 451.5, 483.5[14]	450, 482, 517[14]	429, 462, 495[14]	—	423.5, 451, 483[14]	—	(426), 449.5, 476.5[x,34]	430, 451, 480[138]	421.5, 449.5, 480.5[1a]	—
Zeaxanthin (*trans*)	—	—	—	—	—	—	428, 452, 478[1]	—	429, 451, 478[x,34]	—	—	—
Zeinoxanthin	—	—	—	449, 474, 505[139]	434, 456, 485[139]	—	—	—	422 (428), 445, (463), 474[4] 446.5, 472[x,34]	—	—	—

[a]   Containing 30% acetone.

REFERENCES

1. **Hertzberg, S., Liaaen-Jensen, S., Enzell, C. R., and Francis, G. W.,** *Acta Chem. Scand.,* 23, 3290, 1969.

1a. **Karrer, P. and Jucker, E.,** *Carotinoide,* Verlag Birkhauser, Basel, 1948.

2. **Hager, A. and Stransky, H.,** *Arch. Mikrobiol.,* 73, 77, 1970.

3. **Egger, K., Nitsche, H., and Kleinig, H.,** *Phytochemistry,* 8, 1583, 1969.

4. **Chapman, D. J.,** *Phytochemistry,* 5, 1331, 1966.

5. **Aasen, A. J. and Liaaen-Jensen, S.,** *Acta Chem. Scand.,* 20, 1970, 1966.

6. **Hertzberg, S. and Liaaen-Jensen, S.,** *Acta Chem. Scand.,* 21, 15, 1967.

7. **Karrer, P. and Leumann, E.,** *Helv. Chim. Acta,* 51 (34), 445, 1951.

8. **Ryvarden, L. and Liaaen-Jensen, S.,** *Acta Chem. Scand.,* 18, 643, 1964.

9. **Barber, M. S., Jackman, L. M., Manchand, P. S., and Weedon, B. C. L.,** *J. Chem. Soc.,* C, 2166, 1966.

10. **Surmatis, J. O.,** *Acta Chem. Scand.,* 31, 186, 1966.

11. **Liaaen-Jensen, S.,** *Acta Chem. Scand.,* 17, 500, 1963.

12. **Francis, G. W., Hertzberg, S., Andersen, K., and Liaaen-Jensen, S.,** *Phytochemistry,* 9, 629, 1970.

13. **Toth, G. and Szabolcs, J.,** *Phytochemistry,* 19, 629, 1980.

14. **Frye, A. H.,** *J. Org. Chem.,* 16, 914, 1951.

15. **Stransky, H. and Hager, A.,** *Arch. Mikrobiol.,* 71, 164, 1970.

16. **Goodwin, T. W.,** Carotenoids, in *Modern Methods of Plant Analysis,* Vol. 3, Paech, K. and Tracey, M. V., Eds., Springer-Verlag, Berlin, 1955, 272.

17. **Stewart, J. and Wheaton, T. A.,** *J. Chromatogr.,* 55, 325, 1971.

18. **Jungalwala, F. B. and Cama, H. R.,** *Biochem. J.,* 85, 1, 1962.

19. **Czeczuga, B.,** *Comp. Biochem. Physiol.,* 48B, 349, 1974.

20. **Stransky, H. and Hager, A.,** *Arch. Mikrobiol.,* 72, 84, 1970.

21. **Hertzberg, S. and Liaaen-Jensen, S.,** *Phytochemistry,* 5, 565, 1966.

22. **Aasen, A. J. and Liaaen-Jensen, S.,** *Acta Chem. Scand.,* 21, 970, 1967.

23. **Isler, O. and Schudel, P.,** Carotine und Carotinoide, in *Wiss. Veroff. dt. Ges. Ernähr.,* Vol. 9, Steinkopff, Darmstadt, 1963, 54.

24. **Singh, H., John, J., and Cama, H. R.,** *J. Chromatogr.,* 75, 146, 1973.

25. **Aasen, A. J. and Liaaen-Jensen, S.,** *Acta Chem. Scand.,* 19, 1843, 1965.

26. **Kjøsen, H. and Liaaen-Jensen, S.,** *Phytochemistry,,* 8, 483, 1969.

27. **Egger, K. and Kleinig, H.,** *Phytochemistry,* 6, 903, 1967.

28. **Egger, K.,** *Phytochemistry,* 4, 609, 1965.

29. **Sørensen, N. A., Liaaen-Jensen, S., Børdalen, B., Haug, A., Enzell, C., and Francis, G.,** *Acta Chem. Scand.,* 22, 344, 1968.

30. **Leftwick, A. P. and Weedon, B. C. L.,** *Chem. Commun.,* 1, 49, 1967.

31. **Campbell, S. A., Mallams, A. K., Waight, E. S., and Weedon, B. C. L.,** *Chem. Commun.,* p. 941, 1967.

32. **Kleinig, H. and Egger, K.,** *Phytochemistry,* 6, 611, 1967.

33. **Stobart, A. K., McLaren, J., and Thomas, D. R.,** *Phytochemistry,* 6, 1467, 1967.

34. **Knowles, R. E. and Livingston, A. L.,** *J. Chromatogr.,* 61, 133, 1971.

35. **Liaaen-Jensen, S.,** *Acta Chem. Scand.,* 14, 950, 1960.

36. **Foppen, F. H. and Gribanovski-Sassu, O.,** *Biochim. Biophys. Acta,* 176, 357, 1969.

37. **Merlini, L. and Cardillo, G.,** *Gazz. Chim. Ital.,* 93, 949, 1963.

38. **Grob, E. C. and Pflugshaupt, R. P.,** *Helv. Chim. Acta,* 48, 930, 1965.

39. **Hsieh, L. K., Lee, T.-C., Chichester, C. O., and Simpson, K. L.,** *J. Bacteriol.,* 118, 385, 1974.

40. **Camara, B. and Moneger, R.,** *Phytochemistry,* 17, 91, 1978.

41. **Hiyama, T., Nishimura, M., and Chance, B.,** *Anal. Biochem.,* 29, 339, 1969.

42. **Hertzberg, S. and Liaaen-Jensen, S.,** *Phytochemistry,* 6, 1119, 1967.

43. **Eichenberger, W. and Grob, E. C.,** *Helv. Chim. Acta,* 181, 1556, 1962.

44. **Johansen, J. E., Svec, W. A., and Liaaen-Jensen, S.,** *Phytochemistry,* 13, 2261, 1974.

45. **Toth, G. and Szabolcs, J.,** *Phytochemistry,* 19, 629, 1980.

46. **Braumann, T. and Grimme, H. L.,** *J. Chromatogr.,* 170, 264, 1979.

47. **Nelson, J. W. and Livingston, A. L.,** *J. Chromatogr.,* 28, 465, 1967.

48. **Björnland, T.,** *J. Phycol.,* 15, 457, 1979.

49. **Björnland, T.,** *Phytochemistry,* 21, 1715, 1982.

50. **Smallidge, R. L.,** *Phytochemistry,* 12, 2481, 1973.

51. **Kleinig, H. and Reichenbach, H.,** *Arch. Mikrobiol.,* 68, 210, 1969.

52. **Davies, B. H., Hallett, C. J., London, R. A., and Rees, A. F.,** *Phytochemistry,* 13, 1209, 1974.

53. **Liaaen-Jensen, S.,** *Acta Chem. Scand.,* 18, 1703, 1964.

54. Manchand, P. S., Rüegg, R., Schwieter, U., Siddons, P. T., and Weedon, B. C. L., *J. Chem. Soc.*, p. 2019, 1965.
55. Nakayama, T. O. M., *Arch. Biochem. Biophys.*, 75, 356, 1958.
56. Davis, J. B., Jackman, L. M., Siddons, P. T., and Weedon, B. C. L., *J. Chem. Soc.*, C, 2154, 1966.
57. Sherma, J., *J. Chromatogr.*, 52, 177, 1970.
58. Yokoyama, H. and White, M. J., *J. Org. Chem.*, 30, 2481, 1965.
59. Yamaguchi, M., *Bull. Chem. Soc. Jpn.*, 30, 979, 1957.
60. Pfander, H. and Wittwer, F., *Helv. Chim. Acta*, 58, 1608, 1979.
61. Chapman, D. J., *Phytochemistry*, 5, 1331, 1966.
62. Buckle, K. A. and Rahman, F. M. M., *J. Chromatogr.*, 171, 385, 1979.
63. Cholnoky, L., Szabolcs, J., Cooper, R. D. G., and Weedon, B. C. L., *Tetrahedron Lett.*, 19, 1257, 1963.
64. Cholnoky, L., Szabolcs, J., and Nagy, E., *Liebigs Ann. Chem.*, 616, 207, 1958.
65. Hida, M. and Ida, K., *Bot. Mag. Tokyo*, 77, 458, 1964.
66. Bodea, C. and Tàmas, V., *Ann. Chem.*, 671, 57, 1964.
67. Zechmeister, L. and Wallcave, L., *J. Am. Chem. Soc.*, 75, 5341, 1953.
68. Davies, B. H., Hallett, C. J., London, R. A., and Rees, A. F., *Phytochemistry*, 13, 1209, 1974.
69. Arpin, N. and Liaaen-Jensen, S., *Phytochemistry*, 6, 995, 1967.
70. Isler, O., Montavon, M., Rüegg, R., and Zeller, P., *Helv. Chim. Acta*, 39, 454, 1956.
71. Jackman, L. M. and Liaaen-Jensen, S., *Acta Chem. Scand.*, 15, 2058, 1961.
72. Loeblich, A. R. and Smith, V. E., *Lipids*, 3, 5, 1967.
73. Taylor, R. F. and Davies, B. H., *Biochem. J.*, 139, 751, 1974.
74. Taylor, R. F. and Davies, B. H., *Biochem. J.*, 153, 233, 1976.
75. Allen, M. B., Fries, L., Goodwin, T. W., and Thomas, O., *J. Gen. Microbiol.*, 34, 259, 1964.
76. Hertzberg, S. and Liaaen-Jensen, S., *Phytochemistry*, 8, 1259, 1969.
77. Singh, R. K., Britton, G., and Goodwin, T. W., *Biochem. J.*, 136, 413, 1973.
78. Hertzberg, S. and Liaaen-Jensen, S., *Acta Chem. Scand.*, 20, 1187, 1966.
79. Fiasson, J.-L. and Arpin, N., *Bull. Soc. Chim. Biol.*, 49, 537, 1967.
80. Schwieter, U., Rüegg, R., and Isler, O., *Helv. Chim. Acta*, 49, 992, 1966.
81. Ungers, G. E. and Cocney, J. J., *J. Bacteriol.*, 96, 234, 1968.
82. Petracek, F. J. and Zechmeister, L., *J. Am. Chem. Soc.*, 78, 1427, 1956.
83. Jackman, L. M. and Liaaen-Jensen, S., *Acta Chem. Scand.*, 18, 1403, 1964.
84. Schmidt, K. and Liaaen-Jensen, S., *Acta Chem. Scand.*, 27, 3040, 1973.
85. Krinsky, N. I. and Goldsmith, T. H., *Arch. Biochem. Biophys.*, 91, 271, 1960.
86. Szabolcs, J., *Pure Appl. Chem.*, 47, 147, 1976.
87. Karrer, P. and Solmssen, U., *Helv. Chim. Acta*, 19, 1019, 1936.
88. Karrer, P. and Jucker, E., *Helv. Chim. Acta*, 28, 300, 1945.
89. Eskins, K. and Harris, L., *Photochem. Photobiol.*, 33, 131, 1981.
90. Ricketts, T. R., *Phytochemistry*, 6, 1375, 1967.
91. Arpin, N. and Liaaen-Jensen, S., *Phytochemistry*, 8, 185, 1969.
92. Aasen, A. J., Francis, G. W., and Liaaen-Jensen, S., *Acta Chem. Scand.*, 23, 2605, 1969.
93. Bonnett, R., Spark, A. A., and Weedon, B. C. L., *Acta Chem. Scand.*, 18, 1739, 1964.
94. Liaaen-Jensen, S. and Hertzberg, S., *Acta Chem. Scand.*, 20, 1703, 1966.
95. Liaaen-Jensen, S., *Acta Chem. Scand.*, 19, 1166, 1965.
96. Wallcave, L. and Zechmeister, L., *J. Am. Chem. Soc.*, 75, 4495, 1953.
97. Jensen, A., *Acta Chem. Scand.*, 20, 1728, 1966.
98. Yamaguchi, M., *Bull. Chem. Soc. Jpn.*, 31, 739, 1958.
99. Liaaen-Jensen, S., *Acta Chem. Scand.*, 19, 1025, 1965.
100. Cooper, R. D. G., Davis, J. B., and Weedon, B. C. L., *J. Chem. Soc.*, 10, 5637, 1963.
101. Entschel, R. and Karrer, P., *Helv. Chim. Acta*, 112, 983, 1958.
102. Leftwick, A. P. and Weedon, B. C. L., *Acta Chem. Scand.*, 20, 1195, 1966.
103. Hertzberg, S. and Liaaen-Jensen, S., *Phytochemistry*, 5, 557, 1966.
104. Liaaen-Jensen, S., *Acta Chem. Scand.*, 17, 555, 1963.
105. Jeffrey, S. W., *Biochem. J.*, 80, 336, 1961.
106. Nitsche, H. and Egger, K., *Phytochemistry*, 8, 1577, 1969.
107. Aasen, A. J. and Liaaen-Jensen, S., *Acta Chem. Scand.*, 20, 811, 1966.
108. Conti, S. F. and Benedict, C. R., *J. Bacteriol.*, 83, 929, 1962.
109. Kjøsen, H. and Liaaen-Jensen, S., *Acta Chem. Scand.*, 25, 1500, 1971.
110. Heilbron, I. M. and Lythgoe, B., *J. Chem. Soc.*, p. 1376, 1936.
111. Sherma, J., *Anal. Lett.*, 3, 35, 1970.
112. Sherma, J. and Latta, M., *J. Chromatogr.*, 154, 73, 1978.

113. **Zalokar, M.,** *Arch. Biochem. Biophys.,* 70, 568, 1957.
114. **Liaaen-Jensen, S., Cohen-Bazire, G., Nakayama, T. O. M., and Stanier, R. K.,** *Biochim. Biophys. Acta,* 29, 477, 1958.
115. **Liaaen-Jensen, S.,** *Acta Chem. Scand.,* 17, 303, 1963.
116. **Liaaen-Jensen, S.,** *Acta Chem. Scand.,* 14, 953, 1960.
117. **Pinckard, J. H., Kittredge, J. S., Fox, D. L., Haxo, F. T., and Zechmeister, L.,** *Arch. Biochem. Biophys.,* 44, 189, 1953.
118. **Arpin, N. and Liaaen-Jensen, S.,** *Bull. Soc. Chim. Biol.,* 49, 527, 1967.
119. **Hertzberg, S. and Liaaen-Jensen, S.,** *Acta Chem. Scand.,* 20, 1187, 1966.
120. **Kushwaha, S. C., Pugh, E. L., Kramer, I. K. G., and Kates, M.,** *Biochim. Biophys. Acta,* 260, 492, 1972.
121. **Goodwin, T. W., Land, D. G., and Sissins, M. E.,** *Biochem. J.,* 64, 486, 1956.
122. **Strain, H. H., Manning, W. M., and Hardin, G.,** *Biol. Bull.,* 86, 169, 1944.
123. **Quereshi, A. A., Kim, M., Quereshi, N., and Porter, J. W.,** *Arch. Biochem. Biophys.,* 162, 108, 1974.
124. **Yokoyama, H., White, M. J., and Vandercook, C. E.,** *J. Org. Chem.,* 30, 2482, 1965.
125. **Liaaen-Jensen, S.,** *Acta Chem. Scand.,* 13, 2143, 1959.
126. **Yokoyama, H. and White, M. J.,** *J. Org. Chem.,* 30, 3994, 1965.
127. **Kleinig, H. and Egger, K.,** *Phytochemistry,* 6, 1681, 1967.
128. **Liaaen-Jensen, S.,** *Acta Chem. Scand.,* 17, 489, 1963.
129. **Aasen, A. J. and Liaaen-Jensen, S.,** *Acta Chem. Scand.,* 21, 970, 1967.
130. **Liaaen-Jensen, S.,** *Acta Chem. Scand.,* 17, 500, 1963.
131. **Eugster, C. H. and Karrer, P.,** *Helv. Chim. Acta,* 40, 69, 1957.
132. **Andrewes, A. G. and Liaaen-Jensen, S.,** *Acta Chem. Scand.,* 26, 2194, 1972.
133. **Simpson, K. L., Nakayama, T. O. M., and Chichester, C. O.,** *J. Bacteriol.,* 88, 1688, 1964.
134. **Yokoyama, H. and Guerrero, H. C.,** *J. Org. Chem.,* 35, 2080, 1970.
135. **Kleinig, H. and Egger, K.,** *Z. Naturforsch.,* 22, 868, 1967.
136. **Hertzberg, S. and Liaaen-Jensen, S.,** *Acta Chem. Scand.,* 23, 3290, 1969.
137. **Davies, B. H.,** *Chemistry and Biochemistry of Plant Pigments,* Goodwin, T. W., Eds., Academic Press, London, 1965, 489.
138. **Allen, M. B., Fries, L., Goodwin, T. W., and Thomas, D. M.,** *J. Gen. Microbiol.,* 34, 259, 1934.
139. **Livingston, A. L. and Knowles, R. E.,** *Phytochemistry,* 8, 1511, 1969.

Table I. 6
QUANTITATIVE SPECTROSCOPIC DATA: MOLAR EXTINCTION COEFFICIENTS ($cm^{-1}M^{-1}$) OF CAROTENOIDS

NOTES

Since a number of presumably identical carotenoids have been named differently by different authors, the same pigment may appear under two or even more names. If the desired pigment is not found, refer to row 1 (vertical) of the name list (Table I.3).

The first vertical row contains the pigment names. In the following vertical rows, the first number of each pair gives the wavelength and the second the molar extinction coefficient (with the reference number superscripted to it). If, in the original literature, the A (1%, 1 cm) value is published instead of the molar extinction coefficient, the corresponding ϵ-value is calculated according to the following formula:

$$\epsilon = 0.1 \times A~(1\%, 1~cm) \times mol.~wt.$$

where ϵ = molar extinction coefficient and A (1%, 1 cm) = absorbance of a 1% solution of the carotenoid in a cuvette with a path length of 1 cm. The molecular weight is deduced from Table I.3.

With the values given, it is easily possible to calculate the carotenoid concentration as well as the absolute milligram amounts by using the molecular weight. For the preparation of the present table, the utmost care has been exercised. The available literature has been followed, especially where reviews have been used. Compound names are in strictly alphabetical order.

The amount of carotenoid may be calculated using the formula:

$$mmol~of~carotenoid = volume~(in~m\ell) \times absorbance \times \epsilon^{-1}$$

The amount of carotenoid in milligrams is calculated by multiplication by the molecular weight:

$$amount~of~carotenoid~(in~mg) = mmol \times mol.~wt. \times \epsilon^{-1}$$

Table I. 6 (continued)
QUANTITATIVE SPECTROSCOPIC DATA: MOLAR EXTINCTION COEFFICIENTS (cm$^{-1}M^{-1}$) OF CAROTENOIDS

Pigment	Solvent: Acetone		Benzene		Carbon disulfide		Chloroform		Cyclohexane	
	λmax	ε	λmax	ε	λmax	ε	λmax	ε	λmax	ε
Aleunaxanthin	463	134.9[1]	—	—	—	—	—	—	—	—
Arhydroeschscholtzxanthin	—	—	—	—	—	—	—	—	—	—
Antheraxanthin	—	—	—	—	—	—	—	—	—	—
β-Apo-2'-carotenal	—	—	—	—	—	—	—	—	...	—
β-Apo-8'-carotenal	—	—	—	—	—	—	—	—	—	—
β-Apo-10'-carotenal	—	—	—	—	—	—	—	—	—	—
β-Apo-12'-carotenal (*trans*)	—	—	—	—	—	—	—	—	—	—
β-Apo-4'-carotenoic acid (*trans*)	473	122.7[5]	—	—	—	—	—	—	—	—
β-Apo-8'-carotenoic acid (*trans*)	—	—	—	—	—	—	—	—	—	—
β-Apo-10'-carotenoic acid (*trans*)	—		—	—	—	—			—	—
Apo-3-lycopenal	—	—	488	116.0[6]	—	—			475	118.3[7]
Astacene	—	—	—	—	—	—			—	—
Astaxanthin-diacetate	—	—	482	106.0[8]	—	—			—	—
Aurochrome	—	—	387	73.5[9]	—	—			—	—
			409	116.0[9]						
			434	115.0[9]						
Auroxanthin	—	—	—	—	—	—			—	—
Azafrın	—	—	—	—	—	—			—	—
Bacterioruberin α	499	194.2[10]	—	—	—	—			—	—
3,4,3',4'-bisdehydro-β-carotene	—	—	—	—	—	—			—	—
Bixin	—	—	—	—	—	—			—	—
Canthaxanthin	—	—	480	118.2[1]	—	—			469	124.3[1]
Capsanthin	—	—	518	121.0[2]	—	—				
			483	95.0[2]						
			483							
Capsorubin	—	—	489	132.2[1]	—	—	—	—		
α-Carotene	448	145.0[13]	—	—	477	117.0[1]	456	129.9[3]	—	—
β-Carotene	454	134.4[13]	465	125.5[1]	484	107.8[1]	465	128.6[1]	457	134.5[4]
									485	118.3[4]
β-Carotene-di-epoxide	—	—	426	90.0	—	—	—	—	—	—
			451	136.0						
			481	127.0						
β,β-Carotene	—	—	—	—	—	—	—	—	—	—
γ-Carotene	—	—	—	—	—	—	—	—	—	—
δ-Carotene	—	—	—	—	—	—	—	—	—	—
δ-Carotene (*trans*)	—	—	—	—	—	—	—	—	—	—
ε-Carotene	—	—	—	—	—	—	—	—	—	—
ϵ_1-Carotene	—	—	—	—	—	—	—	—	419	99.9[4]
									444	155.2[4]
									475	151.4[4]
ζ-Carotene	—	—	—	—	—	—	—	—	—	—
β-Carotenone	—	—	—	—	—	—	490	97.7[1]	—	—
Chlorobactene	—	—	—	—	—	—	—	—	—	—

Table I. 6 (continued)
QUANTITATIVE SPECTROSCOPIC DATA: MOLAR EXTINCTION COEFFICIENTS ($cm^{-1}M^{-1}$) OF CAROTENOIDS

Ethanol		Ether		Hexane		Petroleum ether		Methanol		Pyridine	
λmax	ε	λmax	ε	λmax	ε	λmax	ε	λmax	ε	λmax	ε
—	—	—	—	—	—	460	143.8[1]	—	—	—	—
—	—	—	—	499	160.0[2]	—	—	—	—	—	—
446	137.4[3]	—	—	—	—	—	—	—	—	—	—
—	—	—	—	—	—	498	138.9[1]	—	—	—	—
—	—	—	—	—	—	457	110.0[1]	—	—	—	—
—	—	—	—	—	—	435	82.5[1]	—	—	—	—
—	—	—	—	—	—	414	75.7[4]	—	—	—	—
—	—	—	—	—	—	—	—	—	—	—	—
—	—	—	—	—	—	448	108.8[4]	—	—	—	—
—	—	—	—	—	—	430	87.7[4]	—	—	—	—
—	—	—	—	—	—	—	—	—	—	—	—
—	—	—	—	—	—	—	—	—	—	498	100.2[1]
—	—	—	—	—	—	—	—	—	—	—	—
—	—	—	—	—	—	—	—	—	—	—	—
402	111.2[1]	—	—	—	—	—	—	—	—	—	—
—	—	—	—	—	—	409	93.9[1]	—	—	—	—
—	—	—	—	—	—	—	—	—	—	—	—
—	—	—	—	471	127.0[11]	471	127.9[4]	—	—	—	—
—	—	—	—	—	—	456	165.7[1]	—	—	—	—
—	—	—	—	—	—	463	107.3[12]	—	—	—	—
						465—467	124.3[4]				
—	—	—	—	—	—	—	—	—	—	—	—
—	—	—	—	422	98.2[15]	422	102.0[4]	—	—	—	—
				427	96.1[15]	444	150.3[4]				
—	—	—	—	446	145.7[15]	473	135.3[4]	—	—	—	—
				463	98.5[15]						
				475	133.7[15]						
426	118.8[14]	—	—	451	134.5[13]	451	134.2[17]	—	—	—	—
				436	105.2[x,16]	448	139.2[18]				
—	—	—	—	—	—	—	—	—	—	—	—
—	—	—	—	—	—	451	134.5[19]	—	—	—	—
—	—	—	—	462	148.2[1]	433	119.0[20]	—	—	—	—
						459	171.0[20]				
						490	150.0[20]				
—	—	—	—	—	—	280	40.0[20]	—	—	—	—
						431	150.0[20]				
						456	174.0[20]				
						488	165.0[20]				
—	—	—	—	—	—	281	44.0[20]	—	—	—	—
						431	111.0[20]				
						456	174.0[20]				
						489	159.0[20]				
—	—	—	—	—	—	266	36.0[20]	—	—	—	—
						416	104.0[20]				
						440	167.0[20]				
						470	168.0[20]				
—	—	—	—	—	—	—	—	—	—	—	—
—	—	—	—	400	138.2[1]	396	108.2[21]	—	—	—	—
						399	135.2[22]				
—	—	—	—	—	—	—	—	—	—	—	—
—	—	—	—	—	—	435	116.0[6]	—	—	—	—
						461	161.0[6]				
						491	143.0[6]				

Table I. 6 (continued)
QUANTITATIVE SPECTROSCOPIC DATA: MOLAR EXTINCTION COEFFICIENTS (cm$^{-1}M^{-1}$) OF CAROTENOIDS

Pigment	Solvent: Acetone λmax	ε	Benzene λmax	ε	Carbon disulfide λmax	ε	Chloroform λmax	ε	Cyclohexane λmax	ε
Chlorobactene (trans)	—	—	—	—	—	—	—	—	—	—
Chloroxanthin	—	—	—	—	—	—	—	—	—	—
Chrysanthemaxanthin	—	—	—	—	—	—	—	—	—	—
Citranaxanthin	—	—	—	—	—	—	—	—	—	—
Crocetin	—	—	—	—	—	—	—	—	—	—
Crocetin-dialdehyde	—	—	—	--	—	—	—	—	—	—
Crocetin-dimethylester	—	—	—	—	—	—	—	—	—	—
Cryptocapsin	—	—	486	112.0[26]	—	—	—	—	—	—
			520	87.0[26]						
Cryptocapsone	—	—	491	108.0[26]	—	—	—	—	—	—
α-Cryptoxanthin	—	—	277	22.0[15]	—	—	—	—	—	—
			433	86.5[15]						
			439	84.2[15]						
			457	130.0[15]						
			475	86.5[15]						
			488	116.9[15]						
β-Cryptoxanthin	—	—	—	—	—	—	—	—	—	—
Deepoxyneoxanthin	—	—	—	—	—	—	—	—	—	—
Dehydro-β-carotene	—	—	—	—	—	—	—	—	—	—
Dehydro-β-carotene (trans)	—	—	—	—	—	—	—	—	—	—
3,4-Dehydro-β-carotene	—	—	—	—	—	—	—	—	—	—
3,4-Dehydro-lycopene	—	—	—	—	—	—	—	—	—	—
Dehydroretrocarotene	—	—	—	—	—	—	—	—	—	—
3,4-Dehydrorhodopin	—	—	—	—	—	—	—	—	—	—
Deoxyflexixanthin	—	—	—	—	—	—	—	—	—	—
7,7'-Dihydro-β-carotene	—	—	—	—	—	—	—	—	—	—
2,2'-Dihydroxy-β-carotene	452	117.2[1]	—	—	—	—	—	—	—	—
4,4'-Dihydroxy-β-carotene	—	—	—	—	—	—	—	—	—	—
2,2'-Diketobacterioruberin	—	—	—	—	—	—	361	32.2[32]	—	—
							543	145.0[32]		
4,4'-Diketo-β-carotene	—	—	—	—	—	—	—	—	—	—
Echinenone	—	—	472	115.2[1]	—	—	—	—	461	116.2[1]
Eschscholtzxanthin	—	—	—	—	—	—	—	—	—	—
Flavoxanthin	—	—	432	149.1[1]	—	—	—	—	—	—
Fucoxanthinol	—	—	—	—	—	—	—	—	—	—
Gazaniaxanthin	462	142.6[1]	—	—	—	—	—	—	—	—
Helenien	—	—	278	20.7[15]	—	—	—	—	—	—
			433	87.8[15]						
			439	85.8[15]						
			457	130.9[15]						
			474	88.0[15]						
			487	118.1[15]						
2-Hydroxy-α-carotene	447	128.8[1]	—	—	—	—	—	—	—	—
2-Hydroxy-β-carotene	452	126.6[1]	—	—	—	—	—	—	—	—
4-Hydroxy-β-carotene	—	—	—	—	—	—	—	—	457	138.5[4]
									485	121.9[4]

Table I. 6 (continued)
QUANTITATIVE SPECTROSCOPIC DATA: MOLAR EXTINCTION COEFFICIENTS ($cm^{-1}M^{-1}$) OF CAROTENOIDS

Ethanol		Ether		Hexane		Petroleum ether		Methanol		Pyridine	
λmax	ε	λmax	ε	λmax	ε	λmax	ε	λmax	ε	λmax	ε
—	—	—	—	—	—	460	159.9[23]	—	—	—	—
—	—	—	—	—	—	415	91.2[24]	—	—	—	—
						439	137.8[24]				
						469	137.8[24]				
421	122.8[1]	—	—	—	—	—	—	—	—	—	—
—	—	—	—	—	—	463	98.0[25]	—	—	—	—
—	—	—	—	—	—	450	141.9[1]	—	—	—	—
—	—	—	—	—	—	408	86.9[4]	—	—	—	—
						430	155.3[4]				
						458	172.5[4]				
—	—	—	—	—	—	400	83.4[4]	—	—	—	—
						422	138.1[4]				
						450	141.9[4]				
—	—	—	—	—	—	—	—	—	—	—	—
—	—	—	—	—	—	—	—	—	—	—	—
—	—	—	—	—	—	267	25.6[15]	—	—	—	—
						421	97.9[15]				
						427	94.6[15]				
						446	145.5[15]				
						462	96.8[15]				
						475	134.5[15]				
—	—	—	—	—	—	452	131.0[4]	—	—	—	—
						480	109.0[4]				
445	137.4[1]	—	—	—	—	—	—	—	—	—	—
—	—	—	—	471	166.0[27]	—	—	—	—	—	—
—	—	—	—	471	166.0[28]	—	—	—	—	—	—
—	—	—	—	—	—	461	124.2[1]	—	—	—	—
—	—	—	—	—	—	492	160.5[1]	—	—	—	—
—	—	—	—	—	—	446	124.1[29]	—	—	—	—
						472	171.4[29]				
						502	139.6[29]				
—	—	—	—	—	—	483	171.4[30]	—	—	—	—
—	—	—	—	—	—	476.5	150.2[31]	—	—	—	—
—	—	—	—	—	—	382	105.4[4]	—	—	—	—
						405	165.9[4]				
						429	161.5[4]				
—	—	—	—	—	—	—	—	—	—	—	—
—	—	—	—	450	134.0[11]	—	—	—	—	—	—
—	—	—	—	—	—	—	—	—	—	—	—
—	—	—	—	466	114.0[11]	—	—	—	—	—	—
				468	111.0[23]						
—	—	—	—	—	—	458	118.9[1]	—	—	—	—
—	—	—	—	438	136.0[4]	—	—	—	—	—	—
				464	185.9[4]						
				495	147.4[4]						
452	89.6[1]	—	—	—	—	—	—	—	—	—	—
—	—	—	—	—	—	—	—	—	—	—	—
—	—	—	—	267	26.5[15]	—	—	—	—	—	—
				420	98.1[15]						
				427	95.6[15]						
				445	145.5[15]						
				462	99.0[15]						
				475	133.9[15]						
—	—	—	—	—	—	—	—	—	—	—	—
—	—	—	—	—	—	—	—	—	—	—	—

Table I. 6 (continued)
QUANTITATIVE SPECTROSCOPIC DATA: MOLAR EXTINCTION
COEFFICIENTS ($cm^{-1}M^{-1}$) OF CAROTENOIDS

Pigment	Solvent: Acetone		Benzene		Carbon disulfide		Chloroform		Cyclohexane	
	λmax	ε	λmax	ε	λmax	ε	λmax	ε	λmax	ε
1-Hydroxy-1,2-dihydro-γ-carotene	—	—	—	—	—	—	—	—	—	—
4′-Hydroxyechinenone	—	—	—	—	—	—	—	—	—	—
3-Hydroxy-3′-keto-α-carotene	449	137.5[34]	—	—	—	—	—	—	—	—
4-Hydroxy-4′-keto-β-carotene (*trans*)	—	—	—	—	—	—	—	—	—	—
2-Hydroxyplectaniaxanthin	476	143.0[1]	—	—	—	—	—	—	—	—
Isocryptoxanthin	—	—	—	—	—	—	—	—	—	—
Isofucoxantin	—	—	—	—	—	—	—	—	—	—
Isorenieratene	—	—	443	95.0[36]	—	—	—	—	—	—
			465	123.0[36]						
			493	106.0[36]						
β-Isorenieratene	—	—	—	—	456	88.0[36]	—	—	—	—
					487	118.0[36]				
					508	107.0[36]				
Isozeaxanthin	—	—	—	—	—	—	—	—	—	—
4-Keto-α-carotene	—	—	—	—	—	—	—	—	—	—
4-Keto-γ-carotene (*trans*)	471	131.1[38]	—	—	—	—	—	—	—	—
4-Keto-4′-ethoxy-β-carotene	—	—	—	—	—	—	—	—	—	—
4-Keto-4′-hydroxy-β-carotene	—	—	—	—	—	—	—	—	—	—
4-Keto-phleixanthophyll	480	142.0[39]	—	—	—	—	—	—	—	—
Lutein	—	—	458	127.2[1]	475	122.9[1]	—	—	—	—
Lutein-5,6-epoxide	—	—	—	—	—	--	—	—	—	—
Lycopene	474	185.0[41]	487	180.9[1]	—	—	—	—	—	—
Lycopene (*trans*)	—	—	—	—	—	—	—	—	—	—
Lycophyll	508	184.3[1]	—	—	—	—	—	—	—	—
Lycoxanthin	474	170.3[43]	—	—	—	—	—	—	—	—
Lycoxanthin (*trans*)	476	186.9[44]	—	—	—	—	—	—	—	—
		193.5								
Methyl-apo-6′-lycopenoate	—	—	—	—	—	—	—	—	—	—
Methylbixin (*trans*)	—	—	—	—	—	—	—	—	—	—
Methyl-1-hexosyl-1,2-dihydro-3,4-didehydro-apo-8′-lycopenoate	469	122.3[46]	—	—	—	—	—	—	—	—
3,4-Monodehydro-β-carotene	—	—	—	—	—	—	—	—	—	—
Mutatochrome	—	—	416	77.0[9]	—	—	—	—	—	—
			437	113.0[9]						
			463	101.0[9]						
Myxoxanthophyll	878	157.9[48]	—	—	—	—	—	—	—	—
Neochrome	—	—	—	—	—	—	—	—	—	—
Neoxanthin	—	—	—	—	—	—	—	—	—	—
Neurosporaxanthin	—	—	486	110.2[49]	—	—	—	—	—	—
Neurosporaxanthin-methylester (*trans*)	—	—	—	—	—	—	—	—	—	—
Neurosporene	—	—	—	—	—	—	—	—	—	—
OH-Chlorobactene	—	—	—	—	—	—	—	—	—	—
OH-R	—	—	—	—	—	—	—	—	—	—
OH-Spheroidene	—	—	467.5	163.5[52]	—	—	—	—	—	—
OH-Speroidenone	—	—	501	124.4[52]	—	—	—	—	—	—
OH-Y (*trans*)	—	—	467.5	163.5[53]	—	—	—	—	—	—

Table I. 6 (continued)
QUANTITATIVE SPECTROSCOPIC DATA: MOLAR EXTINCTION COEFFICIENTS (cm⁻¹M⁻¹) OF CAROTENOIDS

Ethanol		Ether		Hexane		Petroleum ether		Methanol		Pyridine	
λmax	ε	λmax	ε	λmax	ε	λmax	ε	λmax	ε	λmax	ε
—	—	—	—	—	—	444	128.0[o]	—	—	—	—
						462	156.0[6]				
						494	143.0[6]				
—	—	—	—	—	—	454	127.5[1]	—	—	—	—
—	—	—	—	—	—	—	—	—	—	—	—
—	—	—	—	—	—	454	127.5[12]	—	—	—	—
—	—	—	—	451	134.0[27]	—	—	—	—	—	—
—	—	—	—	—	—	453	105.4[35.]	—	—	—	—
—	—	—	—	—	—	452	137.5[23]	—	—	—	—
—	—	—	—	—	—	—	—	—	—	—	—
—	—	—	—	—	—	451	136.5[4]	—	—	—	—
						478	121.7[4]				
—	—	—	—	452	126.0[37]	—	—	—	—	—	—
				453.5	160.9[4]						
—	—	—	—	—	—	—	—	—	—	—	—
—	—	—	—	459	126.0[33]	—	—	—	—	—	—
—	—	—	—	458	122.0[11]	—	—	—	—	—	—
—	—	—	—	—	—	—	—	—	—	—	—
445	145.1[1]	—	—	—	—	—	—	—	—	—	—
440	140.4[40]	—	—	—	—	—	—	—	—	—	—
—	—	—	—	—	—	446	120.8[4]	—	—	—	—
						472	185.2[4]				
						505	169.1[4]				
—	—	—	—	—	—	474	157.8[42]	—	—	—	—
—	—	—	—	—	—	—	—	—	—	—	—
—	—	—	—	—	—	—	—	—	—	—	—
—	—	—	—	—	—	—	—	—	—	—	—
—	—	—	—	—	—	471	122.9[1]	—	—	—	—
—	—	—	—	—	—	432	108.3[23]	—	—	—	—
						456	165.5[23]				
						490	165.5[23]				
—	—	—	—	—	—	—	—	—	—	—	—
—	—	—	—	—	—	461	124.6[23]	—	—	—	—
—	—	428	125[47]	—	—	—	—	—	—	—	—
—	—	—	—	—	—	—	—	—	—	—	—
424	136.4[1]	—	—	—	—	—	—	—	—	—	—
438	136.4[19]	—	—	—	—	—	—	—	—	—	—
—	—	—	—	472	122.2[49]	477.5	85.5[5]	464	133.7[x,49]	—	—
								470	114.7[y,49]		
—	—	—	—	—	—	473.5	143.6[5]	—	—	—	—
—	—	440	157.3[1]	—	—	413	95.0[50]	—	—	—	—
						437.5	147.0[50]				
						468	147.0[50]				
—	—	—	—	—	—	435	100.0[6]	—	—	—	—
						461	135.0[6]				
						491	115.0[6]				
—	—	—	—	—	—	482	120.2[51]	—	—	—	—
—	—	—	—	—	—	—	—	—	—	—	—
—	—	—	—	—	—	456	146.7[53]	—	—	—	—

Table I. 6 (continued)
QUANTITATIVE SPECTROSCOPIC DATA: MOLAR EXTINCTION COEFFICIENTS (cm$^{-1}M^{-1}$) OF CAROTENOIDS

Pigment	Solvent: Acetone λmax	ε	Benzene λmax	ε	Carbon disulfide λmax	ε	Chloroform λmax	ε	Cyclohexane λmax	ε
Oscillaxanthin	—	—	—	—	—	—	—	—	—	—
P-412	—	—	—	—	—	—	—	—	—	—
P-450	—	—	—	—	—	—	—	—	—	—
P-481	—	—	—	—	—	—	—	—	—	—
P-518	—	—	—	—	561	109.0[54]	360	30.6[32]	—	—
							543	138.1[32]		
Philosamiaxanthin	—	—	—	—	—	—	—	—	—	—
Phleixanthophyll	—	—	—	—	—	—	—	—	—	—
Physalien	—	—	—	—	—	—	—	—	—	—
Phytoene	—	—	—	—	—	—	—	—	—	—
Phytoene (15-*cis*)	—	—	—	—	—	—	—	—	—	—
Phytoene (*trans*)	—	—	—	—	—	—	—	—	—	—
Phytofluene	—	—	—	—	—	—	—	—	—	—
Phytofluene (*trans*)	—	—	—	—	—	—	—	—	—	—
Plectanixanthin	474	142.5[1]	—	—	—	—	—	—	—	—
Pyrenoxanthin	—	—	—	—	—	—	454	103.1[1]	—	—
Renierapurpurin	—	—	—	—	477	83.0[36]	—	—	—	—
					504	109.0[36]				
					544	88.0[36]				
Renieratene	—	—	—	—	467	84.0[36]	—	—	—	—
					497	113.0[36]				
					532	103.0[36]				
Retro-dehydro-β-Carotene	—	—	—	—	—	—	—	—	473	167.4[1]
Rhodopin	—	—	—	—	—	—	—	—	474	165.9[7]
Rhodopin (*trans*)	472	165.4[42]	—	—	—	—	—	—	—	—
Rhodovibrin	—	—	—	—	—	—	—	—	—	—
Rhodoxanthin	—	—	—	—	—	—	—	—	491	134.3[1]
Rubixanthin	462	160.8[1]	—	—	—	—	—	—	—	—
Saproxanthin	478.5	162.0[41]	—	—	—	—	—	—	—	—
Semi-α-carotenone	—	—	—	—	—	—	500	81.4[1]	—	—
Semi-β-carotenone	—	—	—	—	—	—	—	—	—	—
Sintaxanthin	—	—	462	111.5[1]	—	—	—	—	—	—
Spheroidene	—	—	438	104.0[24]	—	—	—	—	—	—
			465	157.0[24]						
			497	147.0[24]						
Spheroidenone	—	—	472	93.4[24]	—	—	—	—	—	—
			533	99.0[24]						
Spheroidenone (*cis*)	—	—	497	127.0[24]	—	—	—	—	—	—
Spirilloxanthin	—	—	—	—	—	—	—	—	—	—
Taraxanthin	—	—	428.5	91.7[56]	—	—	—	—	—	—
			455	138.6[56]						
			485	132.3[56]						
7′,8′,11′,12′-tetrahydro-γ-carotene	—	—	—	—	—	—	—	—	—	—
1,2,1′,2′-Tetrahydro-1,1′-dihydroxylycopene	474	90.8[42]	—	—	—	—	—	—	—	—
5,6,5′,6′-Tetrahydrolycopene	—	—	—	—	—	—	—	—	—	—
7,8,11,12-Tetrahydrolycopene	—	—	—	—	—	—	—	—	—	—
Torularhodin	—	—	—	—	—	—	515	109.1[57]	—	—
Torularhodin-aldehyde	—	—	—	—	—	—	—	—	—	—

Table I. 6 (continued)
QUANTITATIVE SPECTROSCOPIC DATA: MOLAR EXTINCTION COEFFICIENTS (cm$^{-1}M^{-1}$) OF CAROTENOIDS

Ethanol		Ether		Hexane		Petroleum ether		Methanol		Pyridine	
λmax	ε	λmax	ε	λmax	ε	λmax	ε	λmax	ε	λmax	ε
—	—	—	—	—	—	—	—	490	67.0[2,1]	—	—
—	—	—	—	—	—	412	135.3[21]	—	—	—	—
—	—	—	—	—	—	450	176.4[21]	—	—	—	—
—	—	—	—	—	—	482	141.7[21]	—	—	—	—
—	—	—	—	—	—	518	118.7[51]	—	—	—	—
446	144.8[1]	—	—	—	—	—	—	—	—	—	—
—	—	—	—	—	—	—	—	—	—	493	127.0[39]
—	—	—	—	—	—	452	140.1[4]	—	—	—	—
						480	124.4[4]				
—	—	—	—	—	—	276	32.4[55]	—	—	—	—
						286	41.2[55]				
						297	27.8[55]				
—	—	—	—	286	41.3[1]	—	—	—	—	—	—
—	—	—	—	286	49.9[1]	—	—	—	—	—	—
—	—	—	—	347	85.6[1]	347	81.4[21]	—	—	—	—
—	—	—	—	—	—	317	27.0[55]	—	—	—	—
						332	56.0[55]				
						348	88.5[55]				
						367	85.0[55]				
—	—	—	—	—	—	—	—	—	—	—	—
—	—	—	—	—	—	—	—	—	—	—	—
—	—	—	—	—	—	—	—	—	—	—	—
—	—	—	—	—	—	—	—	—	—	—	—
—	—	—	—	—	—	472	171.4[1]	—	—	—	—
—	—	—	—	—	—	473	164.5[6]	—	—	—	—
—	—	—	—	—	—	—	—	—	—	—	—
—	—	—	—	—	—	483	181.3[30]	—	—	—	—
—	—	—	—	490	140.7[4]	—	—	—	—	—	—
				516	104.7[4]						
—	—	—	—	—	—	—	—	—	—	—	—
—	—	—	—	—	—	—	—	—	—	—	—
—	—	—	—	—	—	—	—	—	—	—	—
—	—	—	—	467	105.2[1]	—	—	—	—	—	—
—	—	—	—	—	—	—	—	—	—	—	—
—	—	—	—	—	—	—	—	—	—	—	—
—	—	—	—	—	—	482.5	122.4[51]	—	—	—	—
—	—	—	—	—	—	—	—	—	—	—	—
—	—	—	—	493	151.6[10]	491	140.3[21]	—	—	—	—
442	168.2[1]	—	—	—	—	—	—	—	—	—	—
—	—	—	—	—	—	378	97.4[1]	—	—	—	—
—	—	—	—	—	—	—	—	—	—	—	—
—	—	—	—	—	—	413.5	95.9[4]	—	—	—	—
						437.5	150.4[4]				
						467.5	153.0[4]				
—	—	—	—	395	136.3[1]	—	—	—	—	—	—
—	—	—	—	—	—	497	166.6[4]	—	—	—	—
—	—	—	—	—	—	514	155.9[1]	—	—	—	—

Table I. 6 (continued)
QUANTITATIVE SPECTROSCOPIC DATA: MOLAR EXTINCTION COEFFICIENTS ($cm^{-1}M^{-1}$) OF CAROTENOIDS

Pigment	Solvent: Acetone		Benzene		Carbon disulfide		Chloroform		Cyclohexane	
	λmax	ϵ	λmax	ϵ	λmax	ϵ	λmax	ϵ	λmax	ϵ
Torularhodin-methylester	—	—	—	—	—	—	—	—	—	—
Torulene	—	—	—	—	—	—	—	—	—	—
Triphasiaxanthin	—	—	—	—	—	—	480	96.6[58]	—	—
							510	84.9[58]		
Violaxanthin	—	—	428	88.5[56]	—	—	—	—	—	—
			453.5	134.4[56]						
			483	128.4[56]						
Xanthophyllepoxide	—	—	430	105.1[56]	—	—	—	—	—	—
			456	133.0[56]						
			482.5	100.8[56]						
α-Zeacarotene	—	—	—	—	—	—	—	—	—	—
β-Zeacarotene	—	—	—	—	—	—	—	—	—	—
β₁-Zeacarotene	—	—	—	—	—	—	—	—	—	—
Zeaxanthin	452	133.1[1]	—	—	—	—	—	—	—	—

REFERENCES

1. **Davies, B. H.**, *Chemistry and Biochemistry of Plant Pigments*, Vol. 2, Goodwin, T. W., Eds., Academic Press, London, 1976, 150.
2. **Warren, C. K. and Weedon, B. C. L.**, *J. Chem. Soc.*, p. 3972, 1958a.
3. **Hager, A. and Meyer-Bertenrath, T.**, *Planta*, 69, 198, 1966.
4. **Isler, O. and Schudel, P.**, Carotine und Carotinoide, in *Wiss. Veröff. dt. Ges. Ernähr.*, Vol. 9, Steinkopff, Darmstadt, 1963, 54.
5. **Aasen, A. J. and Liaaen-Jensen, S.**, *Acta Chem. Scand.*, 19, 1843, 1965.
6. **Bonnett, R., Spark, A. A., and Weedon, B. C. L.**, *Acta Chem. Scand.*, 18, 1739, 1964.
7. **Surmatis, J. D.**, *Acta Chem. Scand.*, 31, 186, 1966.
8. **Campbell, S. A., Mallams, A. K., Waight, E. S., Weedon, B. C. L., Barbier, M., Lederer, E., and Salaque, A.**, *Chem. Commun.*, p. 941, 1967.
9. **Barber, M. S., Davis, J. B., Jackman, L. M., and Weedon, B. C. L.**, *J. Chem. Soc.*, p. 2870, 1960.
10. **Liaaen-Jensen, S.**, *Acta Chem. Scand.*, 14, 950, 1960.
11. **Petracek, F. J. and Zechmeister, L.**, *J. Am. Chem. Soc.*, 78, 1427, 1956.
12. **Liaaen-Jensen, S.**, *Acta Chem. Scand.*, 19, 116, 1965.
13. **Hiyama, T., Nishimura, M., and Chance, B.**, *Anal. Biochem.*, 29, 339, 1969.
14. **Braumann, T. and Grimme, H. L.**, *Biochim. Biophys. Acta*, 637, 8, 1981.
15. **Cholnoky, L., Szabolcs, J., and Nagy, E.**, *Liebigs Ann. Chem.*, 616, 207, 1958.
16. **Nelson, J. W. and Livingston, A. L.**, *J. Chromatogr.*, 28, 465, 1967.
17. **Simpson, K. L., Nakayama, T. O. M., and Chichester, C. O.**, *J. Bacteriol.*, 88, 1688, 1964.
18. **Schwieter, U., Bolliger, H. R., Chopard-dit-Jean, L. H., Englert, G., Kofler, M., König, A., von Planta, C., Rüegg, R., Vetter, W., and Isler, O.**, *Chimia (Switz.)*, 19, 294, 1965.
19. **Banthorpe, D. V., Doonan, H. J., and Wirz-Justice, A.**, *J. Chem. Soc. Perkin Trans. I*, p. 1764, 1972.
20. **Manchand, P. S., Rüegg, R., Schwieter, U., Siddons, P. T., and Weedon, B. C. L.**, *J. Chem. Soc.*, p. 2019, 1965.
21. **Liaaen-Jensen, S., Cohen-Bazire, G., Nakayama, T. O. M., and Stanier, R. K.**, *Biochim. Biophys. Acta*, 29, 477, 1958.
22. **Nash, H. A., Quackenbush, F. W., and Porter, J. W.**, *J. Am. Chem. Soc.*, 70, 3613, 1948.
23. **Liaaen-Jensen, S.**, *Acta Chem. Scand.*, 19, 1025, 1965.
24. **Barber, M. S., Jackman, L. M., Manchand, P. S., and Weedon, B. C. L.**, *J. Chem. Soc.*, C, 2166, 1966.
25. **Yokoyama, H. and White, M. J.**, *J. Org. Chem.*, 30, 2481, 1965.
26. **Cholnoky, L., Szabolcs, J., Cooper, R. D. G., and Weedon, B. C. L.**, *Tetrahedron Lett.*, 19, 1257, 1963.
27. **Wallcave, L. and Zechmeister, L.**, *J. Am. Chem. Soc.*, 75, 4495, 1953.

Table I. 6 (continued)
QUANTITATIVE SPECTROSCOPIC DATA: MOLAR EXTINCTION COEFFICIENTS ($cm^{-1}M^{-1}$) OF CAROTENOIDS

Ethanol		Ether		Hexane		Petroleum ether		Methanol		Pyridine	
λmax	ε	λmax	ε	λmax	ε	λmax	ε	λmax	ε	λmax	ε
—	—	—	—	—	—	497	170.8[1]	—	—	—	—
—	—	—	—	—	—	460	123.8[4]	—	—	—	—
						484	173.3[4]				
						518	143.3[4]				
—	—	—	—	—	—	—	—	—	—	—	—
441	150.2[3]	—	—	—	—	441	153.2[59]	—	—	—	—
—	—	—	—	—	—	—	—	—	—	—	—
—	—	—	—	421	99.7[60]	421	132.0[1]	—	—	—	—
—	—	—	—	426	104.5[60]	428	135.8[1]	—	—	—	—
—	—	—	—	—	—	427	97.0[60]	—	—	—	—
450	144.5[1]	—	—	—	—	452	133.7[4]	—	—	—	—
						480	116.6[4]				

REFERENCES

28. **Zechmeister, L. and Wallcave, L.**, *J. Am. Chem. Soc.*, 75, 5341, 1953.
29. **Isler, O., Montavon, M., Rüegg, R., and Zeller, P.**, *Helv. Chim. Acta*, 39, 454, 1956.
30. **Jackman, L. M. and Liaaen-Jensen, S.**, *Acta Chem. Scand.*, 15, 2058, 1961.
31. **Aasen, A. J. and Liaaen-Jensen, S.**, *Acta Chem. Scand.*, 20, 1970, 1966.
32. **Schwieter, U., Rüegg, R., and Isler, O.**, *Helv. Chim. Acta*, 49, 992, 1966.
33. **Entschel, R. and Karrer, P.**, *Helv. Chim. Acta*, 41, 402, 1958.
34. **Liaaen-Jensen, S. and Hertzberg, S.**, *Acta Chem. Scand.*, 20, 1703, 1966.
35. **Jensen, A.**, *Acta Chem. Scand.*, 20, 1728, 1966.
36. **Cooper, R. D. G., Davis, J. B., and Weedon, B. C. L.**, *J. Chem. Soc.*, p. 5637, 1963.
37. **Bush, W. V. and Zechmeister, L.**, *J. Am. Chem. Soc.*, 80, 2991, 1958.
38. **Hertzberg, S. and Liaaen-Jensen, S.**, *Acta Chem. Scand.*, 20, 1187, 1966.
39. **Hertzberg, S. and Liaaen-Jensen, S.**, *Acta Chem. Scand.*, 21, 15, 1967.
40. **Goodwin, T. W.**, Carotenoids, in *Modern Methods of Plant Analysis*, Vol. 3, Paech, K. and Tracey, M. V., Eds., Springer-Verlag, Berlin, 1955, 272.
41. **Aasen, A. J. and Liaaen-Jensen, S.**, *Acta Chem. Scand.*, 20, 811, 1966.
42. **Ryvarden, L. and Liaaen-Jensen, S.**, *Acta Chem. Scand.*, 18, 643, 1964.
43. **Kjøsen, H. and Liaaen-Jensen, S.**, *Acta Chem. Scand.*, 25, 1500, 1971.
44. **Markham, M. C. and Liaaen-Jensen, S.**, *Phytochemistry*, 7, 839, 1968.
45. **Kjøsen, H. and Liaaen-Jensen, S.**, *Phytochemistry*, 8, 483, 1969.
46. **Aasen, A. J., Francis, G. W., and Liaaen-Jensen, S.**, *Acta Chem. Scand.*, 23, 2605, 1969.
47. **Hertzberg, S. and Liaaen-Jensen, S.**, *Phytochemistry*, 6, 1119, 1967.
48. **Hertzberg, S. and Liaaen-Jensen, S.**, *Phytochemistry*, 8, 1259, 1969.
49. **Zalokar, M.**, *Arch. Biochem. Biophys.*, 70, 568, 1957.
50. **Nakayama, T. O. M.**, *Arch. Biochem. Biophys.*, 75, 356, 1958.
51. **Liaaen-Jensen, S.**, *Acta Chem. Scand.*, 17, 303, 1963.
52. **Jackman, L. M. and Liaaen-Jensen, S.**, *Acta Chem. Scand.*, 18, 1403, 1964.
53. **Liaaen-Jensen, S.**, *Acta Chem. Scand.*, 17, 500, 1963.
54. **Liaaen-Jensen, S.**, *Acta Chem. Scand.*, 17, 489, 1963.
55. **Davis, J. B., Jackman, L. M., Siddons, P. T., and Weedon, B. C. L.**, *J. Chem. Soc.*, C, 2154, 1966.
56. **Eugster, C. H. and Karrer, P.**, *Helv. Chim. Acta*, 40, 69, 1957.
57. **Simpson, K. L., Nakayama, T. O. M., and Chichester, C. O.**, *J. Bacteriol.*, 88, 1688, 1964.
58. **Yokoyama, H. and Guerrero, H. C.**, *J. Org. Chem.*, 35, 2080, 1970.
59. **Karrer, P. and Jucker, E.**, *Helv. Chim. Acta*, 26, 626, 1943.
60. **Petzold, E. N., Quackenbush, F. W., and McQuistan, M.**, *Arch. Biochem. Biophys.*, 82, 117, 1959.

PAPER CHROMATOGRAPHY OF CAROTENOIDS

TABLE NOTES

The chromatography of carotenoids has a long tradition dating back to the beginning of the 20th century. (See References 1 and 12 in Methods section.) After PC had been invented, the introduction of adsorbent-loaded paper(s) led to a rapid development in the area of carotenoid chemistry, since a better resolution compared to the columns used previously was possible. Especially useful was the development of circular PC, which even today can be considered as being superior to other methods in some aspects: its advantages are the excellent resolution, the speed of separation, and the minimal apparatus requirements (glass Petri dishes, appropriate solvents, papers, wicks, and — if possible — a nitrogen tank or source). The carotenoid mixture is dissolved (e.g., in acetone) and 10 to 100 μg of carotenoid is applied to the center of the paper; the diameter of the spot should not exceed 1 cm. After the application of each portion of the solution, the solvent is evaporated in a stream of nitrogen. After application, a small drop of acetone is added at the center of the paper to concentrate the carotenoids in the form of a ring (= origin). A wick is inserted at the center and the paper, with the wick protruding from the lower surface, is set horizontally in the lower half of a Petri dish with solvent. When the wick is in contact with the solvent, the apparatus is closed, using the upper Petri dish. Chromatography should be carried out in the dark, but nonitrogen atmosphere is required.

Different solvent systems have been applied by different authors and can be looked up in Table I. PC 1, together with the applicable literature which is, therefore, not explicitly listed at this place. When carried out with paper with a diameter of 12.5 to 15 cm, a separation will take approximately 15 to 20 min and is, therefore, often more rapid than the development of TLC plates.

In the laboratory, papers may be impregnated with a number of adsorbents; however, this poses a certain problem in terms of standardization. Papers have been loaded with the carbonates of calcium, zinc, or magnesium, with magnesium hydroxide or sucrose. Commercially available sorbent-loaded papers seem to give better resolution and reproducibility.

For these tables, references dealing exclusively with papers still commercially available have been included.

Due to the considerable amount of data, two tables (I. PC 1 and I. PC 2) have been composed. Table I. PC 1 deals with papers from Schleicher and Schüll (Düren, F. R. G.) and Table I. PC 2 with Whatman papers. The papers mentioned in Tables I. PC 1 and I. PC 2 are either untreated, reverse phase, or sorbent filled. In many cases, simple cellulose paper is not sufficient for separation. Here, the application of adsorbent-loaded papers has led to excellent results. The papers most often used are those filled with kieselguhr (20%, Schleicher & Schüll No. 287), aluminum oxide (Schleicher & Schüll No. 288), aluminum hydroxide (equ. 7.5% A1203, Whatman AH 21), and silica gel (equ. 22% SiO2, Whatman SG 81). Activation of the papers by heat is possible, but the additional activity (up to a strength equivalent to Brockmann grade II, Schleicher & Schüll paper No. 667) is quickly lost due to the uptake of moisture. Since some papers are no longer available, they are not included in the respective tables.

The most versatile adsorbent-loaded paper is kieselguhr paper, which is considered to be superior for the separation of *cis-trans* isomeric xanthophylls. Alumina-impregnated papers separate ''less polar carotenes'' (α- from β-carotene, for example) and give improved resolution of xanthophylls. Aluminum hydroxide and silica gel-loaded papers have been applied in a number of cases.

It has to be noted that adsorbent-loaded papers should be prewashed with the appropriate solvent system prior to use when a further analysis of pigments is planned. Even then, eluates from kieselguhr paper are unsuitable for mass spectrometric examination.

Table I. PC 1
CELLULOSE PAPERS AND IMPREGNATED CELLULOSE PAPERS — I

Rf × 100

Paper	P1	P2	P2	P2	P2	P2	P2	P2	P2	P2	P2	P2	P2	P2	P2	P2	P2	P2	P2	P2	P3	P3	P4
Solvent	S1	S2	S3	S4	S5	S6	S7	S8	S9	S10	S11	S12	S13	S14	S15	S16	S17	S18	S19	S20	S3	S14	S21
Technique	T1	T2	T2	T2	T2	T2	T2	T2	T2	T2	T2,3	T2	T2,3	T2,3	T2,3	T2,3	T2,3	T2,3	T2,3	T2,3	T2,3	T2,3	T2,3
Detection	D1	D1	D1	D1	D1	D1	D1	D1	D1	D1	D1	D1	D1	D1	D1	D1	D1	D1	D1	D1	D1	D1	D1
Compound																							
Actinioerythrin	—	—	—	—	—	—	—	—	—	—	—	—	—	—	—	52[21]	—	—	—	—	—	—	—
Actinioerythrin-bis-α-ketol	—	—	—	—	—	—	—	—	—	—	—	—	—	—	—	—	78[21]	—	—	—	—	—	—
Aleuriaxanthin	—	—	—	—	—	—	—	—	—	—	—	24[18]	65[13] 33[18]	56[18]	11[21]	—	—	—	—	—	—	—	—
Anhydrorhodovibrin (*trans*)	24[1]	—	—	—	—	—	—	—	—	—	—	—	35[10]	60[10] 65[15]	—	—	—	—	—	—	—	—	—
Aphanizophyll	—	—	—	—	—	—	—	—	—	—	—	—	—	—	—	—	—	—	42[18]	—	—	—	—
Apo-4'-carotenoic acid	—	—	—	—	—	—	—	—	—	—	—	—	—	—	—	26[4]	—	—	—	—	33[4]	—	—
Apo-6'-lycopenal	—	—	—	—	—	—	—	—	—	—	—	—	39[14]	55[14]	—	—	—	—	—	—	—	—	—
Apo-8'-lycopenal	—	—	—	—	—	—	—	—	—	—	—	—	46[14]	77[14]	—	—	—	—	—	—	—	—	—
Astacene	—	—	—	—	—	—	—	—	—	—	—	—	—	40[22]	—	53[21]	—	—	—	—	—	—	46[1]
Astaxanthin (*trans*)	—	—	—	—	—	—	—	—	—	—	—	—	—	—	—	57[10]	—	85[10]	—	—	—	—	—
Asterinic acid	—	—	—	—	—	—	—	—	—	—	—	—	—	—	—	28[29]	—	—	—	—	—	—	—
Auroxanthin epimer 1	—	—	—	—	—	—	—	—	—	70[9]	—	—	—	—	—	—	—	—	—	—	—	—	—
Auroxanthin epimer 2	—	—	—	—	—	—	—	—	—	70[9]	—	—	—	—	—	—	—	—	—	—	—	—	—
Bacterioruberin (*trans*)	—	—	—	—	—	—	—	—	—	—	—	—	—	—	—	—	—	44[10]	—	—	—	—	—
α-Bacterioruberin (*trans*)	35[1]	—	—	—	—	—	—	—	50[9]	—	—	—	—	—	—	—	—	—	—	—	—	—	—
Canthaxanthin	—	—	—	—	—	—	—	—	—	—	—	—	—	—	—	2[10]	—	—	—	96[14]	—	—	42[1]
β-Carotene	—	—	—	—	—	—	—	—	—	—	95[10] 95[13]	—	—	—	—	—	—	—	—	—	—	—	—
β-Carotene (*cis*)	—	—	—	—	—	—	—	—	—	—	88[12]	—	—	—	—	98[10]	—	—	—	—	—	—	—
β-Carotene (*trans*)	—	—	—	—	—	—	—	—	—	—	88[12]	—	98[10]	98[10]	—	—	—	—	—	—	—	—	—
γ-Carotene	—	—	—	—	—	—	—	—	—	—	63[13]	—	—	—	—	—	—	—	—	—	—	—	—
γ-Carotene (*trans*)	—	—	—	—	—	—	—	—	—	—	68[10]	—	—	—	—	—	—	—	—	—	—	—	—
β,β-Carotene-2,3,3'-triol	—	—	—	—	—	—	10[8]	—	—	—	83[11]	—	—	—	—	—	—	—	—	—	—	—	—
Chlorobactene (*trans*)	—	—	—	—	—	—	—	—	—	—	62[11]	—	—	—	—	—	—	—	—	—	—	—	—

Compound	1	2	3	4	5	6	7	8	9	10	11
Chloroxanthin			48[15]								
Chloroxanthin (trans)	58[1]		46[10]	73[10]			90[10]				
Cryptoxanthin		29[10]	62[14]	81[10]	68[20]		91[10]				36[1]
Dehydroadonirubin	30[1]										44[1]
Dehydroadonixanthin	16[1]										55[1]
2'-Dehydroplectaniaxanthin				48[7]	40[23]		80[7]	72[23]			
Deoxyflexixanthin							53[22]				
3,4,-Didehydrolycopene				52[13]							
3,4-Didehydrorhodopin (trans)							70[30]				
1',2'-Dihydro-1'-hydroxy-γ-carotene				61[24]							
1',2'-Dihydro-1'-hydroxy-4-keto-γ-carotene (trans)							54[24]	68[24]			
1',2'-Dihydro-1'-hydroxy-spheroidenone				45[25]							
2'-Dihydrophillipsiaxanthin				63[15]					30[7]		
7,8'-Dihydrorhodovibrin				30[16]		89[16]	60[16]				
1,1'-Dihyroxy-1',2'-tetrahydro-ζ-carotene				8[16]							
Echinenone	90[1]		79[4]								25[1]
Eschscholtzxanthin	9[1]										67[1]
Euglenanone	63[1]						44[22]				27[1]
Flexixanthin										33[34]	
Fucoxanthin							40[10]		81[10]		
Fucoxanthin (trans)											
Gazaniaxanthin			39[19]	22[26]			64[19]				
3-Hydroxy-3'-hydroxy α-carotene (trans)											
4-Hydroxy-4'-keto-β-carotene (trans)				31[40]							
3-Hydroxy-3'-methoxy-α-carotene (trans)				59[26]							
1'-Hydroxy-spheroidene				63[15]							
1'-Hydroxy-spheroidenone (trans)							73[31]				
1'-Hydroxy-spirilloxanthin (trans)							40[32]		80[32]		
Isocryptoxanthin		25[20]									

Table I. PC 1 (continued)
CELLULOSE PAPERS AND IMPREGNATED CELLULOSE PAPERS — I

Paper	P1	P2	P2	P2	P2	P2	P2	P2	P2	P2	P2	P2	P2	P2	P2	P2	P2	P2	P2	P2	P3	P3	P4
Solvent	S1	S2	S3	S4	S5	S6	S7	S8	S9	S10	S11	S12	S13	S14	S15	S16	S17	S18	S19	S20	S3	S14	S21
Technique	T1	T2	T2	T2	T2	T2	T2	T1	T2	T2	T2.3	T2	T2.3	T2.3	T2.3	T2.3	T2.3	T2.3	T2.3	T2.3	T2.3	T2.3	T2.3
Detection	D1	D1	D1	D1	D1	D1	D1	D1	D1	D1	D1	D1	D1	D1	D1	D1	D1	D1	D1	D1	D1	D1	D1
Isofucoxanthin																24[33]							
Isorenieratene (trans)											51[12]												
β-Isorenieratene (trans)											70[12]												
Isozexanthin	10[1]																						65[1]
4-Keto-3'-hydroxy-β-carotene																77[20]							
4-Ketophleixanthophyll			34[3]																				
4-Ketotorulene														53[3]									
Lutein							29[8]																
Lutein (trans)								70[9]						39[10]		72[10] 73[34]		91[10]		80[34]			
Lycopene											40[13] 53[10] 48[14] 48[16]	51[7]		98[27]									
Lycopene (trans)													86[10]										
Lycophyll																15[27]							
Lycoxanthin														30[27]		69[35]							
Methyl-apo-6'-lycopenoate			34[5]											77[14]									
Methyl-1-hexosyl-1,2-dihydro-3,4-didehydroapo-8'-lycopenoate																		1[5]					
Myxol-2'-O-methylmethyl-pentoside			71[2]																				
Myxoxanthophyll	52[2]																		58[18]				
Myxoxanthophyll (trans)																42[36]		70[36]					
Neochrome epimer 1									80[9]														
Neochrome epimer 2									80[9]														
Neochrome (mono-cis)									70[9]														
Neurosporaxanthin (trans)	25[4]																						
Neurosporaxanthinmethylester													63[4]										

Compound	Values
Neurosporene	66[13], 88[15], 89[16]
Okenone (trans)	59[2]
Oscillaxanthin	98[2], 74[2]
Oscillol-2,2'-di-(O-methyl)-methylpentoside	34[28]
Phleixanthophyll—	34[3], 20[6]
Phleixanthophyll (trans)	34[3]
Phillipsiaxanthin	20[7], 36[7], 69[7], 80[7]
Plectianaxanthin	40[23]
Rhodopin (trans)	39[10], 75[10]
Rhodovibrin (trans)	35[39], 54[10]
Rhodoxanthin	33[1]
Rhodoxanthin (trans)	72[10], 27[10], 10[10], 43[1]
Rubixanthin	39[19], 64[19]
Saproxanthin	83[36], 37[2], 42[36]
Saproxanthin (trans)	
Sarcinaxanthin	
Spheroidene (cis)	72[17]
Spheroidene (trans)	52[17]
Spheroidenone	86[15], 66[41]
Spirilloxanthin	74[13], 18[10]
Spirilloxanthin (trans)	76[32], 40[10], 44[15], 18[10]
1,2,1',2'-Tetrahydro-1,1'-dihydroxylycopene (trans)	28[38], 52[13]
Torulene	
Violaxanthin	36[54], 83[10], 44[10], 18[10]
Violaxanthin (trans)	57[21]
Violerythrin	63[9]
Zeaxanthin	66[1], 87[10], 59[10], 30[10], 9[10], 25[8], 10[1]
Zeaxanthin (trans)	60[34]

Table I. PC 1 (continued)

CELLULOSE PAPERS AND IMPREGNATED CELLULOSE PAPERS — I

Paper	P1	= cellulose
	P2	= Schleicher & Schüll No. 287 (silica impregnated)
	P3	= Schleicher & Schüll No. 597
	P4	= cellulose, impregnated with triglyceride
Solvent	S1	= petroleum ether-carbontetrachloride = 3:2
	S2	= acetone-methanol = 49:1
	S3	= benzene-acetone = 7:3
	S4	= benzene-acetone = 1:1
	S5	= benzene-methanol = 99:1
	S6	= benzene-methanol = 49:1
	S7	= hexane-acetone = 19:1
	S8	= hexane-acetone = 9:1
	S9	= hexane-acetone = 7:3
	S10	= hexane-acetone = 4:1
	S11	= petroleum ether
	S12	= petroleum ether-acetone = 99:1
	S13	= petroleum ether-acetone = 49:1
	S14	= petroleum ether-acetone = 19:1
	S15	= petroleum ether-acetone = 17:3
	S16	= petroleum ether-acetone = 9:1
	S17	= petroleum ether-acetone = 7:3
	S18	= petroleum ether-acetone = 4:1
	S19	= petroleum ether-acetone = 1:1
	S20	= petroleum ether-isopropanol = 197:3
	S21	= methanol-acetone = 5:1
Technique	T1	= ambient temperature, chamber saturation, ascending
	T2	= circular paper chromatography
	T3	= for co-chromatograms, the 3-divided paper technique was used (Jensen, A., Aasmuntrud, O., and Eimhjellen, K. E., *Biochim. Biophys. Acta.* 88, 466, 1964)
Detection	D1	= visual observation

REFERENCES

1. Egger, K., *Phytochemistry,* 4, 609, 1965.
2. Francis, G. W., Hertzberg, S., Andersen, K., and Liaaen-Jensen, S., *Phytochemistry,* 9, 629, 1970.
3. Hertzberg, S. and Liaaen-Jensen, S., *Acta Chem. Scand.,* 21, 15, 1967.
4. Aasen, A. J. and Liaaen-Jensen, S., *Acta Chem. Scand.,* 19, 1843, 1965.
5. Aasen, A. J., Francis, G. W., and Liaaen-Jensen, S., *Acta Chem. Scand.,* 23. 2605, 1969.
6. Hertzberg, S. and Liaaen-Jensen, S. *Acta Chem. Scand.,* 20, 1187, 1966.
7. Arpin, N. and Liaaen-Jensen, S., *Bull. Soc. Chim. Biol.,* 49, 527, 1967.
8. Smallidge, R. L. and Quackenbush, F. W., *Phytochemistry,* 12, 2481, 1973.
9. Fiksdahl, A., Mortensen, J. T., and Liaaen-Jensen, S., *J. Chromatogr.,* 157. 111, 1978.
10. Liaaen-Jensen, S. and Jensen, A., *Progress in the Chemistry of Fats or Other Lipids,* Vol. 8 (Part 2), Holman, R. T., Ed., Pergamon Press, New York, 1965, 133.
11. Liaaen-Jensen, S., Hegge, E., and Jackman, L. J., *Acta Chem. Scand.,* 18, 1703. 1964.
12. Liaaen-Jensen, S., *Acta Chem. Scand.,* 19, 1025, 1965.
13. Liaaen-Jensen, S., *Phytochemistry,* 4, 925, 1965.
14. Kjøsen, H. and Liaaen-Jensen, S., *Phytochemistry,* 8, 483, 1969.
15. Liaaen-Jensen, S., *Acta Chem. Scand.,* 17, 500, 1963.
16. Fiasson, J. L. and Arpin, N., *Bull. Soc. Chim. Biol.,* 49, 537, 1967.
17. Jackman, L. M. and Liaaen-Jensen, S., *Acta Chem. Scand.,* 18, 1403, 1964.
18. Arpin, N., Kjøsen, H., Francis, G. W., and Liaaen-Jensen, S., *Phytochemistry,* 12, 2751, 1973.
19. Arpin, N. and Liaaen-Jensen, S., *Phytochemistry,* 8, 185, 1969.
20. Hertzberg, S. and Liaaen-Jensen, S., *Phytochemistry,* 5, 557, 1966.
21. Hertzberg, S., Liaaen-Jensen, S., Enzell, C. R., and Francis, G. W., *Acta Chem. Scand.,* 23, 3290, 1969.
22. Aasen, A. J. and Liaaen-Jensen, S., *Acta Chem. Scand.,* 20, 1970, 1966.
23. Arpin, N. and Liaaen-Jensen, S., *Phytochemistry,* 6, 995, 1967.
24. Hertzberg, S. and Liaaen-Jensen, S., *Acta Chem. Scand.,* 20, 1187, 1966.
25. Liaaen-Jensen, S., *Acta Chem. Scand.,* 17, 489, 1963.

26. **Liaaen-Jensen, S. and Hertzberg, S.,** *Acta Chem. Scand.,* 20, 1703, 1966.
27. **Markham, M. C. and Liaaen-Jensen, S.,** *Phytochemistry,* 7, 839, 1968.
28. **Aasen, A. J. and Liaaen-Jensen, S.,** *Acta Chem. Scand.,* 21, 970, 1967.
29. **Sørensen, N. A., Liaaen-Jensen, S., Børdalen, B., and Haug, A.,** *Acta Chem. Scand.,* 22, 344, 1968.
30. **Jackman, L. M. and Liaaen-Jensen, S.,** *Acta Chem. Scand.,* 15, 2058, 1961.
31. **Liaaen-Jensen, S.,** *Acta Chem. Scand.,* 17, 555, 1963.
32. **Liaaen-Jensen, S.,** *Acta Chem. Scand.,* 14, 953, 1960.
33. **Jensen, A.,** *Acta Chem. Scand.,* 20, 1728, 1966.
34. **Jensen, A. and Liaaen-Jensen, S.,** *Acta Chem. Scand.,* 13, 1863, 1959.
35. **Liaaen-Jensen, S.,** *Acta Chem. Scand.,* 13, 2142, 1959.
36. **Hertzberg, S. and Liaaen-Jensen, S.,** *Phytochemistry,* 8, 1259, 1969.
37. **Aasen, A. J. and Liaaen-Jensen, S.,** *Acta Chem. Scand.,* 20, 811, 1966.
38. **Hertzberg, S. and Liaaen-Jensen, S.,** *Phytochemistry,* 5, 565, 1966.
39. **Surmatis, J. D., Ofner, A., Gibas, J., and Thommen, R.,** *J. Org. Chem.,* 31, 186, 1966.
40. **Liaaen-Jensen, S.,** *Acta Chem. Scand.,* 19, 1166, 1965.
41. **Davies, B. H.,** *Chemistry and Biochemistry of Plant Pigments,* Goodwin, T. W., Eds., Academic Press, London, 1965, 489.

Table I. PC 2
CELLULOSE PAPERS AND IMPREGNATED CELLULOSE PAPERS — II

$R_f \times 100$

Compound	P1 S1 T1 D1 1	P1 S2 T1 D1 1	P1 S3 T1 D1 —	P1 S4 T1 D1 —	P1 S5 T1 D1 1	P2 S1 T1 D1 1	P2 S2 T1 D1 1	P2 S3 T1 D1 1	P2 S4 T1 D1 1	P2 S5 T1 D1 1	P2 S6 T2 D1 2	P3 S7 T1 D1 3	P3 S7 T3a D1 4	P3 S8 T3b D1 4	P3 S9 T3a D1 5	P3 S10 T3b D1 5
Canthaxanthin	—	—	21	100	100	—	—	20	60	100	—	—	—	—	—	—
α-Carotene	73	100	—	—	—	45	—	—	—	—	—	—	—	—	—	—
β-Carotene	50	50	—	—	—	35	100	—	—	—	—	—	—	—	96	96
γ-Carotene	8	100	—	—	—	20	47	—	—	—	—	—	—	—	—	—
ζ-Carotene	—	64	—	—	—	—	100	—	—	—	—	—	—	—	—	—
Cryptoxanthin	—	—	—	—	—	—	29	—	—	—	—	—	—	—	—	—
Diadinoxanthin	—	—	—	—	—	—	—	—	—	—	—	43	54	44	61	23
Diatoxanthin	—	—	—	—	—	—	—	—	—	—	—	59	57	60	62	33
5,6,5',6'-Diepoxy-β-carotene	—	60	—	—	—	—	—	—	—	—	—	—	—	—	—	—
Dinoxanthin	—	—	—	—	—	—	27	—	—	—	—	—	—	—	—	—
Flavochrome	—	00	64	50	—	—	—	65	—	—	—	—	—	—	—	—
Flavoxanthin	—	—	—	—	—	—	00	—	62	—	—	—	54	44	—	—
Fucoxanthin	—	—	—	—	—	—	—	—	—	53	—	27	49	28	60	10
Lutein	11	—	25	—	60	—	—	—	28	—	—	70	74	73	—	—
Lycopene	—	37	—	—	—	10	42	—	—	—	—	—	—	—	—	—
5,6-Monoepoxy-α-carotene	—	31	—	—	—	—	25	—	—	—	—	—	—	—	—	—
5,6-Monoepoxy-β-carotene	—	46	—	—	—	—	38	—	—	—	—	—	—	—	—	—
Mutatochrome	—	00	61	—	00	—	00	61	—	—	—	—	—	—	—	—
Neoxanthin	—	—	—	—	—	—	—	—	—	7	—	5	32	5	—	—
Peridinin and Neoperidinin	—	—	—	—	—	—	—	—	—	—	—	—	51	23	—	—
Phytofluene	78	—	—	—	—	40	—	—	—	—	—	—	—	—	—	—
Rhodoxanthin	8	—	—	—	62	00	—	—	—	60	—	—	—	—	—	—
Rubixanthin	—	41	—	—	—	—	22	—	—	80	49	—	—	—	—	—
Taraxanthin	—	—	—	66	81	—	—	—	71	—	—	—	—	—	—	—
Torulene	—	26	—	—	—	—	32	—	—	—	—	—	—	—	—	—
Violaxanthin	—	—	—	—	32	—	—	—	—	35	—	48	65	48	—	—
Zeaxanthin	—	—	—	—	64	—	—	—	—	85	—	—	—	—	—	—

Paper P1 = Whatman paper Chromedia AH 81 (impregnated with aluminum hydroxide equivalent to 7.5% Al_2O_3)

P2 = Whatman paper Chromedia SG 81 (impregnated with 22% SiO_2)

P3 = Whatman 3 MM chromatography paper

S1 = n-hexane

S2 = n-hexane-acetone = 99:1

S3 = n-hexane-acetone = 95:5

S4 = n-hexane-acetone = 85:15

S5 = n-hexane-acetone = 8:2

S6 = petroleum ether-acetone = 19:1

S7 = petroleum ether (bp 60—80°C)-n-propanol = 24:1

S8 = petroleum ether (bp 60—80°C)-chloroform = 7:3

S9 = petroleum ether (bp 60—110°C)-n-propanol = 24:1

S10 = petroleum ether (bp 60—110°C)-chloroform = 7:3

Technique T1 = ambient temperature, chamber saturation, ascending

T2 = circular paper chromatography; for co-chromatograms the 3-divided paper technique was used (Jensen, A., Aasmuntrud, O., and Eimhjellen, K. E., *Biochim. Biophys. Acta*, 88, 466, 1964)

T3 = two-dimensional chromatography: a, first dimension; b, second dimension.

Detection D1 = visual observation

REFERENCES

1. **Valadon, L. R. G. and Mummery, R. S.**, *Phytochemistry*, 11, 413, 1972.
2. **Arpin, N. and Liaaen-Jensen, S.**, *Phytochemistry*, 8, 185, 1969.
3. **Eskins, K., Scholfield, C. R., and Dutton, H. J.**, *J. Chromatogr.*, 135, 217, 1977.
4. **Jeffrey, S. W.**, *Biochem. J.*, 80, 336, 1961.
5. **Jeffrey, S. W. and Allen, M. B.**, *J. Gen. Microbiol.*, 36, 277, 1964.

THIN LAYER CHROMATOGRAPHY TABLES

TABLE NOTES

Thin-layer chromatography (TLC) of carotenoids is one of the most modern methods of carotenoid separation. Since the basic principles of TLC of carotenoids are already contained in Volumes I and II of the CRC Handbook Series on Chromatography (edited by Zweig and Sherma[23]) and other literature (see "Methods" References 1 to 24 and Tables I. TLC 1 through I. TLC 9), and special precautions have already been discussed in "Carotenoid Handling and Storage" of the Section "Carotenoids" of this handbook, only a few remarks are given here. As already pointed out, carotenoids are extremely sensitive to light and oxygen. It is therefore necessary to chromatograph under subdued light or, better, in the dark. Since carotenoids on TLC plates are much more readily oxidized than on paper, the zones obtained should be rapidly eluted once chromatography has been finished. TLC has been performed on a variety of layers using different developing systems.

Tables I. TLC 1 through I. TLC 4 contain R_f values from diverse one-component layers. Table I. TLC 5 and I. TLC 6 give R_f data obtained on two-component layers containing silica gel G. Tables I. TLC 7 and I. TLC 8 list R_f values from two- and multicomponent layers not containing silica gel G. Table I. TLC 9 contains values from reverse-phase plates.

Table I. TLC 1
TLC ON SILICA GEL G LAYERS

$R_f \times 100$

Compound																											
Layer	L1	L1	L1	L1	L1	L1	L1	L1	L1	L1	L1	L1	L1	L1	L1	L1	L1	L1	L1	L1	L1	L1	L1	L1	L1	L1	L1
Solvent	S1	S1	S1	S1	S1	S2	S2	S3	S4	S5	S6	S7	S8	S9	S10	S11	S12	S13	S14	S15	S16	S17	S17	S18	S18	S18	S19
Technique	T1	T1	T1	T1	T1	T1	T1	T1	T1	T1	T1	T1	T1	T1	T1	T1	T1	T1	T1	T1	T1	T1	T1	T1	T1	T1	T1
Detection	D1,3,4	D1,3	D1	D1	D1,2	D1	D1	D1	D1	D1	D1,3,4	D1,3,4	D5	D1	D1	D1,3,4	D1	D1	D5	D1	D1	D1	D1	D1	D1	D1	D1
Literature	1	2	3	4,5[x]	6	7	8	9	7	10	1	1	11	12	13	1	3	14	11	4	4	14	15	3	15	14	3[x],15
Antheraxanthin	—	—	—	—	—	—	—	—	—	—	—	—	—	—	—	—	—	—	—	—	—	15	—	—	—	32	—
β-Apo-8'-carotenal	—	—	—	—	—	—	—	—	—	—	—	—	—	—	—	—	—	—	—	—	—	—	—	—	—	—	0—7
Astacene	—	—	—	—	—	—	—	—	—	—	—	—	—	—	42	—	—	—	—	—	—	—	—	—	—	—	—
Azafrin	—	—	—	—	—	—	—	—	—	—	—	—	—	—	—	—	—	—	—	—	—	—	—	—	—	—	—
Bixin	—	—	—	—	—	—	—	—	—	—	—	—	—	—	—	—	—	—	—	—	—	—	—	—	—	2	—
Canthaxanthin	—	—	—	—	—	—	—	—	—	—	—	—	—	—	50	—	—	—	—	—	—	—	00	—	—	5	0—7
Capsanthin	—	—	—	—	—	—	—	—	—	—	—	—	—	—	—	—	—	—	—	—	—	—	—	11	—	90	0—7
Capsorubin	—	—	—	—	—	—	—	—	—	—	—	—	—	—	—	—	—	—	—	—	—	—	—	—	—	16	0—7
α-Carotene	—	—	—	—	—	—	—	—	—	—	—	—	89	—	—	—	55	91	—	—	—	—	—	—	—	13	73
β-Carotene	—	1	—	—	—	—	—	—	—	96	—	—	87	76	—	—	87	87	87	—	—	—	82	—	—	100	73[a],75
γ-Carotene	—	—	—	—	—	—	—	—	—	—	—	—	80	—	—	—	—	—	70	—	—	—	—	—	—	100	65
δ-Carotene	—	—	—	—	—	—	—	—	—	—	—	—	—	—	—	—	—	—	—	—	—	—	—	—	100	100	—
ε-Carotene	—	—	—	—	—	—	—	—	—	—	—	—	—	—	—	—	—	—	—	—	—	—	—	—	100	—	—
ζ-Carotene	00	—	—	00	—	—	—	—	—	—	11	28	—	—	—	23	—	—	—	24	—	—	—	—	—	100	—
θ-Carotene	—	—	—	00	—	—	—	—	—	—	—	—	—	—	—	—	—	—	—	—	84	—	—	—	—	—	—
β-Citraurin	—	—	—	—	—	—	—	47	—	—	—	—	53	—	—	—	—	—	—	—	—	—	—	—	—	—	—
Cryptocapsin	—	—	—	—	—	—	—	—	—	—	—	—	—	—	—	—	—	—	—	—	—	—	—	—	—	—	—
Cryptoxanthin	—	—	—	—	—	—	—	—	—	41	—	—	—	85	—	—	—	—	—	—	—	—	—	61	—	75	0—7
Dehydro-retro-β-carotene	—	—	—	—	—	—	—	—	—	—	—	—	—	—	—	—	—	—	—	—	—	—	—	—	—	—	—
Diadinoxanthin	—	—	—	—	—	—	—	—	—	—	—	—	—	—	—	—	—	—	—	—	—	—	—	—	—	—	—
4,4'-Diapo-ζ-carotene	2	—	—	—	—	—	—	—	—	—	19	35	—	—	—	25	—	—	—	—	—	—	—	—	—	—	—
4,4'-Diapolycopen-4-al	—	—	—	—	—	—	52	—	30	—	—	—	—	—	—	—	—	—	—	—	—	—	—	—	—	—	—
4,4'-Diaponeurosporen-4-al	—	—	—	—	—	—	59	—	36	—	—	27	—	—	—	—	—	—	—	—	—	—	—	—	—	—	—
4,4'-Diaponeurosporene	00	—	—	—	—	—	—	—	—	—	14	—	—	—	—	22	—	—	—	—	—	—	—	—	—	—	—
4,4'-Diapophytoene	9	—	—	—	—	—	—	—	—	—	34	55	—	—	—	46	—	—	—	—	—	—	—	—	—	—	—
4,4'-Diapophytofluene	7	—	—	—	—	—	—	—	—	—	25	45	—	—	—	35	—	—	—	—	—	—	—	—	—	—	—

4,4′-Diapo-7,8,11,12-tetrahydrolycopene	—	—	—	—	—	—	—	—	—	—	19	36	—	—	—	26	—	—	—	—	—	—	—	—	—	—	—
1,1′-Dihydroxy-1,2,1′,2′-tetrahydro-ζ-carotene	—	—	—	—	—	—	—	—	—	—	—	—	—	—	—	—	—	—	—	—	00	—	—	—	—	—	—
Dinoxanthin	—	—	—	—	—	—	—	—	—	47	—	—	—	—	—	—	—	—	—	—	—	—	—	—	—	—	—
Echinenone	—	—	—	—	—	—	—	—	—	—	—	—	—	55	82	—	—	—	—	—	—	10	—	—	—	—	—
4-Hydroxy-β-carotene	—	—	—	—	—	—	—	—	—	—	—	—	—	35	—	—	—	—	—	—	—	—	—	—	—	—	—
4-Hydroxy-4,4′-diaponeurosporene	—	—	—	—	—	25	—	—	12	—	—	—	—	—	—	—	—	—	—	—	—	—	—	—	—	—	—
4-Hydroxy-4-keto-β-carotene	—	—	—	—	—	—	—	—	—	—	—	—	—	24	—	—	—	—	—	—	—	—	—	—	—	—	—
Isolutein	—	—	—	—	—	—	—	—	—	—	—	—	—	—	—	—	—	—	—	—	—	—	—	12	—	—	—
Isozeaxanthin	—	—	—	—	—	—	—	—	—	—	—	—	—	16	—	—	—	—	—	—	—	—	—	—	—	—	—
4-Keto-α-carotene	—	—	—	—	—	—	34	—	—	—	—	—	—	—	—	—	—	—	—	—	—	—	—	—	—	—	—
Lutein	—	—	—	—	—	—	—	—	—	—	—	—	—	—	—	—	—	42	—	—	—	—	—	41	35	—	0—7
Lycopene	—	—	—	—	—	—	—	—	—	—	—	—	67	—	—	—	—	—	52	—	72	—	74	—	100	—	56
Lycopersene	—	—	30	—	—	—	—	—	—	—	—	—	—	—	—	—	—	—	—	—	—	—	—	—	100	—	—
Methylbixin	—	—	—	—	—	—	—	—	—	—	—	—	—	—	—	—	—	—	—	—	—	—	—	—	97	—	0—7
Neoxanthin	—	—	—	—	—	—	—	—	—	—	—	—	—	—	—	—	—	23	—	—	—	—	—	—	—	—	—
Neurosporene	00	—	—	—	—	—	—	—	—	—	9	21	—	—	—	18	—	—	—	18	80	—	—	—	—	—	—
Neurosporene *(cis)*	—	—	—	—	—	—	—	—	—	—	—	—	—	—	—	—	—	—	—	25	84	—	—	—	—	—	—
P-457	—	—	—	—	—	—	—	—	—	00	—	—	—	—	—	—	—	—	—	—	—	—	—	—	—	—	—
Peridinin	—	—	—	—	—	—	—	—	—	33	—	—	—	—	—	—	—	—	—	—	—	—	—	—	—	—	—
Peridininol	—	—	—	—	—	—	—	—	—	7	—	—	—	—	—	—	—	—	—	—	—	—	—	—	—	—	—
Phytoene	4	21	—	—	20	—	—	—	—	—	23	47	—	—	—	35	—	—	—	—	—	—	—	—	100	—	—
Phytofluene	1	12	10	100	—	—	—	—	—	—	17	34	—	—	—	28	—	—	—	32	—	—	—	—	100	—	—
Torularhodin-methylester	—	—	—	—	—	—	—	—	—	—	—	—	—	—	—	—	—	—	—	—	—	—	—	—	100	—	0—7
Torulene	—	—	—	—	—	—	—	—	—	—	—	—	—	—	—	—	—	—	—	—	—	—	—	—	100	—	—
Violaxanthin	—	—	—	—	—	—	—	—	—	—	—	—	—	—	—	—	—	37	—	—	—	—	—	19	21	—	0—7
β-Zeacarotene	—	—	—	—	—	—	—	—	—	—	—	—	—	—	—	—	—	—	—	—	—	—	—	—	100	—	—
Zeaxanthin	—	—	—	—	—	—	—	—	—	—	—	—	—	—	—	—	—	—	—	—	—	—	—	—	24	—	0—7

Table I. TLC 1 (continued)
TLC ON SILICA GEL G LAYERS

Layer

L1 = Silica gel G (Merck)

Solvent

S1 = petroleum ether
S2 = petroleum ether-acetone = 85:15
S3 = petroleum ether-acetone = 49:1
S4 = petroleum ether-acetone = 9:1
S5 = petroleum ether-acetone = 7:3
S6 = petroleum ether-benzene = 95:5
S7 = petroleum ether-benzene = 9:1
S8 = petroleum ether-benzene = 3:2
S9 = petroleum ether-benzene-ethanol = 25:25:2
S10 = petroleum ether-benzene-methanol = 49:49:2
S11 = petroleum ether-ether = 99:1
S12 = petroleum ether-ether = 19:1
S13 = petroleum ether-ethyl acetate-diethyl amine = 58:30:12

S14 = petroleum ether-methylene dichloride = 9:1
S15 = petroleum ether (bp 100—120°C)
S16 = petroleum ether (bp 100—120°C)-acetone = 49:1
S17 = petroleum ether (bp 90—110°C)-benzene = 1:1
S18 = methylene chloride-ethyl acetate = 4:1
S19 = undecane-methylene chloride = 4:1

Technique

T1 = ambient temperature, chamber saturation, ascending

Detection

D1 = visual observation
D2 = spraying with Rhodamin 6 G in acetone (1% w/v) and observation under UV light
D3 = brown spots after exposure to J_2 vapor
D4 = observation of fluorescence under UV light
D5 = spraying with $KMnO_4$ (0.5% in 50% sulfonic acid)

REFERENCES

1. **Taylor, R. F. and Davies, B. H.**, *Biochem. J.*, 139, 751, 1974.
2. **Davies, B. H., Goodwin, T. W., and Mercer, E. J.**, *Biochem. J.*, 81, 40P, 1961.
3. **Stobart, A. K., McLaren, J., and Thomas, D. R.**, *Phytochemistry*, 6, 1467, 1967.
4. **Fiasson, J. L. and Arpin, N.**, *Bull. Soc. Chim. Biol.*, 49, 537, 1967.
5. **Davies, B. H., Jones, O., and Goodwin, T. W.**, *Biochem. J.*, 87, 326, 1963.
6. **Bramley, P. M., Davies, B. H., and Rees, A. F.**, *Liq. Scintill. Counting*, 3, 76, 1974.
7. **Taylor, R. F. and Davies, B. H.**, *Biochem. J.*, 153, 233, 1976.
8. **Fox, D. L. and Hopkins, T. S.**, *Comp. Biochem. Physiol.*, 19, 267, 1968.
9. **Cholnoky, L., Szabolcs, J., Cooper, R. D. G., and Weedon, B. C. L.**, *Tetrahedron Lett.*, 5, 1257, 1963.
10. **Johansen, J. E., Svec, W. A., Liaaen-Jensen, S., and Haxon, F. T.**, *Phytochemistry*, 13, 2261, 1974.
11. **Parihar, D. B., Prahash, O. M., Bajaj, J., Tripathi, R., and Verma, K. K.**, *J. Chromatogr.*, 59, 457, 1971.
12. **Grob, E. C. and Pflugshaupt, R. P.**, *Helv. Chim. Acta*, 48, 930, 1965.
13. **Egger, K. and Kleinig, H.**, *Phytochemistry*, 6, 903, 1967.
14. **Bolliger, H. R. and König, A.**, *Dünnschichtchromatographie*, Stahl, E., Ed., Springer-Verlag, Berlin, 1967, 259.
15. **Davies, B. H.**, *Chemistry and Biochemistry of Plant Pigments*, Goodwin, T. W., Ed., Academic Press, London, 1965, 514.

Table I. TLC 2
TLC ON VARIOUS ONE-COMPONENT LAYERS — I

Layer	L1	L2	L3	L3	L4	L5	L6	L7
Solvent	S1	S2	S3	S4	S2	S1	S3	S5
Technique	T1	T1	T1	T1	T1	T2	T1	T1
Detection	D1	D1	D1	D1,2	D1	D1	D1	D1
Literature	1	2	3	4	5	1	6	7
Compound					$R_f \times 100$			
Antheraxanthin	—	—	—	—	16	—	10—11	—
Auroxanthin	36	—	—	—	—	52	—	—
Azafrin	—	—	—	—	—	—	10—11	—
Bixin	—	—	—	—	—	—	10—11	—
Canthaxanthin	—	—	—	—	—	—	63	
Capsanthin	—	—	—	—	—	—	10—11	—
Capsorubin	—	—	—	—	—	—	10—11	—
α-Carotene	—	—	—	—	—	—	100	—
β-Carotene	—	—	—	—	100	—	100	—
γ-Carotene	—	—	—	—	—	—	100	—
ζ-Carotene	—	—	—	—	—	—	100	—
Cryptoxanthin	—	—	—	—	—	—	54	
Cryptoxanthin-monoepoxide	—	—	—	—	68	—	—	—
4,4'-Diapolycopen-4-al	—	—	—	72	—	—	—	—
4,4'-Diaponeurosporen-4-al	—	—	—	72	—	—	—	—
Fucoxanthin	—	15	—	—	—	—	—	95
4-Hydroxy-4,4'-diaponeurosporene	—	—	—	67	—	—	—	—
Lutein	—	22	—	—	27	—	10—11	68
Luteinepoxide	—	—	—	—	17	—	—	77
Lycopene	—	—	—	—	—	—	100	—
Lycopersene	—	—	—	—	—	—	100	—
Methylbixin	—	—	—	—	—	—	81	—
Micronone	—	—	—	—	—	—	—	74
Microxanthin	—	—	—	—	—	—	—	79
Neoxanthin	—	11	—	—	7	—	—	87
Phytoene	—	—	—	—	—	—	100	—
Phytofluene	—	—	—	—	—	—	100	—
Siphonaxanthin	—	9	—	—	—	—	—	87
Siphonaxanthin-monolaurate	—	18	—	—	—	—	—	—
Siphonein	—	18	—	—	—	—	—	73
Torularhodin-methylester	—	—	—	—	—	—	94	—
Torulene	—	—	94	—	—	—	98	—
Trollein	—	—	—	—	—	—	—	76
Vaucheriaxanthin	—	—	—	—	3	—	—	—
Violaxanthin	—	—	—	—	14	—	10—11	86
Xanthophyll K	—	—	—	—	—	—	—	78
Xanthophyll K1	—	—	—	—	—	—	—	72
Xanthophyll K1S	—	—	—	—	—	—	—	87
β-Zeacarotene	—	—	—	—	—	—	100	—
Zeaxanthin	—	—	—	—	27	—	10—11	59

Table I. TLC 2 (continued)
TLC ON VARIOUS ONE-COMPONENT LAYERS — I

Layer	L1	=	alumina
	L2	=	silica gel
	L3	=	silica gel G (Merck)
	L4	=	MN silica gel N (Machery and Nagel)
	L5	=	Kieselgur G (Merck)
	L6	=	$Mg_3(PO_4)_2$
	L7	=	polyamide
Solvent	S1	=	benzene-methanol = 19:1
	S2	=	benzene-methanol = 50:2.5
	S3	=	benzene
	S4	=	benzene-methanol-acetic acid = 87:11:2
	S5	=	butan-2-one-methanol-water = 5:5:1
Technique	T1	=	ambient temperature, chamber saturation, ascending
	T2	=	development of chromatogram under nitrogen
Detection	D1	=	visual observation
	D2	=	brown spots after exposure to J_2 vapor

REFERENCES

1. **Stobart, A. K., McLaren, J., and Thomas, D. R.,** *Phytochemistry,* 6, 1467, 1967.
2. **Kleinig, H. and Egger, K.,** *Phytochemistry,* 6, 1681, 1967.
3. **Cardani, C., Merlini, L., and Mondelli, R.,** *Gazz. Chim. Ital.,* 92, 41, 1962.
4. **Taylor, R. F. and Davies, B. H.,** *Biochem. J.,* 153, 233, 1976.
5. **Kleinig, H. and Egger, K.,** *Z. Naturforsch.,* 22b, 868, 1967.
6. **Davies, B. H.,** *Chemistry and Biochemistry of Plant Pigments,* Goodwin, T. W., Ed., Academic Press, London, 1965, 514.
7. **Ricketts, T. R.,** *Phytochemistry,* 6, 1375, 1967.

Table I. TLC 3
TLC ON VARIOUS ONE-COMPONENT LAYERS — II

$R_f \times 100$

Layer	L1	L2	L2	L2	L2	L2	L2	L3	L3	L4	L4	L4	L4	L4	L4	L4	L5	L5	L5	L5	L5	L6	L7	L7	L8	L9	L9
Solvent	S1	S2	S3	S4	S5	S6	S7	S8	S9	S10	S11	S12	S13	S14	S15	S2	S5	S6	S16	S5	S6	S17	S18	S19	S20	S21	S22
Technique	T1	T2	T1	T2	T2pr.	T2pr.	T2pr.	T1	T1	T1	T1	T1	T1	T1	T1	T1	T2	T2	T2pr.	T2pr.	T2	T3	T1	T1	T1	T1	T1
Detection	D1	D1,4	D1,2,4	D1,4	D1,2,4	D1,2,4	D1,2,4	D1	D1	D1,4	D1	D1	D1	D1	D1	D1	D1,4	D1,4	D1-3	D1-3	D1,4	D1	D1,2	D1,2	D1	D1	D1
Literature	1	2	3	4	4	4	4	5	5	6	7	8	7	7	7	2	4	2	4	4	2	9	10	10	11	11	11
Compound																											
β-Apo-8'-carotenal	—	—	—	—	—	—	—	—	—	—	—	—	—	—	—	—	—	—	—	—	—	—	—	—	00	—	64
β-Apo-10'-carotenal	—	—	—	—	—	—	—	—	—	—	—	—	—	—	—	—	—	—	—	—	—	—	—	—	00	—	53
β-Apo-12'-carotenal	—	—	—	—	—	—	—	—	—	—	—	—	—	—	—	—	—	—	—	—	—	—	—	—	00	—	70
Antheraxanthin	—	—	—	—	—	—	—	—	—	—	—	—	—	—	—	—	—	—	—	—	—	—	—	35	00	10	00
Azafrin	—	—	—	—	—	—	—	—	—	—	—	—	—	—	—	—	—	—	—	—	—	—	—	—	00	00	00
Bixin	—	—	—	—	—	—	—	—	—	—	—	—	—	—	—	—	—	—	—	—	—	—	—	—	00	00	00
Canthaxanthin	—	—	—	—	—	—	—	—	—	—	—	—	26	—	—	—	—	—	—	—	—	—	—	—	—	63	00
Capsanthin	—	—	—	—	—	—	—	—	—	—	—	—	—	—	—	—	—	—	—	—	—	—	51	97	—	5	00
Capsorubin	—	—	—	—	—	—	—	—	—	—	—	—	—	—	—	—	—	—	—	—	—	—	—	66	—	5	00
α-Carotene	52	—	—	—	3	5	16	—	—	—	—	—	—	—	—	—	—	—	40	10	—	—	—	—	—	—	—
β-Carotene	—	—	—	—	5	3	8	—	—	—	—	71	87	87	85	91	—	—	40	10	—	—	74	—	80	100	97
γ-Carotene	—	—	—	—	—	—	—	—	—	—	—	—	—	—	—	—	—	—	—	—	—	—	—	—	41	100	97
δ-Carotene	—	—	—	—	—	—	—	—	—	—	—	—	—	—	—	—	—	—	—	—	—	—	—	—	55	100	97
ε-Carotene	—	—	—	—	—	—	—	—	—	—	—	—	—	—	—	—	—	—	—	—	—	—	—	—	84	100	97
ζ-Carotene	—	—	—	—	35	3	10	—	—	—	—	—	—	—	—	—	—	—	—	—	—	—	—	—	—	—	—
Cryptocapsin	—	—	—	—	—	—	—	—	—	—	—	—	—	—	—	—	—	—	—	—	—	—	9	97	—	—	—
Cryptoxanthin	—	—	—	37	—	—	—	—	—	—	—	—	—	—	—	—	—	—	—	—	—	83	—	—	—	54	00
Cryptoxanthin-5,6-epoxide	—	—	—	—	—	—	—	—	—	—	—	—	—	—	—	—	—	—	—	—	—	83	—	—	—	—	—
4,4'-Diapo-ζ-carotene	—	—	—	—	37	7	16	—	—	70	—	—	—	—	—	—	—	—	—	—	—	—	—	—	—	—	—
4,4'-Diapolycopen-4-al	—	—	—	—	—	—	—	—	—	70	—	—	—	—	—	—	—	—	—	—	—	—	—	—	—	—	—
4,4'-Diaponeurosporen-4-al	—	—	—	—	—	—	—	—	—	31	—	—	—	—	—	—	—	—	—	—	—	—	—	—	—	—	—
4,4'-Diaponeurosporene	—	—	—	—	21	00	4	—	—	—	—	—	—	—	—	—	—	—	—	—	—	—	—	—	—	—	—
4,4'-Diapophytoene	—	—	—	—	67	45	61	—	—	—	—	—	—	—	—	—	—	—	—	—	—	—	—	—	—	—	—
4,4'-Diapophytofluene	—	—	—	—	58	19	38	—	—	—	—	—	—	—	—	—	—	—	—	—	—	—	—	—	—	—	—
4,4'-Diapo-7,8,11,12-tetrahydrolycopene	—	—	—	—	37	7	18	—	—	—	—	—	—	—	—	—	—	—	—	—	—	—	—	—	—	—	—

Table I. TLC 3 (continued)
TLC ON VARIOUS ONE-COMPONENT LAYERS — II

Rf × 100

Compound	L1 S1 T1 D1,4 [1]	L2 S2 T2 D1,2,4 [2]	L2 S3 T1 D1,4 [3]	L2 S4 T2 D1,2,4 [2]	L2 S5 T2pr. D1,2,4 [3]	L2 S5 T1 D1,2,4 [3]	L2 S6 T2pr. D1,2,4 [4]	L2 S6 T1 D1,2,4 [3]	L2 S7 T2pr. D1,2,4 [4]	L3 S8 12T1 D1 [5]	L3 S9 T1 D1 [5]	L4 S10 T1 D1,4 [6]	L4 S11 T1 D1 [8]	L4 S12 T1 D1 [7]	L4 S13 T1 D1 [8]	L4 S14 T1 D1 [7]	L4 S15 T1 D1 [7]	L5 S2 T2 D1,4 [2]	L5 S5 T2pr. D1-3 [4]	L5 S6 T2 D1,4 [2]	L5 S16 T2pr. D1-3 [4]	L6 S17 T3 D1 [9]	L7 S18 T1 D1,2 [10]	L7 S19 T1 D1,2 [10]	L8 S20 T1 D1 [11]	L9 S21 T1 D1 [11]	L9 S22 T1 D1 [11]
Dihydroxy-α-carotene										20																	
4,4'-Diketo-3-hydroxy-β-carotene																											
Echinenone											100																
Hydroxy-α-carotene															69								92				
4-Hydroxy-4,4'-diaponeurosporene																											
Isozeaxanthin															10												
Lutein																							77	29		10	00
Luteoxanthin a																								25			
Luteoxanthin b																											
Lycopenal																											
Lycopene	14	98		34														84		23							43
Lycopersene		98		88														83		63							97
Methylbixin																											13
Mutatoxanthin																							81	82	00		
Neoxanthin																						69		20	00		
Neurosporene			18			00		00																			
Phytoene			62		35		46	56	67										10		40						
Phytofluene			51		11		17	32	40																		
Phytofluene (cis)																			10		40						
Phytofluene (trans)																			10		40						
Torularhodin-methylester																											
Torulene													64	57											00	94	83
Violaxanthin																								38		98	
Xanthophyll													30	37												5	00
β-Zeacarotene															84												
Zeaxanthin																							19	97		5	00

Layer   L1 = Alox 25-22 (Brinkman Instruments)
         L2 = Alumina G (type E, Merck)
         L3 = CaCO$_3$ (specified as low in alkalinity)
         L4 = Silica gel G (Merck)
         L5 = Silica gel H (Merck)
         L6 = Kieselguhr G (Merck)
         L7 = Cellulose MN 300 (Machery and Nagel)
         L8 = MgO
         L9 = Mg$_3$(PO$_4$)$_2$

Solvent S1   = hexane-acetone = 99:1
         S2   = chloroform-ether = 99:1
         S3   = hexane-benzene = 95:5
         S4   = hexane-benzene = 1:1
         S5   = hexane-ether = 99.75:0.25
         S6   = hexane-ether = 99:1
         S7   = hexane-ether = 97:3
         S8   = ether
         S9   = ether-petroleum ether = 1:1
         S10 = chloroform
         S11 = ether-hexane = 9:1

         S12 = hexane-benzene = 4:1
         S13 = hexane-ether = 1:1
         S14 = hexane-ethyl acetate = 9:1
         S15 = hexane-ethyl acetate = 2:1
         S16 = hexane-ether = 99.5:0.5
         S17 = hexane-ethyl acetate = 19:1
         S18 = hexane-*n*-propanol = 99.9:0.1
         S19 = hexane-*n*-propanol = 99:1
         S20 = benzene-petroleum ether = 9:1
         S21 = benzene
         S22 = carbon tetrachloride

Technique T1 = ambient temperature, chamber saturation, ascending
           T2 = previous to the actual use, the plates are washed with chloroform-methanol = 1:1, then with acetone; pr. indicates layer thickness of 1 mm
           T3 = development of chromatograms under nitrogen

Detection D1 = visual observation
           D2 = observation under UV light
           D3 = spraying with 0.5% aqueous KMnO$_4$
           D4 = iodine vapor

REFERENCES

1. **David, H. L.**, *J. Bacteriol.*, 119, 527, 1974.
2. **Kushwaha, S. C. and Kates, M.**, *Biochim. Biophys. Acta*, 316, 235, 1973.
3. **Taylor, R. F. and Davies, B. H.**, *Biochem. J.*, 139, 751, 1974.
4. **Kushwaha, S. C., Pugh, E. L., Kramer, J. K. G., and Kates, M.**, *Biochim. Biophys. Acta*, 260, 492, 1972.
5. **Ungers, G. E. and Cooney, J. J.**, *J. Bacteriol.*, 96, 234, 1968.
6. **Taylor, R. F. and Davies, B. H.**, *Biochem. J.*, 153, 233, 1976.
7. **Cardani, C., Merlini, L., and Mondelli, R.**, *Gazz. Chim. Ital.*, 92, 41, 1962.
8. **Merlini, L. and Cardillo, G.**, *Gazz. Chim. Ital.* 93, 949, 1963.
9. **Stobart, A. K., McLaren, I., and Thomas, D. R.**, *Phytochemistry*, 6, 1467, 1967.
10. **Buckle, K. A. and Rahman, F. M. M.**, *J. Chromatogr.*, 171, 385, 1979.
11. **Bolliger, H. R. and König, A.**, *Dünnschichtchromatographie*, Stahl, E., Ed., Springer-Verlag, Berlin, 1967, 253.

Table I. TLC 4
TLC ON VARIOUS ONE-COMPONENT LAYERS — III

Layer	L1	L2	L3	L4	L4	L4	L4	L5	L5	L5	L6	L6	L7	L8	L8	L8	L8	L8	L8	L8	L9
Solvent	S1	S2	S3	S4	S5	S6	S7	S8	S9	S10	S11	S12	S13	S14	S14	S14	S14	S14	S2	S15	S16
Technique	T1	T1	T1	T2	T1	T2	T1	T1	T1	T1	T1	T1	T1	T1	T1	T1	T1	T1	T1	T1	T1
Detection	D1	D1	D1	D1,2	D1	D1,2	D1	D1	D1	D1	D1	D1	D1	D1	D1	D1	D1	D1	D1	D1	D1
Literature	1	2	3	4	5	4	5	6	7	5	8	5	3	9	10	11	12	13	2	14	15
Compound ($R_f \times 100$)																					
Alloxanthin (*trans*)	—	—	—	—	—	—	—	—	—	—	—	—	72	—	—	—	—	—	—	—	—
Antheraxanthin	—	—	00	50	—	—	—	—	—	—	—	—	65	—	—	—	—	—	—	25	—
β-Apo-8'-carotenal	—	—	—	—	—	—	—	—	—	—	—	00	—	—	—	—	—	—	—	—	—
β-Apo-10'-carotenal	—	—	—	—	—	—	—	—	—	—	—	00	—	—	—	—	—	—	—	—	—
β-Apo-12'-carotenal	—	—	—	—	—	—	—	—	—	—	—	00	—	—	—	—	—	—	—	—	—
Astacene	—	—	—	—	—	—	—	—	—	—	—	—	—	32	—	28	—	—	—	—	—
Auroxanthin	—	—	00	25	—	—	—	—	—	—	—	00	—	—	—	—	—	—	—	—	—
Azafrin	—	—	00	—	—	—	—	—	—	—	—	00	—	—	—	—	—	—	—	—	—
Bixin	—	—	00	—	—	—	—	—	—	—	—	00	—	—	—	—	—	—	—	—	—
Canthaxanthin	—	—	00	—	35	—	—	—	—	—	—	00	—	58	—	60	65	—	—	—	—
Capsanthin	—	—	00	—	—	—	—	—	—	—	—	00	—	—	—	—	—	—	—	—	—
Capsorubin	—	—	00	—	—	—	—	—	—	—	—	00	—	—	—	—	—	—	—	—	—
α-Carotene	—	—	84	—	—	—	—	—	—	—	—	80	—	—	—	—	—	—	—	—	—
β-Carotene	—	—	69	—	97	—	98	—	—	99	—	74	—	—	—	—	—	—	—	100	—
γ-Carotene	—	—	—	—	—	—	—	—	—	—	25—30	41	—	—	—	—	—	—	—	—	—
δ-Carotene	—	—	—	—	—	—	—	—	—	—	—	55	—	—	—	—	—	—	—	—	—
ε-Carotene	—	—	—	—	58	—	—	—	—	—	—	84	—	—	—	—	—	—	—	—	—
Cryptoxanthin	—	—	00	—	—	—	—	—	—	—	—	00	—	—	—	—	—	—	—	—	—
Dehydroadonirubin	—	—	—	—	—	—	—	—	—	—	—	—	—	—	—	—	—	—	—	—	—
Diadinoxanthin	—	—	—	—	—	—	—	—	—	—	—	—	63	45	—	43	—	—	—	—	—
Diatoxanthin	—	—	—	—	—	—	—	—	—	—	—	—	75	—	—	—	—	—	—	—	—
Dihydroxy-α-carotene	—	—	—	—	—	—	—	—	—	—	—	—	—	—	—	—	—	—	00	—	—
Dihydroxy-3,4-dehydro-α-carotene	—	100	—	—	—	—	—	—	—	—	—	—	—	—	—	—	—	—	—	—	—

Compound	1	2	3	4	5	6	7	8	9	10	11	12	13	14	15	16	17	18	19
3,4-Diketo-α-carotene	20	—	—	—	—	—	—	—	—	—	—	—	—	—	—	—	—	—	—
Echinenone	—	—	—	—	—	—	—	—	—	—	—	—	—	87	98	—	—	—	—
Fucoxanthin	78	—	—	90	—	—	—	—	—	—	—	—	28	—	—	24	—	—	—
Isorenieratene	—	00	75	—	76	—	—	83	—	00	—	—	—	—	—	—	—	—	—
Lutein	—	—	—	—	—	—	—	—	—	—	—	—	35	—	—	26	—	35	—
Lutein-5,6-epoxide	—	—	—	—	—	—	—	—	—	—	—	—	33	—	—	—	—	—	—
Lycopenal	—	00	—	—	—	—	—	—	—	00	—	—	—	—	—	—	—	—	—
Lycopene	—	—	—	—	—	—	—	—	—	13	—	—	—	—	—	—	—	—	30
Methylbixin	—	00	—	—	—	—	—	—	—	00	—	—	—	—	—	—	—	—	—
Micronone	—	—	—	—	—	—	—	—	—	—	—	—	44	—	—	—	—	—	—
Microxanthin	—	—	—	—	—	—	—	—	—	—	—	—	27	—	—	—	—	—	—
Neoxanthin	—	—	20	—	35	10	—	00	—	—	—	—	25	—	—	13	—	22	—
Prolycopene	—	—	—	—	—	—	—	—	—	—	—	—	—	—	—	—	—	—	53
Renierapurpurin	50	—	—	—	—	—	—	—	—	—	—	—	—	—	—	—	—	—	—
Renieratene	69	—	—	—	—	—	—	—	—	—	—	—	—	—	—	—	—	—	—
Rhodoxanthin	—	—	—	33	—	—	—	—	—	—	—	—	23	—	—	—	—	—	—
Siphonaxanthin	—	—	—	—	—	—	—	—	—	—	—	—	—	—	—	10	—	—	—
Siphonaxanthin-monolaurate	—	—	—	—	—	—	—	—	—	—	—	—	—	—	—	27	—	—	—
Siphonein	—	—	—	—	—	—	—	—	—	—	—	—	34	—	—	27	—	—	—
Spheroidene	—	00	—	—	—	—	36	—	—	00	—	—	—	—	—	—	—	—	—
Torularhodin-dimethylester	—	—	—	—	—	—	—	—	—	00	—	—	—	—	—	—	—	—	—
Trollein	—	—	—	—	—	—	—	—	—	—	—	—	16	—	—	—	—	—	—
Vaucheriaxanthin	—	00	—	—	—	—	—	—	—	—	—	—	—	—	—	—	—	10	—
Violaxanthin	—	00	62	—	70	—	—	65	—	00	—	—	33	—	—	—	—	28	—
Xanthophyll K	—	—	—	—	—	—	—	—	—	—	—	—	25	—	—	—	—	—	—
Xanthophyll K1	—	—	—	—	—	—	—	—	—	—	—	—	34	—	—	—	—	—	—
Xanthophyll K1S	—	—	—	—	—	—	—	—	—	—	—	—	22	—	—	—	—	—	—
Zeaxanthin	—	00	—	10	—	—	—	—	—	00	—	—	32	—	29	—	—	31	—

Table I. TLC 4 (continued)
TLC ON VARIOUS ONE-COMPONENT LAYERS — III

Layer   L1   = Alumina G
        L2   = CaCO$_3$
        L3   = Ca(OH)$_2$
        L4   = Cellulose MN 300
        L5   = kieselguhr
        L6   = MgO
        L7   = Mg(OH)$_2$·MgCO$_3$
        L8   = polyamide
        L9   = Silica gel H

Solvent   S1   = petroleum ether-acetone = 199:1
          S2   = petroleum ether
          S3   = methylene chloride-petroleum ether = 95:5
          S4   = petroleum ether-acetone-*n*-propanol = 90:10:0.25
          S5   = petroleum ether-carbon tetrachloride = 6:4
          S6   = petroleum ether-benzene-chloroform-acetone = 50:35:10:5
          S7   = petroleum ether-acetone-*n*-propanol = 90:10:0.45
          S8   = petroleum ether-acetone = 200:1
          S9   = petroleum ether-benzene = 1:1
          S10 = petroleum ether-*n*-propanol = 99:1
          S11 = petroleum ether-benzene = 3:1
          S12 = petroleum ether (bp 100—140°C)-benzene = 8:1:1
          S13 = petroleum ether-acetone = 25:3
          S14 = petroleum ether-methanol-butanone-2 = 8:1:1
          S15 = petroleum ether (bp 100—140°C)-methanol-butanone-2 = 8:1:1
          S16 = petroleum ether-benzene = 9:1

Technique   T1   = ambient temperature, chamber saturation, ascending
            T2   = development of chromatogram under nitrogen

Detection   D1   = visual observation
            D2   = observation under UV light (λ = 366 nm)

REFERENCES

1. **Cooper, R. D. G., Davis, J. B., and Weedon, B. C. L.,** *J. Chem. Soc.,* p. 5637, 1963.
2. **Ungers, G. E. and Cooney, J. J.,** *J. Bacteriol.,* 96, 234, 1968.
3. **Davies, B. H.,** *Chemistry and Biochemistry of Plant Pigments,* Goodwin, T. W., Ed., Academic Press, London, 1965, 489
4. **Buckle, K. A. and Rahman, F. M. M.,** *J. Chromatogr.,* 171, 385, 1979.
5. **Bolliger, H. R. and König, A.,** *Dünnschichtchromatographie,* Stahl, E., Ed., Springer-Verlag, Berlin, 1967, 253.
6. **Stobart, A. K., McLaren, J., and Thomas, D. R.,** *Phytochemistry,* 6, 1467, 1967.
7. **Barber, M. X., Jackman, L. M., Manchand, P. S., and Weedon, B. C. L.,** *J. Chem. Soc.,* p. 2166, 1966.
8. **Bramley, P. M. and Davies, B. H.,** *Phytochemistry,* 14, 463, 1974.
9. **Egger, K.,** *Phytochemistry,* 4, 609, 1965.
10. **Ricketts, T. R.,** *Phytochemistry,* 6, 1375, 1967.
11. **Egger, K. and Kleinig, H.,** *Phytochemistry,* 6, 903, 1967.
12. **Kleinig, H. and Egger, K.,** *Phytochemistry,* 6, 611, 1967.
13. **Kleinig, H. and Egger, K.,** *Phytochemistry,* 6, 1681, 1967.
14. **Kleinig, H. and Egger, K.,** *Z. Naturforsch. B,* 22, 868, 1967.
15. **Qureshi, A. A., Manok, K., Qureshi, N., and Porter, J. W.,** *Arch. Biochem. Biophys.,* 162, 108, 1974.

Table I. TLC 5
TLC ON VARIOUS TWO-COMPONENT LAYERS (CONTAINING SILICA GEL G) — I

	L1	L2	L3	L4	L4	L5	L6	L7
Layer	L1	L2	L3	L4	L4	L5	L6	L7
Solvent	S1	S2	S2	S3	S4	S5	S6	S7
Technique	T1	T1	T1	T1	T1	T1	T1	T1
Detection	D1	D1	D1	D1	D1	D1	D1	D1
Literature	1	2	2	3	3	4	5	3

Compound	$R_f \times 100$							
β-Apo-8'-carotenal	—	—	—	—	—	—	48	—
β-Apo-10'-carotenal	—	—	—	—	—	—	36	—
β-Apo-12'-carotenal	—	—	—	—	—	—	50	—
β-Apo-8'-carotenoic acid	—	—	—	—	—	—	00	—
β-Apo-10'-carotenoic acid	—	—	—	—	—	—	00	—
β-Apo-12'-carotenoic acid	—	—	—	—	—	—	00	—
Bixin	—	—	—	—	—	—	—	51
Canthaxanthin	50	—	—	43	—	—	—	38
Canthaxanthin *(cis)*	48	—	—	—	—	—	—	—
β-Carotene	—	—	—	—	—	90	100	96
Cryptoxanthin	—	—	—	34	74	—	—	—
Echinenone	72	—	—	82	—	—	—	—
4-Hydroxy-4'-keto-β-carotene	40	—	—	—	—	—	—	—
Isocryptoxanthin	55	—	—	—	—	—	—	—
Isozeaxanthin	26	—	—	—	—	—	—	63
Lutein	—	—	—	—	55	—	—	—
Lycopene	—	—	—	—	—	6	—	—
Rhodoxanthin	—	—	—	16	94	—	—	—
Zeaxanthin	—	—	—	—	57	—	—	17
Zeinoxanthin	—	48	65	—	—	—	—	—

Layer	L1 =	Silica gel G with fluorescence indicator (Eastman chromagram sheets 6060)
	L2 =	Silica gel G-lime = 1:4
	L3 =	Silica gel G-lime = 1:6
	L4 =	Silica gel G-Ca(OH)$_2$ = 1:6
	L5 =	Silica gel G-MgO = 1:1
	L6 =	Silica gel G-plaster of Paris = 8:1
	L7 =	Silica gel G-rice starch = 98:2
Solvents	S1 =	benzene-methanol = 97:3
	S2 =	benzene-*n*-butanol = 49:1
	S3 =	benzene
	S4 =	benzene-methanol = 49:1
	S5 =	benzene-petroleum ether = 1:1
	S6 =	cyclohexane-acetone = 47:3
	S7 =	ether-*n*-hexane = 7:3
Technique	T1 =	ambient temperature, chamber saturation, ascending
Detection	D1 =	visual observation

REFERENCES

1. **Hsieh, L. K., Lee, T. Ch., Chichester, C. O., and Simpson, K. L.,** *J. Bacteriol.*, 118, 385, 1974.
2. **Livingstone, A. L. and Knowles, R. E.,** *Phytochemistry*, 8, 1311, 1969.
3. **Bollinger, H. R. and König, A.,** *Dünnschichtchromatographie*, Stahl, E., Ed., Springer-Verlag, Berlin, 1967, 253.
4. **Bramley, P. M. and Davies, B. H.,** *Phytochemistry*, 14, 463, 1974.
5. **Singh, H., John, H., and Cama, H. R.,** *J. Chromatogr.*, 75, 146, 1973.

Table I. TLC 6
TLC ON VARIOUS TWO-COMPONENT LAYERS (CONTAINING SILICA GEL G) — II

$R_f \times 100$

	L1	L2	L3	L3	L3	L4	L4	L5	L6	L7	L7	L8	L9	L10	L11	L11	L11
Solvent	S1	S2	S3	S4	S5	S6	S7	S8	S9	S10	S11	S12	S12	S13	S14	S15	S16
Technique	T1	T2	T1	T1	T1	T1	T1	T1	T1	T1	T1	T1	T1	T1	T1	T1	T1
Detection	D1	D2,3	D1	D1	D1	D1	D1	D1	D1	D1	D1	D1	D1	D1	D1	D1	D1
Literature	1	2	3	3	3	4	4	5	6	7	7	8	8	6	9	9	9
Compound																	
Alloxanthin	—	—	—	—	—	—	—	—	—	—	—	—	—	—	—	—	—
β-Apo-8′-carotenal	—	—	—	42	—	—	—	—	—	—	—	—	—	—	—	—	—
β-Apo-10′-carotenal	—	—	—	—	—	—	—	—	—	—	—	—	—	—	—	67	67
β-Apo-12′-carotenal	—	—	—	—	—	—	—	—	—	—	—	—	—	—	—	58	53
β-Apo-8′-carotenoic acid	—	—	—	—	—	—	—	—	—	—	—	—	—	—	22	71	63
β-Apo-10′-carotenoic acid	—	—	—	—	—	—	—	—	—	—	—	—	—	—	12	00	00
β-Apo-12′-carotinoic acid	—	—	—	—	—	—	—	—	—	—	—	—	—	—	28	00	00
Astaxanthin	—	—	—	—	—	66	—	—	—	—	—	—	—	—	—	00	00
Canthaxanthin	—	—	—	—	—	82	—	—	—	—	—	—	—	—	—	—	00
α-Carotene	86	—	—	—	—	—	—	—	88	92	96	—	—	—	—	—	—
β-Carotene	—	—	—	—	—	—	—	—	84	92	92	—	—	81	—	100	100
β,β-Carotene	—	—	—	—	—	100	—	—	—	—	—	—	—	—	—	—	—
β,ε-Carotene	—	—	—	—	—	100	—	—	—	—	—	—	—	—	—	—	—
γ-Carotene	60	—	—	—	—	—	—	—	40—50	83	75	—	—	—	—	—	—
ζ-Carotene	—	—	—	—	—	—	—	—	—	—	—	—	—	62	—	—	—
β-Citraurin	—	—	—	—	—	—	—	—	—	—	—	—	—	—	—	—	—
Crocoxanthin	—	—	57	—	—	—	—	—	—	58	52	—	—	—	—	—	—
α-Cryptoxanthin	—	—	61	—	—	82	—	—	—	—	—	—	—	—	—	—	—
β-Cryptoxanthin	—	—	—	—	—	59	6	—	—	—	—	—	—	—	—	—	—
Diadinoxanthin	—	—	—	—	—	67	—	—	—	—	—	—	—	—	—	—	—
Diatoxanthin	—	—	—	—	—	—	38	—	—	—	—	—	—	—	—	—	—
Echinenone	—	—	—	—	—	96	—	71	—	—	—	—	—	—	—	—	—
4-Hydroxy-α-carotene	—	—	77	—	—	—	—	—	—	—	—	—	—	—	—	—	—
Isorenieratene	—	—	—	—	41	—	—	—	—	—	—	—	—	—	—	—	—
Lutein	—	—	—	—	—	—	—	—	10—20	72	57	—	—	—	—	—	—
Lycopene	10	—	—	—	—	—	—	—	—	—	—	—	—	—	—	—	—

Monadoxanthin	—	—	—	—	38	—	25	—	—	—	—	—	—	—	—	—	—
Neoxanthin	—	—	—	—	—	—	—	—	—	—	—	—	—	—	—	—	—
Phytoene	35—40	—	—	—	—	—	—	—	—	—	—	—	—	—	—	—	—
β-Zeacarotene	—	—	—	—	—	—	—	—	—	—	—	—	75	—	—	—	—
Zeaxanthin	—	48	—	—	—	—	—	—	27	—	—	—	—	—	—	—	—
Zeinoxanthin	—	—	—	—	—	—	—	—	47	—	—	—	—	—	—	—	—

Layer
L1 = Silica gel G with fluorescence indicator (Merck)
L2 = Silica gel G (Merck) impregnated with 3% AgNO₃ (w/w)
L3 = Silica gel G (Merck)-aluminum oxide = 1:1
L4 = Silica gel G (Merck)-calcium carbonate = 1:1
L5 = Silica gel G (Merck)-calcium hydroxide = 1:4
L6 = Silica gel G (Merck)-calcium hydroxide = 1:6
L7 = Silica gel G (Merck)-Kieselguhr = 3:1
L8 = Silica gel G (Merck)-lime = 1:4
L9 = Silica gel G (Merck)-lime = 1:6
L10 = Silica gel G (Merck)-magnesium oxide = 1:1
L11 = Silica gel G (Merck)-plaster of Paris = 8:1

Solvent
S1 = petroleum ether-acetone = 85:15
S2 = petroleum ether-ether = 7:3
S3 = hexane-acetone = 3:1
S4 = hexane-acetone = 13:7
S5 = hexane-acetone = 7:3
S6 = petroleum ether-acetone-benzene-isopropanol = 69.5:25:4:1.5
S7 = petroleum ether-acetone = 12:1
S8 = petroleum ether-benzene = 4:1
S9 = petroleum ether-benzene = 49:1
S10 = petroleum ether-benzene = 3:2
S11 = petroleum ether-methylene chloride = 9:1
S12 = hexane-acetone = 49:1
S13 = petroleum ether-benzene = 9:1
S14 = petroleum ether-acetone = 75:25
S15 = petroleum ether-acetone = 9:1
S16 = petroleum ether-diethyl ether = 75:25

Technique
T1 = ambient temperature, chamber saturation, ascending

Detection
D1 = visual observation
D2 = spraying with a solution of Rhodamin 6G in acetone (1% w/v)
D3 = observation under UV light

REFERENCES

1. **Hsieh, L. K., Lee, T. C., Chichester, C. O., and Simpson, K. L.,** *J. Bacteriol.,* 118, 385, 1974.
2. **Bramley, P. M. and Davies, B. H.,** *Phytochemistry,* 14, 463, 1974.
3. **Chapman, D. J.,** *Phytochemistry,* 5, 1331, 1966.
4. **Bjørnland, T.,** *Phytochemistry,* 21, 1715, 1982.
5. **Cooper, R. D. G., Davis, J. B., and Weedon, B. C. L.,** *J. Chem. Soc. London,* p. 5637, 1963.
6. **Davies, B. H.,** *Chemistry and Biochemistry of Plant Pigments,* Goodwin, T. W., Ed., Academic Press, London, 1965, 489.
7. **Parihar, D. B., Prahash, D. M., Bajaj, J., Tripathi, R. P., and Verma, K. K.,** *J. Chromatogr.,* 59, 457, 1971.
8. **Livingstone, A. L. and Knowles, R. E.,** *Phytochemistry,* 8, 1311, 1969.
9. **Singh, H., John, J., and Cama, H. R.,** *J. Chromatogr.,* 75, 146, 1973.

Table I. TLC 7
TLC ON TWO- AND MULTICOMPONENT LAYERS

$R_f \times 100$

Layer	L1	L1	L1	L1	L1	L1	L1	L2	L3	L3	L3	L3	L3	L3	L3	L4
Solvent	S1	S1	S1	S1	S1	S2	S3	S4	S5	S6	S7	S8	S9	S10	S11	S12
Technique	T1	T1	T1	T1	T1	T1	T1	T1	T1	T1	T1	T1	T1	T1	T1	T2
Detection	D1	D1	D1	D1	D1	D1	D1	D1	D1	D1	D1	D1	D1	D1	D1	D1
Literature	1	2	3	4	5	1	1	6	7	7	7	7	7	7	7	8
Compound																
Alloxanthin	—	20	—	—	—	21	—	—	—	—	—	—	—	—	—	—
Anhydroeschscholtzxanthin	76	—	—	—	—	—	—	—	—	—	—	—	—	—	—	—
Antheraxanthin	52	52	52	—	52	—	—	—	—	—	—	—	—	—	—	—
Aphanizophyll	—	—	—	5	—	—	—	—	—	—	—	—	—	—	—	—
β-Apo-β-carotenoic acid	—	—	—	—	—	—	—	—	28	38	30	15	5	00	00	—
Astacene	—	—	—	—	—	—	—	—	34	50	69	72	42	25	7	—
Auroxanthin	20	—	—	—	—	—	13	—	—	—	—	—	—	—	—	—
Caloxanthin	83	—	—	24	—	32	—	—	—	—	—	—	—	—	—	—
Canthaxanthin	—	—	—	83	—	—	77	—	58	65	79	80	55	37	20	—
Capsanthin	—	—	—	—	—	—	—	—	24	42	79	81	62	42	15	—
Capsorubin	—	—	—	—	—	—	—	—	19	37	74	76	60	42	15	—
α-Carotene	—	97	—	—	—	—	—	—	—	—	—	—	—	—	—	—
β-Carotene	—	92	—	—	92	—	92	—	100	100	100	80	25	10	00	—
γ-Carotene	—	—	71	—	—	—	—	—	—	—	—	—	—	—	—	—
ε-Carotene	—	96	—	—	—	—	—	—	—	—	—	—	—	—	—	—
β-Carotene-5,6,5',6'-diepoxide	—	—	—	—	86	—	—	—	—	—	—	—	—	—	—	—
Crocoxanthin	—	75	—	—	—	—	—	—	—	—	—	—	—	—	—	—
Cryptoxanthin	—	77	—	77	—	—	—	—	62	70	76	74	39	21	4	—
Cryptoxanthin-5,6,5',6'-diepoxide	—	—	—	—	82	—	—	—	—	—	—	—	—	—	—	—
Cryptoxanthin-5',6'-monoepoxide	—	79	—	—	79	—	—	—	—	—	—	—	—	—	—	—
Diadinoxanthin	41	—	—	—	41	—	—	—	—	—	—	—	—	—	—	—
Diatoxanthin	28	—	—	—	28	48	—	—	—	—	—	—	—	—	—	—
Echinenone	—	—	—	—	—	—	—	—	91	92	90	72	35	18	2	—
Eschscholtzxanthin	6	88	—	88	—	—	—	—	12	22	25	22	8	1	00	—
Euglenanone	—	—	—	—	—	—	—	—	62	68	81	80	54	34	9	—
Fucoxanthin	—	72	—	—	—	—	—	—	27	44	84	98	95	85	55	—
Heteroxanthin	—	—	—	—	12	—	—	—	—	—	—	—	—	—	—	—
3'-Hydroxyechinenone	—	—	—	80	—	—	—	—	—	—	—	—	—	—	—	—

3-Hydroxy-3'-keto-α-carotene	—	—	—	—	—	—	—	—	—	50	—	—	—
4-Hydroxy-4'-keto-β-carotene	—	63	—	—	—	—	—	—	—	52	—	—	—
Isocryptoxanthin	—	—	—	—	—	—	—	—	—	67	—	—	—
Isozeaxanthin	65	—	—	34	56	92	91	57	36	10	54	31	—
Lutein	—	—	—	35	57	93	95	68	45	14	—	—	70
Lutein (*trans*)	—	—	—	—	—	—	—	—	—	44	—	—	—
Luteindipalmitate	—	—	75	100	100	95	30	00	00	00	—	—	—
Lutein-5,6-epoxide	—	—	—	—	—	—	—	—	—	—	—	—	—
Luteoxanthin	—	—	—	—	—	—	—	—	—	22	—	—	—
Lycopene	—	—	—	95	90	75	60	5	00	00	—	—	83
Lycophyll	—	—	—	8	20	22	20	7	00	00	—	—	—
Lycoxanthin	—	—	—	29	37	40	32	8	00	00	—	—	—
Myxoxanthophyll	—	3	—	—	—	—	—	—	—	—	—	—	—
Neoxanthin	49	—	49	22	42	82	96	90	78	40	35	—	83
Nostoxanthin	10	—	—	—	—	—	—	—	—	—	12	—	—
Oscillaxanthin	—	6	—	—	—	—	—	—	—	—	—	—	—
Retroanhydro-β-cryptoxanthin	61	—	—	—	—	—	—	—	—	56	—	—	—
Rhodoxanthin	—	—	—	28	42	43	40	14	7	1	—	—	—
Rubixanthin	—	—	—	45	60	45	64	15	4	00	—	—	—
Torularhodin	—	—	—	6	10	9	2	1	00	00	—	—	—
Trihydroxy-α-carotene	—	—	36	—	—	—	—	—	—	—	—	—	—
Vaucheriaxanthin	19	—	—	—	—	—	—	—	—	—	—	—	—
Vaucheriaxanthinester	69	—	—	—	—	—	—	—	—	—	—	—	—
Violaxanthin	74	—	74	30	52	76	93	88	73	35	63	—	83
Zeaxanthin	39	39	39	30	54	78	82	55	35	10	59	—	—

Layer
L1 = CaCO₃-MgO-Ca(OH)₂ = 15:3:2
L2 = MgO-Kieselguhr G (Merck) = 1:1
L3 = polyamide-Cellulose MN 300 = 85:15
L4 = powdered polyethylene-Cellulose MN 300 = 23:2; 25 g of the mixture are suspended in 75 mℓ methanol-chloroform = 4:1

Solvent
S1 = "benzine"-acetone-chloroform-methanol = 50:50:40:1
S2 = "benzine"-acetone-chloroform-methanol = 50:50:40:8
S3 = "benzine"-chloroform-acetone = 50:40:30
S4 = benzene-acetone = 3:2
S5 = petroleum ether (bp. 100—120°C)-methanol-butan-2-one = 8:1:1
S6 = petroleum ether (bp. 100—120°C)-methanol-butan-2-one = 4:1:1
S7 = petroleum ether (bp. 100—120°C)-methanol-butan-2-one = 2:1:1
S8 = methanol-butan-2-one = 1:1
S9 = water-methanol-butan-2-one = 1:5:5
S10 = water-methanol-butan-2-one = 1:3:3
S11 = water-methanol-butan-2-one = 2:3:3
S12 = methanol-water-benzene = 90:12:4

Technique
T1 = ambient temperature, chamber saturation, ascending
T2 = two runs: the plate is developed first with chloroform, then with solvent system indicated

Detection
D1 = visual observation

Table I. TLC 7 (continued)
TLC ON TWO- AND MULTICOMPONENT LAYERS

REFERENCES

1. **Hager, A. and Stransky, H.**, *Arch. Mikrobiol.*, 71, 132, 1979.
2. **Hager, A. and Stransky, H.**, *Arch. Mikrobiol.*, 73, 77, 1970.
3. **Hager, A. and Stransky, H.**, *Arch. Mikrobiol.*, 72, 68, 1970.
4. **Stransky, H. and Hager, A.**, *Arch. Mikrobiol.*, 72, 84, 1970.
5. **Stransky, H. and Hager, A.**, *Arch. Mikrobiol.*, 71, 164, 1970.
6. **Cyronak, M. J., Britton, G., and Simpson, K. L.**, *Phytochemistry*, 16, 612, 1977.
7. **Bolliger, H. R. and König, A.**, *Dünnschichtchromatographie*, Stahl, E., Ed., Springer-Verlag, Berlin, 1967, 253.
8. **Schenk, J. and Dussler, H. G.**, *Pharmazie*, 24, 116, 1969.

Table I. TLC 8
TLC ON "THINLAYER A"

Layer	L1	L1	L1	L1
Solvent	S1	S2	S2	S3
Technique	T1	T1	T1	T1
Literature	1	1	1	1

Compound[a]

Antheraxanthin	3	5	4	—
α-Carotene	1	—	—	1
β-Carotene	1	1	—	2
γ-Carotene	—	—	—	3
ξ-Carotene	—	—	—	4
Lutein	2	4	3	—
Lutein-5,6-epoxide	3	2	1	—
Lycopene	—	—	—	5
Neoxanthin	5	6	5	—
Rhodoxanthin	—	—	7	—
Violaxanthin	4	3	2	—
Zeaxanthin	2	7	6	—

Layer	L1	=	"thinlayer A": 12 g Kieselguhr G (Merck 8129), 3 g silica gel "unter 0.08 mm" for chromatography (Merck 7729), 3 g CaCO₃ p.A. (Merck 2066), 0.02 g Ca(OH)₂ p.A. (Merck 2047), 50 mℓ ascorbic acid $(8 \times 10^{-3} \text{ mol} \times \ell^{-1})$
Solvent	S1	=	"benzine" (bp 100—140°)-isopropanol-dist. water = 100:12:0.25
	S2	=	"benzine" (bp 100—140°)-acetone-chloroform = 50:50:40
	S3	=	"benzine" (bp 100—140°)-benzene-acetone = 40:10:1
Technique	T1	=	layer thickness 0.125 mm, ascending; the plates are dried for 1.5 hr at 50—60°C; good ventilation is necessary; the plates are to be used immediately after preparation

[a] Order of separation; no R_f values. 1 = top pigment.

REFERENCE

1. **Hager, A. and Meyer-Bertenrath, T.**, *Planta (Berlin)*, 69, 198, 1966.

Table I. TLC 9
TLC ON VARIOUS REVERSED-PHASE LAYERS

Layer	L1	L1	L1	L1	L1	L2	L3	L4	L4	L5	L5	L5	L6
Solvent	S1	S2	S3	S3	S4	S5	S6	S5	S6	S5	S7	S8	S9
Technique	T1	T1	T1	T1	T1	T1	T1	T1	T1	T1	T2	T2	T3
Detection	D1	D1	D1	D1	D1	D1	D1	D1	D1	D1	D1	D1	D1
Literature	1	2	3	4	5	6	6	7	7	7	8	8	8
Compound							$R_f \times 100$						
Alloxanthin *(trans)*	—	62	—	—	—	—	—	—	—	—	—	—	—
β-Apo-8′-carotenal	—	—	—	—	—	—	—	—	—	—	80	65	—
Antheraxanthin	—	—	45	—	—	—	—	—	—	—	—	—	—
Antheraxanthin *(trans)*	—	66	—	—	—	—	—	—	—	—	—	—	—
Astacene	—	—	—	—	45	—	—	—	—	—	—	—	—
Capsanthin	—	—	—	—	—	—	74	—	—	—	—	—	—
Canthaxanthin	38	—	—	—	38	—	—	—	—	—	94	86	—
β-Carotene	—	—	00	—	—	10	00	—	—	8	22	3	00
γ-Carotene	—	—	—	—	—	15	00	—	—	—	—	—	—
Cryptoxanthin	—	—	—	—	—	90	7	11	—	94	91	80	7
Cryptoxanthinepoxide	—	—	14	—	—	—	—	—	—	—	—	—	—
Dehydroadonirubin	—	—	—	—	43	—	—	—	—	—	—	—	—
Diadinoxanthin *(trans)*	—	70	—	—	—	—	—	—	—	—	—	—	—
Diatoxanthin *(trans)*	—	59	—	—	—	—	—	—	—	—	—	—	—
Echinenone	23	—	—	—	23	61	—	—	—	—	69	42	—
Fucoxanthin	—	—	—	62	—	—	—	—	—	—	—	—	—
Isozeaxanthin	—	—	—	—	—	—	49	—	—	—	—	—	—
Lutein	—	—	36	39	—	—	56	—	5	—	100	97	56
Lutein-dipalmitate	—	—	—	—	—	2	—	—	—	—	11	00	00
Luteinepoxide	—	—	—	—	—	—	72	—	—	—	—	—	72
Neoxanthin	—	—	67	68	—	—	95	—	89	—	—	—	95
Rhodoxanthin	—	—	—	—	—	—	26	—	—	—	—	—	26
Siphonaxanthin	—	—	—	72	—	—	—	—	—	—	—	—	—
Siphonein	—	—	—	36	—	—	—	—	—	—	—	—	—
Torularhodin-methyl-ester	—	—	—	—	—	48	—	—	—	—	57	25	—
Vaucheriaxanthin	—	—	85	—	—	—	—	—	—	—	—	—	—
Violaxanthin	—	—	52	—	—	—	84	—	—	—	—	—	84
Zeaxanthin	53	—	37	—	—	—	54	—	—	—	—	—	55
Zeaxanthin *(trans)*	—	56	—	—	—	—	—	—	—	—	—	—	—

Layer L1 = cellulose, impregnated with triglyceride
　　　L2 = kieselguhr, impregnated with paraffin
　　　L3 = kieselguhr, impregnated with vegetable oil
　　　L4 = Kieselguhr G (Merck), impregnated with coconut butter
　　　L5 = Kieselguhr G (Merck), impregnated with paraffin
　　　L6 = Kieselguhr G (Merck), impregnated with triglyceride
Solvent S1 = acetone-methanol-water = 15:5:2
　　　S2 = methanol-acetone-water = 40:10:3
　　　S3 = methanol-acetone-water = 30:10:3
　　　S4 = methanol-acetone-water = 15:5:2
　　　S5 = methanol-acetone = 5:2
　　　S6 = methanol-acetone-water = 20:4:3
　　　S7 = acetone-methanol-water = 50:47:3
　　　S8 = methanol-acetone-water = 76:20:4
　　　S9 = methanol-acetone-water = 75:15:10

Table I. TLC 9 (continued)
TLC ON VARIOUS REVERSED-PHASE LAYERS

Technique   T1 = ambient temperature, chamber saturation, ascending

             T2 = a slurry is prepared from Kieselguhr G (Merck) and dioxane-water = 3:1 and distributed equally onto glass plates (10 × 20 cm, 20 × 20 cm); the layer is dried for 4 hr at 100°C and thereafter partially impregnated with paraffin oil-petroleum ether (bp 100—140°C) = 8:92 to about 3—4 cm from the upper rim (ascending technique); for solvent removal, the plates are kept at 70°C for 1 hr

             T3 = the layer is prepared as described under T2; partial impregnation with an 8% solution of a triglyceride with low acid content (for example, olive oil [e.g., Livio®])

Detection   D1 = visual observation

REFERENCES

1. **Kleinig, H. and Egger, K.,** *Phytochemistry,* 6, 611, 1967.
2. **Egger, K., Nitsche, H., and Kleinig, H.,** *Phytochemistry,* 8, 1583, 1969.
3. **Kleinig, H. and Egger, K.,** *Z. Naturforsch. B,* 22, 868, 1967.
4. **Kleinig, H. and Egger, K.,** *Phytochemistry,* 6, 1681, 1967.
5. **Egger, K. and Kleinig, H.,** *Phytochemistry,* 6, 903, 1967.
6. **Davies, B. H.,** in *Chemistry and Biochemistry of Plant Pigments,* Goodwin, T. W., Ed., Academic Press, London, 1965, 489.
7. **Stobart, A. K., McLaren, J., and Thomas, D. R.,** *Phytochemistry,* 6, 1467, 1967.
8. **Bolliger, H. R. and König, A.,** *Dünnschichtchromatographie,* Stahl, E., Ed., Springer-Verlag, Berlin, 1967, 253.

LIQUID CHROMATOGRAPHY (LC) TABLES

NOTES ON TABLE I. LC 1

A more "classical" way to separate fat-soluble chloroplast pigments is column chromatography. A diversity of stationary phases has been used. Much of the early work (see Reference 12 in "Methods") and a number of recent publications still focus on liquid chromatography. Table I. LC 1 deals with the separation of carotenoids on columns filled with various sorbents. Since often in the original literature small figures (drawings) instead of retention volumes, or other data have been published, no numbers are given. Also, the resolution of zones and order of zones eluted is strongly dependent on the experimental conditions. The user is therefore advised to consult the original literature, when pigment name, sorbent, and solvent (= eluent) have been found. The amount of pigment present in a band can be roughly estimated by judging from the width of this band and the depth of color.

As stationary phases, mixtures often are used. One example for such a mixture is $Ca(OH)_2$-Hyflo-Super-Gel with toluene-petrol ether in various compositions as the mobile phase. The stationary phase is more polar than the mobile phase. Prior to the application to the column, the plant extract is evaporated to a small volume or brought to dryness and taken up in a small volume of hexane. In this way, a number of interfering pigments, such as flavins, are separated. For working on a semipreparative scale, a variety of column sizes is suitable. A more general rule from the chemical point of view is that the larger the molecule of the carotenoid applied, the more firm is the linkage to the column material. The distance between bands should be such that the bands appear well separated, especially the main bands: for example, for *Neurospora crassa*, γ-carotene and neurosporene.

Sterols tend to interfere with carotenoid separation; they also absorb in the UV region. For their removal they may be frozen out or digitonized and removed by subsequent centrifugation.

For literature sources, the reader is referred to Table I. LC 1.

Table I. LC 1
SURVEY OF DIFFERENT ADSORBENTS AND SOLVENTS USED FOR CAROTENOIDS

Pigment	Solvent	Ref.
Alumina[a]		
Antheraxanthin	S1	1a
Astaxanthin	S2	1a
	S3	1a
β-Carotene	S5	1a
	S6	2
γ-Carotene	S7	1a
Cryptoxanthin	S8	1a
Lycopene	S6	2
Lycophyll	S9	2
Lycoxanthin	S10	2
Mutatoxanthin	S9	1a
	S11	1a
Neoxanthin	S12	1a
P-481	S13	2
Spirilloxanthin	S13	2
Violaxanthin	S14	1a
Alumina (act. grade 0 — II)		
Aleuriaxanthin	S15	3
Anhydrorhodovibrin	S7	4
	S7	5
Aphanicin	S8	6
Aphanin	S13	6
Bacterioruberin	S16	7[h]
Canthaxanthin	S17	8
α-Carotene	S18	9[h]
β-Carotene	S5	5
	S6	11
	S15	3
	S15	6
	S15	9[h]
	S15	10
	S15	12
	S19	7[h]

Pigment	Solvent	Ref.
β-Carotene (*trans*)	S15	13
γ-Carotene	S7	14
	S10	33
	S13	3
δ-Carotene	S15	10
	S20	15
ε-Carotene	S20	15
	S21	15
ζ-Carotene	S5	21
	S15	3
	S15	10
	S15	17
	S15	21
ζ-Carotene (*cis*)	S15	16
ζ-Carotene (*trans*)	S15	16
Chlorobactene	S7	10
	S19	18[u]
Chloroxanthin	S17	4
Cryptoxanthin	S1	11
3,4-Dehydrolycopene	S35	3
2'-Dehydroplectaniaxanthin	S30	12
2'-Dehydroplectaniaxanthin-1'-ester	S17	27
3,4-Dehydrotorulene	S7	19
Deoxyflexixanthin	S22	20
1',2'-Dihydro-1'-hydroxy-γ-carotene	S8	14
1',2'-Dihydro-1'-hydroxy-4-keto-γ-carotene	S8	14
Dihydroxy-ζ-carotene	S23	21
Dihydroxylycopene	S24	21
1,1'-Dihydroxy-1,2,1',2'-tetrahydro-ζ-carotene	S25	17
Echinenone	S17	11
Gazaniaxanthin	S1	22
3-Hydroxy-3'-hydroxy-α-carotene	S9	23
3-Hydroxy-3'-keto-α-carotene	S11	23
4-Hydroxy-4'-keto-β-carotene	S8	8

Isorenieratene (*trans*)	S13	13
β-Isorenieratene (*trans*)	S15	13
Isozeaxanthin	S11	8
4-Keto-γ-carotene	S17	14
4-Keto-3'-hydroxy-β-carotene	S11	11
4-Ketorulene	S17	19
Lutein	S9	23
Lycopene	S7	5
	S7	10
	S7	17
	S7	27
	S10	3
	S13	21
Lycopene	S15	24
	S15	25
	S19	26
	S26	7[a]
	S35	33
Lycopersene	S8	7[d]
	S19	7[b]
	S29	7[a]
Lycophyll	S27	25
Lycoxanthin	S28	25
Methyl-apo-6'-lycopenoate	S13	24
Myxoxanthin	S17	11
Neurosporaxanthin-methylester	S13	28
Neurosporene	S7	4
	S7	29
	S13	3
	S13	17
	S15	21
Neurosporene (*cis*)	S5	17
OH-ζ-Carotene	S17	10
	S17	21
OH-Chlorobactene	S8	10
OH-Lycopene	S17	21
OH-Neurosporene	S17	21

OH-P-481	S8	21
OH-Phytofluene	S17	10
	S17	21
OH-R	S8	4
	S8	29
	S8	21
OH-Spirilloxanthin	S17	4
OH-Y	S15	21
P-412	S13	21
P-450	S7	21
P-481	S17	29
P-518	S5	10
Phytoene	S5	21
	S19	7[b]
	S29	9[b]
Phytofluene	S5	10
	S5	17
	S5	21
	S15	3
	S19	7[b]
	S29	9
Phytofluene (*cis*)	S29	9[b]
Phytofluene (*trans*)	S29	9[b]
Plectaniaxanthin	S30	12
Rhodopin	S8	5
	S8	10
Rhodovibrin	S8	4
	S8	5
Rubixanthin	S1	22
Rubixanthin (*cis*)	S8	10
Rubixanthin (*trans*)	S1	10
Saproxanthin	S9	32
	S23	32
	S31	30
S.g. 434	S32	31
S.g. 460	S7	32
	S33	32

Table I. LC 1 (continued)
SURVEY OF DIFFERENT ADSORBENTS AND SOLVENTS USED FOR CAROTENOIDS

Pigment	Solvent	Ref.	Pigment	Solvent	Ref.
S.g. 500	S8	32	Chrysanthemaxanthin	S8	34g
Spheroidenone	S7	29		S35	33k
Spirilloxanthin	S17	4	Crocetindial	S10	41
	S17	5	Cryptoxanthin	S7	35e
	S17	21	Cryptoxanthin (trans)	S8	34e
1,2,1',2'-Tetrahydro-1,1'-dihydroxylylycopene	S11	5	3,4-Dehydrolycopene	S15	16i
7,8,11,12-Tetrahydrolycopene (cis)	S15	16		S15	37h
7,8,11,12-Tetrahydrolycopene (all trans)	S15	16	5,6-Diepoxy-β-carotene (trans)	S15	34e
Torulene	S10	3	7,8-Dihydrosarcinaxanthin	S35	40j
Zeaxanthin	S14	10	Echinenone	S7	35e
	S34	11	Euglenanone	S8	35e
Alumina (act. grade III — IV)			Flavochrome	S15	33k
Antheraxanthin	S9	35e	Flavoxanthin	S8	34g
	S11	34g		S35	33k
Apo-3-lycopenal	S19	18j	1-Hydroxy-1,2-dihydro-γ-carotene	S19	18j
Auroxanthin	S9	34g	Lutein	S8	34g
β-Carotene	S5	16i		S10	33k
	S5	35e	Lycopene	S5	40j
	S5	39m		S6	16i
	S6	33l		S10	33l
	S6	37h		S15	37h
	S6	38		S15	38
β-Carotene (trans)	S6	34f	Lycopene (trans)	S19	18j
γ-Carotene	S6	16i	5,6-Monoepoxy-β-carotene (trans)	S13	34e
	S7	35e	5,6-Monoepoxylutein	S8	34g
	S19	18j	Mutatochrome	S13	33k
γ-Carotene (trans)	S15	34f	Mutatochrome (trans)	S15	34e
δ-Carotene (trans)	S15	34f	Neoxanthin	S36	34g
ζ-Carotene	S6	37h		S37	35e
ζ-Carotene (trans)	S15	34f	Neurosporene	S6	16i
ϑ-Carotene	S6	16i		S6	37h

OH-Chlorobactene	S19	18[i]
Phytoene	S5	16[i]
	S5	34[f]
	S5	37[h]
	S5	38
	S5	39[m]
Phytofluene	S5	16[i]
	S5	37[h]
	S5	39[m]
	S6	34[f]
Rhodopin	S19	18[i]
Rubixanthin (*cis*)	S13	34[e]
Rubixanthin (*trans*)	S10	34[e]
Sarcinaxanthin	S13	40[i]
Torulene	S6	37[h]
Violaxanthin	S11	34[g]
β-Zeacarotene	S6	16[i]
Zeaxanthin	S8	34[g]
CaCO₃		
α-Carotene	S8	42
β-Carotene	S8	42
Cryptoxanthin	S8	42
Lutein	S8	42
Neoxanthin	S8	42
Violaxanthin	S8	42
Zeaxanthin	S8	42
CaCO₃-Ca(OH)₂-Diatomaceous earth = 2:2:1 (w/w)		
Alloxanthin (*trans*)	S38	43
Crocoxanthin	S38	43
α-Cryptoxanthin	S38	43
Diatoxanthin	S38	43
4-Hydroxy-α-carotene	S38	43
Lutein	S38	43
Monadoxanthin	S38	43
Zeinoxanthin	S38	43
Ca(OH)₂		
α-Carotene	S5	45
	S7	44

β-Carotene	S5	45
	S7	44
β-Carotene-monoepoxide	S5	45
γ-Carotene	S7	44
	S8	44
	S8	44
Cryptoxanthin	S8	44
α-Cryptoxanthin	S5	45
α-Cryptoxanthin (*cis*)	S5	45
β-Cryptoxanthin	S5	45
	S8	44
β-Cryptoxanthin (*cis*)	S5	45
Lycopene	S8	44
Phytofluene	S7	44
Phytofluenol	S8	44
Zeaxanthin	S8	45
Ca(OH)₂-Celite 535 = 4:1		
β-Carotene	S21	46
4-Keto-α-carotene	S39	46
Retrodehydrocarotene	S39	46
Cellulose		
Deoxyflexixanthin	S41	20
	S42	20
Flexixanthin	S35	20
	S42	20
1-Hexosyl-1,2-dihydro-3,4-didehydro-apo-8'-lycopenol	S8	47
4-Ketophleixanthophyll	S1	19
Lycopene	S5	40
Methyl-1-hexosyl-1,2-dihydro-3,4-didehydro-apo-8'-lycopenoate	S9	47
Myxoxanthophyll	S1	7
	S1	11
	S1	31
Neurosporaxanthin	S5	28
Oscillaxanthin	S43	11
P-476	S8	30
P-496	S44	30

Table 1. LC 1 (continued)
SURVEY OF DIFFERENT ADSORBENTS AND SOLVENTS USED FOR CAROTENOIDS

Pigment	Solvent	Ref.
Phyleixanthophyll	S1	15
	S1	19
Sarcinaxanthin	S8	40
Lime Superfine (Sierrahydrated)-Celite 545 = 2:1		
β-Carotene	S45	48
	S47	48
β-Carotene (*cis*)	S46	48
Isocryptoxanthin	S47	48
Isocryptoxanthin (*cis*)	S47	48
MgO		
Fucoxanthin	S17	50
MgO-Celite = 3:2		
Anhydroeschscholtzxanthin	S19	51
Antheraxanthin	S1	34
Auroxanthin	S9	34
Chrysanthemaxanthin	S11	34
Eschscholtzxanthin	S19	51
Flavoxanthin	S11	34
Lutein	S8	34
5,6-Monoepoxylutein (*trans*)	S17	34
Neoxanthin	S11	34
Violaxanthin	S8	34
Xanthophyll	S19	51
Zeaxanthin	S1	34
	S19	51
MgO-Celite 503 = 2:1 (w/w)		
β-Carotene	S45	52
γ-Carotene	S46	52
3,4,3',4'-Bisdehydro-β-carotene	S48	52
3,4-Dehydro-β-carotene	S47	52
Lycopene	S46	52
Rodoxanthin	S47	52
MgO-Celite 545 = 1:1		
β-Carotene	S49	38
ζ-Carotene	S49	38
Echinenone	S17	35
Euglenanone	S8	35
Phytofluene	S49	38
Magnesium silicate		
Neurosporaxanthin (*cis*)	S51	28
Neurosporaxanthin (*trans*)	S52	28
S.g. 434	S15	32
S.g. 460	S9	32
Saproxanthin	S23	32
Microcel C	S10	32
Alloxanthin (*trans*)	S53	43
Crocoxanthin	S53	43
α-Cryptoxanthin	S53	43
Diatoxanthin	S53	43
4-Hydroxy-α-carotene	S53	43
Lutein	S53	43
Monadoxanthin	S53	43
Zeinoxanthin	S53	43
Polyamide		
(0.2 — 0.8 mesh)		
β-Carotene	S45	53
Sea Sorb 43 Magnesia-Celite 545 = 1:2		
α-Carotene	S54	42
	S55	42
	S56	42
β-Carotene	S54	42
	S55	42
	S56	42

Compound	System	Ref.
Cryptoxanthin	S54	42
	S55	42
	S56	42
Lutein	S54	42
	S55	42
	S56	42
Neoxanthin	S54	42
	S55	42
	S56	42
Violaxanthin	S54	42
	S55	42
	S56	42
Zeaxanthin	S54	42
	S55	42
	S56	42

Sephadex-LH-20

Compound	System	Ref.
Auroxanthin	S4	55
α-Carotene	S57	36
β-Carotene	S57	36
ζ-Carotene	S57	36
Chrysanthemaxanthin	S4	55
Lutein	S4	55
	S57	36
Lycopene	S57	36
Phytoene	S57	36
Phytofluene	S57	36

Starch

Compound	System	Ref.
α-Carotene	S58	42
β-Carotene	S58	42
Cryptoxanthin	S58	42
Lutein	S58	42
Neoxanthin	S58	42
Violaxanthin	S58	42
Zeaxanthin	S58	42

Sugar

Compound	System	Ref.
α-Carotene	S19	42
β-Carotene	S19	42
Cryptoxanthin	S19	42
Lutein	S19	42
Neoxanthin	S19	42
Violaxanthin	S19	42
Zeaxanthin	S19	42

ZnCO₃-Celite = 3:1 (w/w)

Compound	System	Ref.
α-Carotene	S59	54
β-Carotene	S59	54
Diatoxanthin	S59	54
Lutein	S59	54
Zeaxanthin	S59	54

ZnCO₂-Diatomaceous earth = 3:1 (w/w)

Compound	System	Ref.
Alloxanthin	S49	43
Crocoxanthin	S49	43
α-Cryptoxanthin	S49	43
Diatoxanthin	S49	43
4-Hydroxy-α-carotene	S49	43
Lutein	S49	43
Monadoxanthin	S49	43
Zeinoxanthin	S49	43

a Activity grade not available.
b 3% (v/w) water-deactivated alumina (i.e., Brockmann grade II).
c Alumina (Peter Spence, Type H); Brockmann grade II.
d 2% (v/w) water-deactivated alumina.
e Alumina, deactivated by treatment with 5% of water.
f Partially deactivated by treatment with 1% (v/w) of water.
g Methanol-deactivated alumina.
h Deactivated neutral alumina (80 — 200 mesh), 6% water.
i Woelm neutral (Brockmann grade III).
j Alumina (Peter Spence, Type H); Brockmann grade IV.
k Deactivated alumina made by treating grade "O" with methanol.
l 3:2 Mixture of activated and deactivated (Brockmann grade III) alumina.

Table I. LC 1 (continued)

SURVEY OF DIFFERENT ADSORBENTS AND SOLVENTS USED FOR CAROTENOIDS

Solvent S1 = 20—30% acetone in petroleum ether
S2 = ethyl ether-cold acetic acid = 20:1
S3 = 15% KOH in 90% methanol
S4 = chloroform
S5 = petroleum ether
S6 = 0—5% ether in petroleum ether
S7 = 0—5% acetone in petroleum ether
S8 = 10—25% acetone in petroleum ether
S9 = 40—60% acetone in petroleum ether
S10 = 30—50% ether in petroleum ether
S11 = 25—40% acetone in petroleum ether
S12 = acetone-*n*-propanol = 9:1
S13 = 15—25% ether in petroleum ether
S14 = acetone
S15 = 5—15% ethyl ether in petroleum ether
S16 = benzene-methanol = 9:1
S17 = 5—10% acetone in petroleum ether
S18 = 7—10% ethyl ether in hexane
S19 = benzene
S20 = petroleum ether-benzene = 85:15
S21 = petroleum ether-benzene = 1:1
S22 = 99% ether-1% methanol
S23 = 1—5% methanol in petroleum ether
S24 = petroleum ether-methanol = 75:25
S25 = benzene-methanol = 95:5
S26 = benzene-ethyl ether = 75:25
S27 = ether-methanol = 98:2
S28 = ether
S29 = 0—4% ethyl ether in hexane
S30 = 1—2% methanol in benzene

S31 = benzene-acetone = 6:4
S32 = 0.5—1% methanol in ethyl ether
S33 = 100% methanol
S34 = from 50% acetone in petroleum ether to 2% methanol in petroleum ether
S35 = 50% ether in petroleum ether
S36 = 60% acetone in petroleum ether
S37 = ethanol
S38 = 0—20% acetone in benzene
S39 = 20—30% benzene in petroleum ether
S40 = petroleum ether-benzene = 95:5
S41 = 0—30% ether in petroleum ether
S42 = 10—50% ether in petroleum ether
S43 = pyridine
S44 = 30% acetone in benzene
S45 = hexane
S46 = 1—5% acetone in hexane
S47 = 6—15% acetone in hexane
S48 = hexane-acetone = 75:25
S49 = 0—25% acetone in petroleum ether
S50 = benzene-ether-ethanol = 2:3:1
S51 = methanol
S52 = 1% acetic acid in methanol
S53 = 10—80% ethyl ether in petroleum ether
S54 = 1,2-dichlorethane
S55 = petroleum ether + 10—50% acetone
S56 = petroleum ether + 6—10% *n*-propanol
S57 = chloroform-methanol-*n*-hexane = 65:5:30
S58 = petroleum either + 5% *n*-propanol
S59 = petroleum ether + ethyl ether (conc unspecified)

REFERENCES

1. **Czeczuga, B.**, *Comp. Biochem. Physiol.*, 48B, 349, 1974.
1a. **Czeczuga, B.**, *Comp. Biochem. Physiol.*, 39B, 945, 1971; *J. Insect. Physiol.*, 17, 2017, 1971.
2. **Conti, S. F. and Benedict, C. R.**, *J. Bacteriol.*, 83, 929, 1962.
3. **Liaaen-Jensen, S.**, *Phytochemistry*, 4, 925, 1965.
4. **Liaaen-Jensen, S.**, *Acta Chem. Scand.*, 17, 500, 1963.
5. **Ryvarden, L. and Liaaen-Jensen, S.**, *Acta Chem. Scand.*, 18, 643, 1964.
6. **Hertzberg, S. and Liaaen-Jensen, S.**, *Phytochemistry*, 5, 565, 1966.
7. **Kushwaha, S. C. and Kates, M.**, *Biochim. Biophys. Acta*, 316, 235, 1973.
8. **Liaaen-Jensen, S.**, *Acta Chem. Scand.*, 19, 1166, 1965.
9. **Kushwaha, S. C., Pugh, E. L., Kramer, J. K. G., and Kates, M.**, *Biochim. Biophys. Acta*, 260, 492, 1972.
10. **Liaaen-Jensen, S., Hegge, E., and Jackman, L. M.**, *Acta Chem. Scand.*, 18, 1703, 1964.
11. **Hertzberg, S. and Liaaen-Jensen, S.**, *Phytochemistry*, 5, 557, 1966.
12. **Arpin, N. and Liaaen-Jensen, S.**, *Phytochemistry*, 6, 995, 1967.
13. **Liaaen-Jensen, S.**, *Acta Chem. Scand.*, 19, 1025, 1965.
14. **Hertzberg, S. and Liaaen-Jensen, S.**, *Acta Chem. Scand.*, 20, 1187, 1966.
15. **Manchand, P. S., Rüegg, R., Schwieter, U., Siddons, P. T., and Weedon, B. C. L.**, *J. Chem. Soc.*, p. 2019, 1965.
16. **Davies, B. H., Hallett, C. J., London, R. A., and Rees, A. F.**, *Phytochemistry*, 13, 1209, 1974.
17. **Fiasson, J. L. and Arpin, N.**, *Bull. Soc. Chim. Biol.*, 49, 537, 1967.
18. **Bonnett, R., Spark, A. A., and Weedon, B. C. L.**, *Acta Chem. Scand.*, 18, 1739, 1964.
19. **Hertzberg, S. and Liaaen-Jensen, S.**, *Acta Chem. Scand.*, 21, 15, 1967.
20. **Aasen, A. J. and Liaaen-Jensen, S.**, *Acta Chem. Scand.*, 20, 1970, 1966.
21. **Liaaen-Jensen, S., Cohen-Bazire, G., Nakayama, T. O. M., and Stanier, R. Y.**, *Biochim. Biophys. Acta*, 29, 477, 1958.
22. **Arpin, N. and Liaaen-Jensen, S.**, *Phytochemistry*, 8, 185, 1969.
23. **Liaaen-Jensen, S. and Hertzberg, S.**, *Acta Chem. Scand.*, 20, 1703, 1966.
24. **Kjøsen, H. and Liaaen-Jensen, S.**, *Phytochemistry*, 8, 483, 1969.
25. **Markham, M. C. and Liaaen-Jensen, S.**, *Phytochemistry*, 7, 839, 1968.
26. **Surmatis, J. D., Ofner, A., Gibas, J., and Thommen, R.**, *J. Org. Chem.*, 31, 186, 1966.
27. **Arpin, N. and Liaaen-Jensen, S.**, *Bull. Soc. Chim. Biol.*, 49(5), 527, 1967.
28. **Aasen, A. J. and Liaaen-Jensen, S.**, *Acta Chem. Scand.*, 19, 1843, 1965.
29. **Liaaen-Jensen, S.**, *Acta Chem. Scand.*, 17, 489, 1963.
30. **Francis, G. W., Hertzberg, S., Andersen, K., and Liaaen-Jensen, S.**, *Phytochemistry*, 9, 629, 1970.
31. **Hertzberg, S. and Liaaen-Jensen, S.**, *Phytochemistry*, 8, 1259, 1969.
32. **Aasen, A. J. and Liaaen-Jensen, S.**, *Acta Chem. Scand.*, 20, 811, 1966.
33. **Goodwin, T. W.**, *Biochem. J.*, 58, 90, 1954.
34. **Jungalwala, F. B. and Cama, H. R.**, *Biochem. J.*, 85, 1, 1962.
35. **Krinsky, N. I. and Goldsmith, T. H.**, *Arch. Biochem. Biophys.*, 91, 271, 1960.
36. **Hasegawa, K.**, *Methods Enzymol.*, 67, 261, 1980.
37. **Goldie, A. H. and Subden, R. E.**, *Biochem. Genet.*, 10(3), 275, 1973.
38. **Bramley, P. M. and Davies, B. H.**, *Phytochemistry*, 14, 463, 1975.
39. **Mitzka-Schnabel, U.**, dissertation University of Munich, Munich, 1978.
40. **Arpin, N., Norgård, S., Francis, G. W., and Liaaen-Jensen, S.**, *Acta Chem. Scand.*, 27, 2321, 1973.
41. **Aasen, A. J. and Liaaen-Jensen, S.**, *Acta Chem. Scand.*, 21, 970, 1967.
42. **Strain, H. H. and Svec, W. A.**, *Adv. Chromatogr.*, 8, 119, 1969.
43. **Chapman, D. J.**, *Phytochemistry*, 5, 1331, 1966.
44. **Zechmeister, L. and Pinckard, J. H.**, *Experientia*, 4(12), 474, 1948.
45. **Cholnoky, L., Szabolcs, J., and Nagy, E.**, *Justus Liebigs Ann. Chem.*, 616, 207, 1958.
46. **Entschel, R. and Karrer, P.**, *Helv. Chim. Acta*, 41, 112, 1958.
47. **Aasen, A. J., Francis, G. W., and Liaaen-Jensen, S.**, *Acta Chem. Scand.*, 23, 2605, 1969.
48. **Wallcave, L. and Zechmeister, L.**, *J. Am. Chem. Soc.*, 75, 4495, 1953.
49. **Zechmeister, L. and Wallcave, L.**, *J. Am. Chem. Soc.*, 75, 5341, 1953.
50. **Jensen, A.**, *Acta Chem. Scand.*, 20, 1728, 1966.
51. **Karrer, P. and Leumann, E.**, *Helv. Chim. Acta*, 50/51, 445, 1951
52. **Foppen, F. H. and Gribanovski-Sassu**, *Biochim. Biophys. Acta*, 176, 357, 1969.
53. **Fric, F. and Haspel-Horvatovic, E.**, *J. Chromatogr.*, 68, 264, 1972.
54. **Allen, M. B., Fries, L., Goodwin, T. W., and Thomas, D. M.**, *J. Gen. Microbiol.*, 34, 259, 1964.
55. **Shimizu, S.**, *J. Chromatogr.*, 59, 440, 1971.

HIGH PERFORMANCE LIQUID CHROMATOGRAPHY (HPLC) TABLES

TABLE NOTES

In the last few years, the application of high-performance liquid chromatography has greatly increased. We, therefore, have included tables on HPLC which is basically a special application of liquid chromatography.

Numbers for retention times (t_R) are given only in relatively few publications. From other publications, however, it is possible to deduce approximate values from figures. The data thus obtained are compiled in Table I.HPLC 1, together with the applicable chromatographic parameters. The other relevant literature is combined in Tables I.HLPC 2 and I.HLPC 3, which are a complementary survey of the separation of carotenoids in HPLC systems.

Table I. HLPC 1
CAROTENOID (RETENTION TIMES)

Packing	P1	P1	P1	P1	P2	P2	P3	P3	P4	P4	P5	P6
Column length (mm)	250	250	250	250	135	135	2 × 610	2 × 610	250	250	500	300
diameter (mm)	4.6	4.6	4.6	4.6	6.35(OD)	6.35(OD)	7(ID)	7(ID)	4.6	4.6	2(ID)	3(ID)
form	st	st	st	st	st	st	st	st	st	st	st	st
material	n.a.	n.a.	n.a.	n.a.	SS	SS	SS	SS	SS	SS	SS	G
Solvent	S1	S2	S3	S4	S5	S5	S6	S7	S8	S9	S10	S11
Flow rate (mℓ/min)	1.4	1.4	1.4	1.4	0.5	<0.5	2.5	4.0	2.0	2.0	0.8	2.0
Pressure (psi)	300	300	300	300	<1600	<800	600—250	600—250	n.a.	n.a.	500	n.a.
Temperature (°C)	n.a.	n.a.	n.a.	n.a.	15	15	28	18	amb.	amb.	amb.	amb.
Detection	D1	D1	D1	D1	D2	D2	D3	D3	D4	D4	D5	D6
Technique	T1	T1	T1	T1	T2	T2	T3	T3	T4	T4	T5	T6
Literature	1	1	1	1	2	2	3	3	4	4	5	6
Compound	t_R(min)	t_R(min)	t_R(min)	t_R(min)	t_R(min)[a]	t_R(min)[b]	t_R(min)	t_R(min)	t_R(min)[c]	t_R(min)[d]	t_R(min)[d]	t_R(min)[e]
Antheraxanthin	—	—	—	—	173	—	—	—	—	—	—	13
Antheraxanthin(cis)	—	—	—	—	215	—	—	—	—	—	—	—
Auroxanthin epimer 1	—	25.5[f]	—	—	—	—	—	—	—	—	—	—
Auroxanthin epimer 2	—	25.9[f]	—	—	—	—	—	—	—	—	—	—
Bacterioruberin all *trans*	—	—	14.6[g]	—	—	—	—	—	—	—	—	—
Bacterioruberin neo A	—	—	14.1[g]	—	—	—	—	—	—	—	—	—
Bacterioruberin neo U	—	—	15.1[g]	—	—	—	—	—	—	—	—	—
Bacterioruberin neo V	—	—	16.0[g]	—	—	—	—	—	—	—	—	—
Bacterioruberin neo W	—	—	16.3[g]	—	—	—	—	—	—	—	—	—
β-Carotene	4.2[h]	—	—	—	—	—	—	—	19	—	—	38
β,β-Carotene	4.2[h]	—	—	—	—	—	—	—	—	16	—	—
β,ε-Carotene	4.2[h]	—	—	—	—	—	—	—	—	—	4.9	—
β,ψ-Carotene	4.6[h]	—	—	—	—	—	—	—	—	—	—	—
β-Citraurin	—	—	—	—	84	—	—	—	—	—	—	—
Cryptoxanthin	—	—	—	—	—	46	—	—	—	—	—	—
Diadinoxanthin	—	—	—	—	—	—	135.2	—	16	—	—	—
Diatoxanthin	—	—	—	—	—	—	157.2	—	—	—	—	—
Dinoxanthin	—	—	—	—	—	—	—	—	—	17	—	—
2,2'-Diol*	8.0[a]	—	—	—	—	—	—	—	—	—	—	—

Compound										
Fucoxanthin	—	—	—	—	—	—	—	—	—	—
Lutein	9.5[i]	—	124	145	79.5	53	13	13	—	25
Lutein all-*trans*	—	14.0[g]	—	—	—	—	—	18	—	—
Lutein neo A	—	13.1[g]	—	—	—	—	—	—	—	—
Lutein neo B	—	12.4[g]	—	—	—	—	—	—	—	—
Lutein neo U	—	14.8[g]	—	—	—	—	—	—	—	—
Lutein neo V	—	15.4[g]	—	—	—	—	—	—	—	25
Lutein 5,6-epoxide	—	—	—	—	—	—	—	—	—	—
Lutein-3'-ether epimer 1	17.5[f]	—	—	—	—	—	—	—	—	—
Lutein-3'-ether epimer 2	17.9[f]	—	—	—	—	—	—	—	—	—
Lycopene	5.0[h]	—	—	—	—	—	—	—	—	—
Neochrome epimer 1	—	17.5[g]	—	—	—	—	—	—	—	—
Neochrome epimer 2	—	17.8[g]	—	—	—	—	—	—	—	—
Neochrome mono-*cis*	—	18.4[g]	—	—	—	—	—	—	—	—
Neofucoxanthin A	—	—	—	—	93.2	—	—	—	—	—
Neofucoxanthin B	—	—	—	—	93.2	—	—	—	—	—
Neoperidinin	—	—	—	—	—	—	—	18	—	—
Neoxanthin	—	—	268	—	—	23	—	—	—	11
Neoxanthin neo A	—	—	—	—	—	—	—	—	—	9
Peridinin	—	—	—	—	—	—	—	12	—	—
Phytoene	—	—	—	—	—	—	—	—	2.5	—
Violaxanthin	—	—	176	196	—	30	—	—	—	15
Violaxanthin (*cis*)	—	—	—	240	—	—	—	—	—	—
Violeoxanthin	—	—	—	—	—	—	—	—	—	22
Zeaxanthin	9.7[i]	13.6[f]	—	137	—	—	—	—	—	—

a   Values deduced from Figure 3 of Reference 2.
b   Values deduced from Figure 4 of Reference 2.
c   Values deduced from Figure 1 of Reference 4.
d   Values deduced from Figure 2 of Reference 4.
e   Values deduced from Figure 1 of Reference 6.

f   Gradient rate (%/min) = 1.
g   Gradient rate (%/min) = 3.
h   Gradient rate (%/min) = 0.
i   Gradient rate (%/min) = 10.

Packing   P1   =   Spherisorb; 5μm
P2   = $ZnCO_3$ (Fisher Scientific Co., Z-29)
P3   =   Bondapak® C_{18}-Porasil B; 37 — 75 μm (Waters Associates)
P4   =   Partisil 10 ODS C_{18} (Whatman)
P5   =   1% ODS Permaphase, chemically bonded on Zipax (DuPont)
P6   =   Sorb-Sil 60-D 10 C_{18} (Macherey & Nagel)

Table I. HLPC 1 (continued)
CAROTENOID (RETENTION TIMES)

Solvent (= eluent)

S1 = hexane-methanol = 99.9:0.1

S2 = gradient 0—30% acetone in solvent S1

S3 = gradient 0—40% acetone in solvent S1

S4 = gradient 20—60% acetone in solvent S1

S5 = hexane-TPA-gradient (see T2)

    chamber 1: hexane-BHT (butylated hydroxytoluene: Eastman Chemical Products, Tenox BHT) = 90 mℓ:1 g

    chamber 2:hexane-TPA (tert. pentylalcohol; Baker Chemical Co., 9046) = 89:1

    chamber 3: hexane-TPA = 65:20

S6 = solvent program: (1) aqueous methanol: 80% (from 0—19 min), 90% (from 19—66 min), 95% (from 66—108 min), 97.5% (from 108—174 min), and 100% (from 174—216 min); (2) ether in methanol: 10% (from 216—246 min), 50% (from 246—270 min), and 75% (from 270 min to end of separation)

S7 = 98% methanol (from 0—77 min); then 50% ethyl acteate in methanol

S8 = gradient methanol-water from 70:30—95:5

S9 = gradient methanol-water from 70:30—95:5; water containing 5 m*M* tetrabutylammonium phosphate (''Pic A'')

S10 = methanol-water = 95:5

S11 = stepwise gradient elution: (1) aqueous methanol: 85% (from 0—18 min), 95% (from 18—27 min), 100% (from 27—33 min): (2) methanol in ethanol: 50% (from 33—40 min)

Detector

D1 = Varian series 634 double-beam spectrometer set monochromatically at 400—490 nm; the spectrometer was equipped with an 8-μℓ flow-through cell, and during separation, absorption spectra in the range of 350—600 nm were recorded for the components of each HPLC peak

D2 = column effluent was monitored with either a Beckman DB spectrophotometer (set at 440 nm) with connected microflow cell (light path: 10 nm) and a Sargeant SRL recorder (chart speeds 0.1 or 0.2 in./min) or a Technicon colorimeter with 440 nm interference filter and tubular flow cell (light path: 15 mm) and recorder

D3 = UV detector set at 440 nm

D4 = a variable wavelength detector set at 440 nm

D5 = a model 836 multiwavelength UV and fluorescence detector (operated in the UV mode at 280 nm) was used

D6 = LC-55 UV-visible light detector (set at 440 nm) and a digital scan unit (Perkin Elmer)

Technique

T1 = DuPont 830 liquid chromatograph equipped with gradient-elution accessories and thermostat-regulated oven (temperature not given)

T2 = gradient device was patterned after the Contigrad® (Metaloglass), but machined from a block of aluminum 15 × 20 × 10 cm high, with nine interconnected chambers, each holding 90 mℓ of solvent; the gradient was mixed in each chamber by a metal paddle; only the first three chambers were used; chamber 1 fitted with outlet. Pump: Milton Roy Chromatographic Mini-Pump Model 196-31 (Milton Roy Co.) at a pressure up to 1600 psi; tert. pentylalcohol (TPA) distilled over KOH prior to use

T3 = Waters Assoc. ALC 202 liquid chromatograph; pump pressure varied from 600 — 250 psi, depending on the solvent used; the residue of pigment extract in ether was taken up in 6 mℓ of methanol with several drops of acetone

T4 = Waters analytical high pressure liquid chromatograph equipped with model 660, a solvent program accessory (Waters Associates)

T5 = DuPont 830 high pressure liquid chromatograph; standard solution: 25—100 nm/10 μℓ

T6 = Altex Model 100A liquid chromatograph (Altex Scientific Instruments); the column was self-packed according to the balanced-density method and protected by a stainless-steel mantle according to Stahl[7] (Riedel-deHaen); the samples were introduced into the column by the stop-flow technique from a 10-μℓ precision syringe via a septum inlet port

REFERENCES

1. **Fikskahl, A., Mortensen, J. T., and Liaaen-Jensen, S.,** *J. Chromatogr.,* 157, 111, 1978.
2. **Stewart, J. and Wheaton, T. A.,** *J. Chromatogr.,* 55, 325, 1971.
3. **Eskins, K., Scholfield, C. R., and Dutton, H. J.,** *J. Chromatogr.,* 135, 217, 1977.
4. **Davies, D. and Holdsworth, E. S.,** *J. Liquid Chromatogr.,* 3(1), 123, 1980.
5. **Puglisi, C. V. and de Silva, J. A. F.,** *J. Chromatogr.,* 120, 457, 1976.
6. **Braumann, T. and Grimme, L. H.,** *J. Chromatogr.,* 170, 264, 1979.
7. **Stahl, K. W., Schuppe, E., and Potthast, H.,** *GIT-Fachz. Lab.,* p. 536, 1973.

Table I. HPLC 2
CAROTENOIDS: SURVEY OF SEPARATED PIGMENTS AND SEPARATION CONDITIONS — I

Packing	P1	P2	P3	P4	P5	P6	P6	P7	P8	P8	P8	P9	P9	P10
Solvent	S1	S2	S3	S4	S5	S6	S7	S8	S9	S10	S11	S9	S10	S12
Literature	1	2	2	3	4	5	6	7	8	8	8	8	8	9

Compound	XX = Carotenoids separated in solvent system given below													
Antheraxanthin	—	—	—	—	—	XX	—	—	—	—	—	—	—	—
Canthaxanthin	—	XX	XX	—	—	—	—	—	—	—	—	—	—	—
α-Carotene	—	—	—	—	—	—	—	—	XX	XX	XX	XX	XX	XX
β-Carotene	XX	XX	XX	—	XX	XX	XX	XX	XX	XX	XX	XX	XX	XX
β-Carotene, *cis* isomer 1	—	—	—	—	—	XX	—	—	—	—	—	—	—	—
β-Carotene, *cis* isomer 2	—	—	—	—	—	XX	—	—	—	—	—	—	—	—
γ-Carotene	—	—	—	—	—	—	—	—	—	—	—	—	—	XX
Cryptoxanthinester	—	—	—	—	XX	—	—	—	—	—	—	—	—	—
3,4-Dehydrorhodopin	—	—	XX	—	—	—	—	—	—	—	—	—	—	—
Echinenone	—	XX	XX	—	—	—	—	—	—	—	—	—	—	—
Lutein	XX	XX	XX	XX	—	XX	—	XX	—	—	—	—	—	—
Luteinepoxide	—	—	—	—	—	—	—	XX	—	—	—	—	—	—
Lycopene	—	—	XX	—	—	—	XX	—	XX	XX	XX	XX	XX	XX
Neolutein A	—	—	—	—	—	XX	—	—	—	—	—	—	—	—
Neolutein B	—	—	—	—	—	XX	—	—	—	—	—	—	—	—
Neoxanthin	XX	—	XX	XX	—	XX	—	XX	—	—	—	—	—	—
Neoxanthin X	—	—	—	—	—	XX	—	—	—	—	—	—	—	—
Neurosporene	—	—	—	—	—	—	XX	—	—	—	—	—	—	—
Okenone	—	—	XX	—	—	—	—	—	—	—	—	—	—	—
Phytoene	—	—	—	—	—	—	XX	—	—	—	—	—	—	—
Phytofluene	—	—	—	—	—	—	XX	—	—	—	—	—	—	—
Rhodopin	—	—	XX	—	—	—	—	—	—	—	—	—	—	—
Spheroidenone	—	—	XX	—	—	—	—	—	—	—	—	—	—	—
Spirilloxanthin	—	—	XX	—	—	—	—	—	—	—	—	—	—	—
Tetrahydrospirilloxanthin	—	—	XX	—	—	—	—	—	—	—	—	—	—	—
Torulene	—	—	XX	—	—	—	—	—	—	—	—	—	—	—
Violaxanthin	XX	XX	XX	XX	—	XX	—	XX	—	—	—	—	—	—
Violaxanthin 9-*cis*	—	—	—	—	—	XX	—	—	—	—	—	—	—	—
Violaxanthin 13-*cis*	—	—	—	—	—	XX	—	—	—	—	—	—	—	—
Xanthophyllester	—	—	—	—	XX	—	—	—	—	—	—	—	—	—
Zeaxanthin	—	XX	XX	—	—	—	—	—	—	—	—	—	—	—

Packing
P1 = silica gel-column (trade name Lobar; Merck)
P2 = 10 μm Spherisorb; spherical silica packed as a slurry in methanol
P3 = 5 μm Partisil; irregular silica particles (Whatman)
P4 = silica gel powder SS-05 (particle size: 0.5 μm; Japan Spectroscopic)
P5 = Merck RP 8, No. 750 682
P6 = Nucleosil 10 C_{18} (Macherey & Nagel)
P7 = μ-C_{18} reversed phase
P8 = Partisil-PXS-5/ODS
P9 = Partisil-PXS-10/25-ODS-2
P10 = Spherisorb ODS (5 μm)

Table I. HPLC 2 (continued)
CAROTENOIDS: SURVEY OF SEPARATED PIGMENTS AND SEPARATION CONDITIONS — I

Solvent (= eluent) S1 = three-component solvent mixture of heptane, diethyl-ether, and acetone (HEA), changed at unequal intervals in a five-step gradient

S2 = gradient 2 — 50% acetone in hexane (concave over 20 min)

S3 = gradient 1 —75% acetone in hexane (concave over 30 min)

S4 = step gradient of isopropyl alcohol in hexane (1, 2, 5, and 10%) at unequal intervals

S5 = gradient: methanol-water (90:10)/methanol

S6 = gradient of acetone-water (100:40 — 100:5) (v/v)

S7 = gradient of (methanol-water = 3:1)-(tetrahydrofurane-acetonitrile = 1:1) from 100/0 — 30/70

S8 = gradient of 80% methanol-ethyl acetate (100:0 — 50:50) (v/v); linear solvent program

S9 = acetonitrile-chloroform = 88.4:11.6

S10 = acetonitrile-chloroform = 91.5:8.5

S11 = acetonitrile-chloroform = 92.0:8.0

REFERENCES

1. **DeJong, D. W. and Woodlief, W. G.,** *J. Agric. Food. Chem.,* 26(6), 1281, 1978.
2. **Hajibrahim, S. K., Tibbetts, P. J. C., Watts, C. D., Maxwell, J. R., Eglinton, G., Colin, H., and Guiochon, G.,** *Anal. Chem.,* 50(4), 549, 1978.
3. **Iriyama, K., Yoshiura, M., and Shiraki, M.,** *J. Chromatogr.,* 154, 302, 1978.
4. **Benk, E., Treiber, H., and Bergmann, R.,** *Riechstoffe, Aromen, Körperpflegemittel,* 10, 216, 1976.
5. **Tóth, G. and Szabolcs, J.,** *Phytochemistry,* 19, 629, 1980.
6. **Beyer, P., Kreuz, K., and Kleinig, H.,** *Planta,* 150, 435, 1980.
7. **Eskins, K. and Harris, L.,** *Photochem. Photobiol.,* 33, 131, 1981.
8. **Zakaria, M., Simpson, K., Brown, P. R., and Krstulovic, A.,** *J. Chromatogr.,* 176, 109, 1979.
9. **Pfander, H., Schurtenberger, H., and Meyer, V. R.,** *Chimia,* 34(4), 179, 1980.

Table I. HPLC 3
CAROTENOIDS: SURVEY OF SEPARATED PIGMENTS AND SEPARATION CONDITIONS — II

Packing	P1	P1	P1	P1	P1	P2	P2	P2	P2
Solvent	S1	S2	S3	S4	S5	S6	S7	S7	S7
Flow rate (mℓ/min)	0.5	1.2	1.5	1.5	1.5	1.0	0.8	0.8	1.0
Detection	D1	D2	D3	D4	D4	D4		D5	D2
Literature	1	1	1	1	1	1	1	1	1

Compound	XX = Carotenoids separated in solvent system given below								
β-Carotene	—	—	—	—	—	—	XX	—	—
(Z-)β-Carotene	—	—	—	—	—	—	XX	—	—
β,ψ-Carotene	—	—	—	—	—	—	XX	—	XX
(Z-)β,ψ-Carotene	—	—	—	—	—	—	—	—	XX
ε,ψ-Carotene	—	—	—	—	—	—	XX	—	XX
(Z-)ε,ψ-Carotene	—	—	—	—	—	—	—	—	XX
ξ-Carotene	—	—	—	—	—	—	XX	XX	—
(Z-)ξ-Carotene	—	—	—	—	—	—	—	XX	—
all-(E)-β-Cryptoxanthin	—	XX	—	—	—	—	—	—	—
(Z-)β-Cryptoxanthin	—	XX	—	—	—	—	—	—	—
Lutein	—	—	—	XX	—	—	—	—	—
Lycopene	XX	—	—	—	—	—	XX	—	—
all-(E)Lycopene	XX	—	—	—	—	—	—	—	—
(8R-)Mutatoxanthin	—	—	—	XX	—	—	—	—	—
(8S-)Mutatoxanthin	—	—	—	XX	—	—	—	—	—
Neochrome	—	—	XX	—	—	—	—	—	—
Neoxanthin	—	—	XX	—	—	—	—	—	—
Neurosporene	—	—	—	—	—	XX	—	—	—
all-(E)-Neurosporene	—	—	—	—	—	XX	—	—	—
(5Z-)Neurosporene	—	—	—	—	—	XX	—	—	—
Phytofluene	—	—	—	—	—	—	XX	—	—
all-(E)-Phytofluene	—	—	—	—	—	—	XX	—	—
all-(E)-Rubixanthin	—	XX	—	—	—	—	—	—	—
(5'Z-)Rubixanthin	—	XX	—	—	—	—	—	—	—
(9'Z-)Rubixanthin	—	XX	—	—	—	—	—	—	—
(13Z-)Rubixanthin	—	XX	—	—	—	—	—	—	—
(13'Z-)Rubixanthin	—	XX	—	—	—	—	—	—	—
(5'Z,13Z-) or (5'Z,13' Z-)Rubixanthin	—	XX	—	—	—	—	—	—	—
3,5,6,3'-Tetrol	—	—	—	—	XX	—	—	—	—
Tetrolfuranoxyd	—	—	—	—	XX	—	—	—	—
Zeaxanthin	—	—	—	XX	—	—	—	—	—
(9Z-)Zeaxanthin	—	—	—	XX	—	—	—	—	—
(13Z-)Zeaxanthin	—	—	—	XX	—	—	—	—	—

Packing	P1	= Spherisorb CN 5 μm
	P2	= Spherisorb ODS 5 μm
Solvent (= eluent)	S1	= hexane with 0.1% ethyldiisopropylamine
	S2	= hexane-methylenchloride-methanol-ethyldiisopropylamine = 95:5:0.05:0.1
	S3	= hexane-methylenchloride-methanol-ethyldiisopropylamine = 60:40:0.5:0.1
	S4	= hexane-methylenchloride-methanol-ethyldiisopropylamine = 60:40:0.1:0.1
	S5	= hexane-methylenchloride-methanol-ethyldiisopropylamine = 50:50:0.1:0.1
	S6	= CH₃CN-CH₃OH = 93:7
	S7	= CH₃CN-CH₃OH-2-propanol = 85:10:5
Detection	D1	= detector set at 465 nm
	D2	= detector set at 455 nm
	D3	= detector set at 425 nm
	D4	= detector set at 450 nm
	D5	= detector set at 500 nm

REFERENCE

1. Märki-Fischer, E., Marti, U., Buchecker, R., and Eugster, C. H., Helv. Chim. Acta, 66, 494, 1983.

GAS-LIQUID CHROMATOGRAPHY (GLC) TABLES

TABLE NOTES

The gas-liquid chromatography of carotenoids is hampered by their large molecular weight and, consequently, low volatility. In addition, carotenoids possess an extended system of conjugated double bonds that tend to rearrange, especially at the relatively high temperatures needed for GC. To prevent double-bond rearrangement and decomposition, perhydrogenated cartotenoids usually are used. Perhydrogenation, however, deletes any information on double bonds that could be furnished by GC. In spite of all these and some other drawbacks, e.g., the required instrumentation, a number of fine results have been obtained; they are summarized in Tables I.GC 1 and I.GC 2. (For literature, see Tables I.GC 1 and I.GC 2.)

Table I. GC 1
GC OF CAROTENOIDS — I

	t_R^a	t_R^b	t_R^c	t_R^d	t_R^e	t_R^f	t_R^g	t_R^h	t_R^i	t_R^j	t_R^k	t_R^l	t_R^m
Column packing	P1	P1	P1	P1	P2	P2	P2	P2	P3	P3	P3	P3	P4
Temperature (°C) column	240	275	300	225—300	240	275	300	225—300	240	275	300	225—300	225—300
injector	265—290	300—325	325—350	325	265—290	300—325	325—350	325	265—290	300—325	325—350	325	325
detector	265—290	300—325	325—350	350	265—290	300—325	325—350	350	265—290	300—325	325—350	350	350
Gas flow rate (mℓ/min) pressure (kp/cm²)	N₂; 60	N₂; 60	N₂; 60	N₂; 60	N₂; 60	N₂; 60	N₂; 60	N₂; 60	N₂; 60	N₂; 60	N₂; 60	N₂; 60	N₂; 60
Column length (cm)	150	150	150	150	150	150	150	150	150	150	150	150	150
diameter (mm [ID])	4	4	4	4	4	4	4	4	4	4	4	4	4
form	Spiral	Spiral	Spiral	Spiral	Spiral	Spiral	Spiral	Spiral	Spiral	Spiral	Spiral	Spiral	Spiral
material	G	G	G	G	G	G	G	G	G	G	G	G	G
solvent	S1	S1	S1	S1	S1	S1	S1	S1	S1	S1	S1	S1	S1
Detector	D1	D1	D1	D1	D1	D1	D1	D1	D1	D1	D1	D1	D1
Technique	T1	T1	T1	T1	T1	T1	T1	T1	T1	T1	T1	T1	T1
Literature	1	1	1	1	1	1	1	1	1	1	1	1	2
Compound													
Lycopersene	—	1.03	1.06	3.95	—	1.02	1.00	3.15	—	1.38	1.02	2.76	2.86
Phytol	0.14	—n	—n	0.18	—n	—n	—n	0.17	0.05	0.1 5	—n	0.10	0.09
Squalene	1.00	0.14	0.16	1.00	1.00	0.14	0.17	1.00	1.00	0.1 5	0.18	1.00	1.00
Hydrogenated Carotenoids													
H₂-β-apo-4′-carotenal	2.55	0.30	0.35	2.10	2.55	0.33	0.35	1.84	—	0.26	0.29	1.44	1.71
H₂-β-apo-8′-carotenal	0.79	0.11	—n	0.82	0.83	0.10	0.15	0.85	0.54	0.10	0.12	0.70	0.76
H₂-β-apo-10′-carotenal	0.41	0.06	—n	0.42	0.42	0.07	0.09	0.52	0.27	0.05	0.05	0.41	0.45
H₂-β-apo-8′-carotenoic acid	1.65	0.23	0.27	1.57	1.80	0.23	0.27	1.48	1.77	0.24	0.27	1.40	1.53
H₂-β-apo-8′-carotenoic acid ethyl ester	2.02	0.26	0.30	1.77	2.05	0.26	0.30	1.59	1.96	0.28	0.30	1.50	1.60
H₂-β-apo-8′-carotenoic acid methyl ester	1.80	0.24	0.26	1.61	1.88	0.24	0.26	1.54	1.85	0.26	0.28	1.46	1.56
H₂-astacene	—	1.70	1.63	5.79	—	1.64	1.55	3.58	—	1.46	1.34	3.00	4.66
H₂-azafrin	2.61	0.27	0.32	2.08	1.97	0.27	0.30	1.75	1.63	0.45	0.49	1.80	1.55
H₂-bixin	1.13	0.15	0.18	1.14	1.12	0.15	0.18	1.09	1.33	0.20	0.21	1.20	0.95
H₂-canthaxanthin	—	2.85	2.58	5.90	—	2.73	2.47	4.20	—	3.01	2.06	3.66	5.43
H₂-capsanthin	—	1.87	2.03	5.13	—	1.17	1.11	3.22	—	1.49	1.40	2.81	4.16
H₂-α-carotene	—	1.00	1.00	3.81	—	1.00	1.00	2.87	—	1.00	1.00	2.43	3.02
H₂-β-carotene	—	1.00	1.00	3.81	—	1.00	1.00	2.88	—	1.00	1.00	2.43	3.03
H₂-γ-carotene	6.33	0.78	0.79	3.66	6.83	0.81	0.81	2.83	—	0.67	0.68	2.24	2.71
H₂-ζ-carotene	0.61	0.62	3.13		0.63	0.64	2.66		0.46	0.46	1.78	2.41	
H₂-β-carotenone	—	2.69	2.53	5.33	—	2.64	2.25	4.12	—	2.55	1.74	3.09	5.33

H₂-carotinin	—	1.00	1.00	—	3.81	1.00	1.00	2.88	—	1.00	1.00	2.42	3.03
H₂-crocetin	0.07	—[n]	—[n]	0.05	0.13	—[n]	—[n]	0.11	0.28	0.02	—[n]	0.13	0.10
H₂-cryptoxanthin	—	1.27	1.25	—	4.14	1.34	1.25	3.33	—	1.43	1.38	2.55	3.40
H₂-cryptoxanthin,Ac	—	1.23	1.21	—	4.10	1.33	1.25	3.28	—	1.76	1.76	2.53	3.29
H₂-cryptoxanthin,TMS	—	1.47	1.41	—	4.12	1.40	1.45	3.31	—	1.82	1.76	2.54	3.35
H₂-decapreno-β-carotene	0.79	5.04	—	0.83	7.54	4.86	6.12	5.58	0.54	6.87	5.45	5.32	5.89
H₂-3,4-dehydro-β-apo-8′-carotenal	—	—	—	—	0.82	0.15	0.10	0.85	—	0.10	0.12	0.69	0.75
H₂-dehydro-β-carotene	—	1.00	1.00	—	3.81	1.00	1.00	2.88	—	1.00	1.00	2.43	3.02
H₂-4,4′-diapo-ζ-carotene	0.61	—	—	0.66	0.67	0.12	0.09	0.73	0.35	0.07	0.08	0.53	0.73
H₂-4,4′-diaponeurosporene	0.61	—	—	0.67	0.67	0.12	0.09	0.73	0.36	0.07	0.09	0.53	0.73
H₂-4,4′-diaponeurosporen-4-oate methyl ester	—	—	—	—	1.41	—	—	1.33	—	—	—	1.27	1.32
H₂-4,4′-diaponeurosporen-4-oic acid	0.61	—	—	0.67	2.40	0.12	0.09	2.27	0.37	0.07	—	2.16	1.97
H₂-4,4′-diapophytoene	0.60	—	—	0.67	0.67	0.11	0.05	0.73	0.37	0.06	0.09	0.53	0.71
H₂-4,4′-diapophytofluene	0.39	—	—	0.39	0.67	—[n]	0.07	0.74	0.43	0.07	0.09	0.54	0.72
H₂-diethylcrocetin	—	—	—	—	0.45	—[n]	—[n]	0.45	—	—	—[n]	0.50	0.39
H₂-dimethoxyisozeaxanthin	—	1.23	1.05	—	4.02	1.03	1.10	2.88	—	1.46	1.34	2.37	3.01
H₂-dimethoxyzeaxanthin	0.29	2.02	1.84	0.21	5.15	1.78	1.96	3.67	0.32	2.68	2.22	3.44	3.93
H₂-dimethylcrocetin	—	0.03	—[n]	—	0.34	0.03	0.03	0.33	—	0.05	—[n]	0.40	0.30
H₂-echinenone	—	1.71	1.61	—	4.65	1.54	1.61	3.59	—	2.25	1.38	3.19	4.64
H₂-fucoxanthin	—	2.24	2.22	—	6.24	1.62	1.70	3.49	—	2.56	2.35	3.42	4.51
H₂-4-hydroxy-4,4′-diaponeurosporene	—	—	—	6.85	0.75	0.21	0.21	0.80	—	1.40	—	0.60	0.77
H₂-isocryptoxanthin	6.32	1.18	1.21	6.80	3.92	1.17	1.22	2.91	0.37	1.74	1.37	2.53	2.99
H₂-isocryptoxanthin, Ac	6.30	1.21	1.23	6.80	3.78	1.23	1.23	2.81	0.37	1.80	1.74	2.44	2.75
H₂-isocryptoxanthin, TMS	1.99	1.36	1.44	1.74	4.07	1.44	1.44	3.02	0.43	1.90	1.76	2.51	3.01
H₂-isozeaxanthin	6.32	1.24	1.24	6.79	4.16	1.32	1.32	3.00	—	1.64	1.74	2.59	3.15
H₂-isozeaxanthin, diAc	6.32	1.25	1.25	6.85	4.31	1.37	1.37	3.14	—	1.58	1.57	2.47	3.22
H₂-isozeaxanthin, diTMS	6.32	1.25	1.25	6.85	4.35	1.41	1.41	3.17	—	1.58	1.55	2.53	3.31
H₂-lycopene	0.15	0.62	0.64	—	3.15	0.65	0.60	2.66	—	0.46	0.46	1.75	2.40
H₂-lycopersene	0.15	0.62	0.64	—	3.15	0.64	0.59	2.67	—	0.46	0.46	1.76	2.40
H₂-methylazafrin	—	0.30	0.27	—	1.70	0.23	0.21	1.49	1.35	0.37	0.28	1.73	1.44
H₂-methylbixin	0.61	0.17	0.14	0.67	1.08	0.14	0.14	1.08	1.22	0.19	0.18	1.18	1.05
H₂-neurosporene	—	0.62	0.65	—	3.14	0.65	0.60	2.68	—	0.46	0.45	1.77	2.41
H₂-physalien; C₄₀ fragment	—	1.43	1.35	—	3.90	1.42	1.46	2.92	—	1.88	1.73	2.42	3.67
H₂-phytoene	—	0.62	0.65	—	3.14	0.65	0.60	2.64	—	0.45	0.46	1.77	2.40
H₂-phytofluene	—	0.61	0.65	—	3.15	0.65	0.60	2.66	—	0.46	0.44	1.75	2.42
H₂-retinaldehyde	—	—	—	—	0.10	—	—	0.10	0.04	—[n]	—[n]	0.06	0.08
H₂-retinol	—	—	—	—	0.11	—	—	0.13	0.04	—[n]	—[n]	0.07	0.09
H₂-rubixanthin	—	0.09	0.09	—	4.15	0.09	0.09	2.94	—	0.07	0.09	2.55	2.99
H₂-squalene	—	1.05	1.00	—	0.66	1.00	0.82	0.73	0.37	1.26	1.13	0.54	0.72
H₂-torularhodin	—	0.80	0.81	0.67	4.23	0.81	0.79	2.83	—	0.67	0.68	2.58	4.02
H₂-β-zeacarotene	—	1.41	1.50	—	3.66	1.50	1.68	3.45	—	2.21	1.79	2.24	2.71
H₂-zeaxanthin	—	1.39	1.40	—	4.68	1.40	1.63	3.50	—	1.93	1.74	3.30	3.41
H₂-zeaxanthin, diAc	—	1.39	1.44	—	5.04	1.44	1.66	3.55	—	1.91	1.71	2.98	3.47
H₂-zeaxanthin, diTMS	—	1.38	—	—	5.07	1.44	1.71	3.58	—	1.91	1.71	3.05	3.58

Table I. GC 1 (continued)
GC OF CAROTENOIDS — I

a   Retention time relative to squalene (t_R = 2.95 min).
b   Retention time relative to perhydro-β-carotene (t_R = 0.70 min).
c   Retention time relative to perhydro-β-carotene (t_R = 0.29 min).
d   Retention time relative to squalene (t_R = 5.50 min).
e   Retention time relative to squalene (t_R = 5.96 min).
f   Retention time relative to perhydro-β-carotene (t_R = 1.44 min).
g   Retention time relative to perhydro-β-carotene (t_R = 0.68 min).

h   Retention time relative to squalene (t_R = 8.65 min).
i   Retention time relative to squalene (t_R = 11.23 min).
j   Retention time relavtive to perhydro-β-carotene (t_R = 2.48 min).
k   Retention time relative to perhydro-β-carotene (t_R = 1.06 min).
l   Retention time relative to squalene (t_R = 11.15 min).
m   Retention time relative to squalene (t_R = 10.69 min).
n   Compound is eluted with solvent.

Packing   P1 = silicone gum rubber SE-52: 2% SE-52 on Gas-Chrom Q (80 — 100 mesh)
           P2 = 2% Dow-Corning high-vacuum grease (HVG) on Chromosorb W: AW-DMCS (85 — 100 mesh)
           P3 = 3% OV-17 on Universal B (85 — 100 mesh)
           P4 = 2% Dexsil 300 GC on Gas Chrom Q (100 — 120 mesh)

Solvent   S1 = chloroform
Detection   D1 = dual flame ionization detector system
Technique   T1 = type of gas chromatograph: Pye-Unicam (Cambridge, England) series 104
               G = glass

Packed columns were conditioned for at least 72 hr at 325°C with a nitrogen carrier gas flow rate of 40 mℓ/min prior to use; 1 — 3 µℓ of 1 — 5 mg/mℓ solutions in chloroform were injected directly onto silylated glass wool plugs in the columns; blocking of any remaining or newly appearing active sites on the columns was accomplished by periodic injections of Silyl-8 (Pierce, Rockford, Il.)

ABBREVIATIONS

Ac   = Acetate of perhydro-monohydroxycarotenoid
diAc = Diacetate of perhydro-dihydroxycarotenoid
TMS   = TMS ether of perhydro-monohydroxycarotenoid

diTMS = TMS diether of perhydro-dihydroxycarotenoid
AW   = Acid washed
DMCS = Dimethyldichlorosilane

REFERENCES

1. **Taylor, R. F. and Davies, B. H.,** *J. Chromatogr.*, 103, 327, 1975.

2. **Taylor, R. F. and Ikawa, M.,** *Methods Enzymol.*, 67, 233, 1980.

Table I. GC 2
GC OF CAROTENOIDS — II

Column Packing		P1		P2		P3		P3		P4		P4		P4
Temperature (°C)	column	40 — 180		245		293		293		275		290		225 — 290
	injector	250		—		—		—		300		315		315
	detector	250		—		—		—		300		315		315
Gas	flow rate (mℓ/min)	H_2; 100		N_2; 100		N_2; 100		N_2; 100		N_2; 75		N_2; 75		N_2; 75
Column	pressure (kp/cm²)	—		—		1		1		—		—		—
	length (cm)	1500		244		60		60		183		183		183
	diameter (mm [ID])	n.a.		6		3.5		3.5		4		4		4
	form	Capillary		n.a.		n.a.		n.a.		U-tube		U-tube		U-tube
	material	G		n.a.		SS		SS		G		G		G
Solvent		S1		S2		S3		S3		S4		S4		S4
Detector		D1		D2		D1		D1		D1		D1		D1
Technique		T1		T2		T3		T3		T4		T4		T5
Literature		1		2		3		3		4		4		4
Compound		t_R (min)		t_R^a		t_R (min)		t_R^b		t_R^c		t_R^d		t_R^e
6-But-2-enylidene-1,5,5-trimethylcyclohex-1-ene		7.2[f]		—		—		—		—		—		—
Echinenone				—		—		—		0.55		0.62		0.62
β-Ionone		9.8[f]		—		—		—		—		—		—
Lycopene				—		—		—		0.51		0.59		0.61
Phytoene				—		—		—		5.15		4.46		2.04
Squalene				—		—		—		1.00		1.00		1.00
Triacontane				1.00		3.23[g]		1.00[g]		—		—		—
Hydrogenated Carotenoids														
H_2-β-apo-8'-carotenenal				—		—		—		0.92		0.91		0.91
H_2-canthaxanthin				—		—		—		7.23		5.93		2.33
H_2-β-carotene		10.40		—		14.58[g]		4.51[g]		11.80		10.30		2.95
H_2-γ-carotene		8.21		—		11.65		3.58[g]		7.24		5.98		2.33
H_2-ζ-carotene		6.58		—		—		—		—		—		—

Table I. GC 2 (continued)
GC OF CAROTENOIDS — II

Compound	t_R (min)	t_R^a	t_R (min)	t_R^b	t_R^c	t_R^d	t_R^e
H$_2$-dehydrosqualene	—	—	1.59	0.49	—	—	—
H$_2$-dihydrosqualene	—	—	1.59	0.48	—	—	—
H$_2$-echinenone	—	—	—	—	7.23	5.93	2.33
H$_2$-lycopene	—	6.58	—	—	4.66	3.81	1.93
H$_2$-lycopersene	—	—	9.26	2.87^g	—	—	—
H$_2$-neo-α-carotene	—	—	14.62	4.53	—	—	—
H$_2$-neo-β-carotene	—	—	14.61	4.52	—	—	—
H$_2$-neurosporene	—	6.58	—	—	—	—	—
H$_2$-phytoene	—	6.58	9.27	2.87	—	—	—
H$_2$-phytofluene	—	6.58	—	—	—	—	—
H$_2$-phytofluene (cis)	—	—	9.23	2.89	—	—	—
H$_2$-phytofluene (trans)	—	—	9.32	2.89	—	—	—
H$_2$-phytol	—	—	—	—	0.16	0.24	0.11
H$_2$-retinol	—	—	—	—	0.19	0.24	0.13
H$_2$-squalene	—	—	1.61^g	0.50^g	0.75	0.77	0.82
H$_2$-tetrahydrosqualene	—	—	1.59	0.48	—	—	—

a Retention time relative to triacontane (t_R = 2.32 min).
b Retention time relative to triacontane (t_R = 3.23 min).
c Retention time relative to squalene (t_R = 5.88 min).
d Retention time relative to squalene (t_R = 3.88 min).

e Retention time relative to squalene (t_R = 21.5 min).
f Values deduced from Figure 1 of Reference 1.
g Standard.

Packing   P1 = OV-101 (methyl silicone) on capillary
          P2 = 1% SE-30 (methyl silicone) on Gas-Chrom Q
          P3 = 5% SE-30 (methyl silicone) on chromosorb G; AW-DMCS (100 — 120 mesh)
          P4 = 5% Dow-Corning High Vacuum Grease on Chromosorb W; AW-DMCS (80 — 100 mesh)

Solvent   S1 = ether
          S2 = petroleum ether
          S3 = hexane
          S4 = chloroform

Detection D1 = flame ionization detector
          D2 = detection through radioactivity of effluent (see T2)

Technique   T1 = type of gas chromatograph: Carlo Erba 180 with an integrator (Supergrator-2, Columbia Scientific Industries) filled with a WCOT glass capillary column (split ratio of carrier gas: 1:30; temperature program: 5°C/min; volume injected: 2 µℓ)

T2 = type of gas chromatograph: Barber-Colman model 10;[5] fractions emerging from the column were collected on glass wool in a Packard fraction collector; each effluent fraction, trapped on leaving the column, was eluted with the toluene solution of scintillators and assayed for radioactivity; cocktail mixture consisted of 0.03% PO-POP and 0.5% PPO in toluene

T3 = type of gas chromatograph: not available

T4 = type of gas chromatograph: Barber-Colman model 5000; coating of the support was carried out by dissolving the liquid phase in chloroform, adding the resultant solution to the support, and concentrating to dryness in vacuo; for further details see Reference 4

T5 = see T4; however, the temperature was increased from 225 to 290°C at a rate of 2°C/min after an initial isothermic period of 3 min; samples (1 — 3 µℓ of 5- to 10-mg.mℓ solutions in glass redistilled chloroform) were injected directly onto siliconized wool plugs in the column

ABBREVIATIONS

AW = Acid washed
DMCS = Dimethyldichlorosilane
G = glass

REFERENCES

1. **Jüttner, F.,** *Z. Naturforsch.,* 34 c, 186, 1979.
2. **Kushwaha, S. C., Kates, M., and Porter, J. W.,** *Can. J. Biochem.,* 54, 816, 1976.
3. **Kushwaha, S. C., Pugh, E. L., Kramer, J. K. G., and Kates, M.,** *Biochim. Biophys. Acta,* 260, 492, 1972.
4. **Taylor, R. F. and Ikawa, M.,** *Anal. Biochem.,* 44, 623, 1971.
5. **Kushwaha, S. C., Suzue, G., Subbarayan, C., and Porter, J. W.,** *J. Biol. Chem.,* 245, 4708, 1970.

Part II: Porphyrins (Exclusive of Chlorophylls)

Bruce Burnham and Hans-P. Köst

Chromatographic Methods for the Separation of Porphyrins and Metalloporphyrins

STRUCTURE, FUNCTION, OCCURRENCE, AND BIOSYNTHESIS
OF PORPHYRINS

Introductory Remarks

Porphyrins are a class of ubiquitously occurring compounds which are found in plants, animals, and microorganisms.[1-7] Most porphyrins which occur in the plant kingdom occur elsewhere as well. Therefore, we decided to take into account porphyrins from all sources (including animal and bacterial) to give the maximum possible information. Actually, porphyrins were described first in animal systems.[5] From this early work, separation techniques — mostly chromatographic — have been developed.

Chemically, porphyrins are macrocyclic systems consisting of four pyrrole nuclei which are linked by four methine bridges. Naturally occurring porphyrins (listed in Table II.1) generally are substituted at the porphyrin ground skeleton, but not at the methine bridges. Except for the state of a biosynthetic intermediate, biologically functional prophyrins are always present as metal complexes:[6,7] chlorophylls contain magnesium as central ligand, whereas iron is found in hemes and cobalt in the porphyrin-derived corrin ring system of cobalamines.[1] In nature, especially in geological deposits, other metals have been incorporated such as zinc, copper, nickel, and vanadium.[8-10] Thus far, few of those have been found to be functional.[11] It is generally believed that in those deposits the incorporation of metals is nonenzymatical and simply represents the tendency of porphyrins to combine with metals. The massive appearance of free porphyrins in organisms usually indicates an aberration of metabolism and may lead to severe disease (e.g., porphyria[5,12]), although today porphyrins are even used in cancer phototherapy.[13] Normally, functional metalloporphyrins are conjugated with proteins, either covalently as in cytochromes (e.g., cytochrome c[6]) or noncovalently as in hemoglobin,[14] often after a preceeding modification of the porphyrin ring (e.g., esterification, hydrogenation; see Part III: Chlorophylls).

The function of the resulting metalloporphyrin-protein complexes is manifold. The complex-bound iron in hemoglobins, myoglobins, catalases, and peroxidases enables the transfer of oxygen or electrons. In this way, redox processes are catalyzed whose specificity is directed by the apoprotein(s).[15] On the other hand, porphyrins are strongly colored, due to their light absorption: this latter property is used by nature in chlorophylls and bacteriochlorophylls which contain divalent magnesium[7] that does not undergo redox changes.

Ring-opening of the iron-porphyrins (hemes) leads to metal-free open-chained tetrapyrroles (bile pigments),[16,17] which serve in cyanobacteria, red algae, and cryptophyta in combination with proteins as antenna and sensor pigments (phycocyanin, phycoerythrin, phycochrome; see References 18 to 20). Some microorganisms obviously perform ring-splitting on porphyrins other than (iron)-protoporphyrin. For example, Brumm et al.[21] isolated a blue bile pigment "bactobilin", a compound that also has been obtained semisynthetically from urohemin III.[22] A number of changes at the side chains of the intact porphyrin ring skeleton are also possible — for example, reactions leading to "petroporphyrins" contained in crude oil shales.[8-10]

Biosynthesis[23-26]

The first steps in tetrapyrrole biosynthesis are the conversion of 5-aminolevulinic acid (5-ALA) (1) to the monopyrrole porphobilinogen (PGB) (2) and its further condensation to the porphyrin skeleton which is still hydrogenated at the bridge carbons ("porphyrinogens") (Figure 1).

The first porphyrinogen synthesized is uroporphyrinogen (uro'gen). Although numerous isomeric forms are possible, only two — uro'gen I (4b) and uro'gen III (4a) — play an important role. For synthesis of the natural uro'gen III, two enzymes, uroporphyrinogen synthase and uroporphyrinogen cosynthase, are required.[25] If the synthase exclusively is active, the pathological uroporphyrinogen I will be formed, indicating metabolic malfunction.

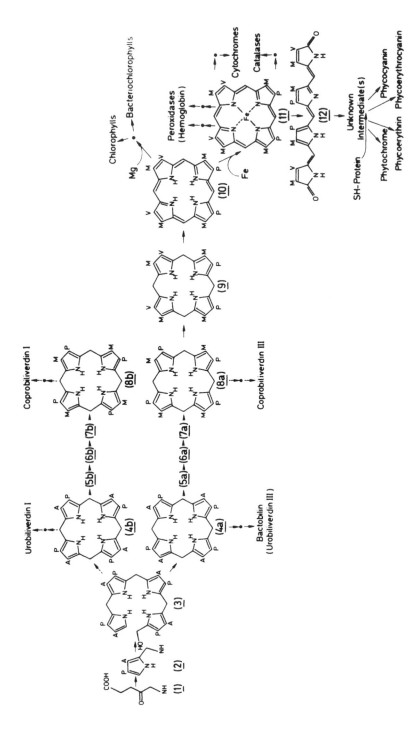

FIGURE 1. Tetrapyrrole biosynthetic pathway (see text). (From Scheer, H. and Inhoffen, H. H., in *The Porphyrins*, Vol. 2, Dolphin, D., Ed., Academic Press, New York, 1978, 45. With permission.)

The substrate of the uroporphyrinogen III cosynthase is the linear intermediary tetrapyrrole hydroxymethylbilane (3).

By action of a decarboxylase,[24] four of the carboxyl groups at the acetic acid side chains of uro'gen are subsequently split off to yield 7-carboxylic porphyrinogen (5), 6-carboxylic porphyrinogen (6), 5-carboxylic porphyrinogen (7), and 4-carboxylic porphyrinogen (co-proporphyrinogen) (8). Coproporphyrinogen III is finally decarboxylated to protopophyrinogen IX (9) which exhibits two vinyl side chains. The latter is enzymatically oxidized in vivo to protoporphyrin IX (10), which is a key intermediate in the synthesis of porphyrins and bile pigments (see Reference 26; oxidation in vitro is easily achieved with oxidants like hydrogen peroxide or chloramin T). At the point of protoporphyrin the biosynthetic pathway branches: magnesium insertion leads to the chlorophyll of plants and certain bacteria (see H. Scheer, Part III), iron insertion yields the heme component of cytochromes, catalases, and peroxidases (see Figure 1, [11]).

CHARACTERIZATION OF PORPHYRINS

Several physical and chemical techniques are available for the characterization, separation, and study of porphyrins.[27] For a detailed study of the various aspects of the different techniques, the reader is referred to the book by Smith, *Porphyrins and Metalloporphyrins,*[1] and the series *The Porphyrins* edited by Dolphin, starting in 1978.[2] Some techniques are sufficient by themselves, but frequently they are used in concert. Chromatography, the subject of this book, is useful in two respects. First, in characterizing a molecule it is necessary to have a pure material. Chromatography frequently is the most powerful technique to accomplish this task. Second, a compound, once pure, may often be identified by its behavior under analytical chromatographic conditions with the aid of appropriate standards. In addition, the following techniques are used along with chromatography.

Spectroscopic Methods
UV-Vis Spectroscopy
Coupled with chromatography, this is probably one of the most useful methods for characterizing an unknown porphyrin. The UV-vis absorption spectra of porphyrins are highly characteristic for this class of compounds. Frequently, a spectrum by itself is sufficient to identify a pigment as a porphyrin. Metal-free porphyrins exhibit five sharp absorption bands.

All porphyrins, whether they contain a metal or not, show a very strong absorbance in the region around 400 nm, the so-called Soret band[28] (see, also, Reference 29). Metal-free prophyrins exhibit four additional absorption bands in the visible region of the spectrum, i.e., from 500 to 650 nm. Spectral data of representative porphyrins are given in Table II.1. The intensity of these various bands is informative about the symmetry of substitution of the macrocycle as shown in Figure 2, a through d.

Acidification of the porphyrin causes a significant change in the visible spectrum with the appearance of only two principal bands (Figure 3a). The two-banded spectrum resembles somewhat the spectrum of the metal complexes of most porphyrins. The metal complexes exhibit two absorption bands which vary in intensity as a function of the bound metal. Two examples are given in Figure 3b and 3c.

Porphyrins bind different metals with significantly different affinities. Magnesium is the metal least tightly bound of all encountered in biological systems, and it is displaced by protons under even the mildest of acidic conditions. Zinc is somewhat more tightly bound, but again it is displaced by protons in the presence of even low concentrations of mineral acids. Other metals, such as iron, copper, cobalt, nickel, and vanadium, are displaced only under special conditions using concentrated mineral acids.

Additional techniques that have proven useful for the characterization and identification

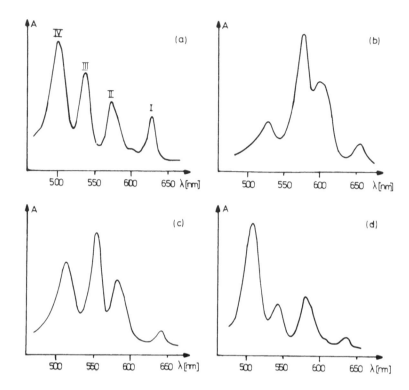

FIGURE 2. (a) The "etio-type" spectrum is found in all porphyrin in which six or more peripheral positions carry side chains such as methyl-, ethyl-, acetic acid, or propionic acid groups, the remaining positions being unsubstituted. The commonly occurring uro-, copro-, proto-, and deuteroporphyrins all exhibit the etiotype spectra. (b) two "rhodofying" groups on opposite (diagonal) of ring result in the oxorhodo-type spectrum. (c) As single carbonyl, a carboxylic acid substitution results in the "rhodo-type" spectrum. (d) The "phyllo-type" spectrum is most commonly seen when there is a substitution on one of the bridge (meso) carbons, as it occurs in some chlorophylls. (From Scheer, H. and Inhoffen, H. H., in *The Porphyrins*, Vol. 2, Dolphin, D., Ed., Academic Press, New York, 1978, 45. With permission.)

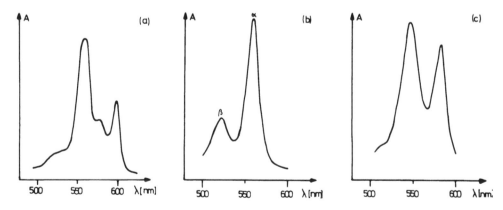

FIGURE 3. (a) Typical visible absorption spectra (Soret band omitted) of a porphyrin di-cation in chloroform containing 0.5% TFA (according to Smith[1]). (b and c) Visible absorption spectra (Soret band omitted) of typical square planar metalloporphyrins: (b) nickel(II) octaethylporphyrin (strong binding); (c) zinc(II) octaethylporphyrin (weak binding). The β-band is often less intense than the α-band when the metal is strongly bound (b). Spectra of porphyrins which bind metals only weakly frequently exhibit a more intense (stronger) β- than α-band (c). (From Scheer, H. and Inhoffen, H. H., in *The Porphyrins*, Vol. 2, Dolphin, D., Ed., Academic Press, New York, 1978, 45. With permission.)

of porphyrins are indicated below. Except for melting points and HCl numbers, the techniques involve instrumentation which is generally very expensive and whose use is quite specialized. Thus, only a very limited discussion is given for each method. We have chosen to give reference to recent work in each area for the reader to obtain information if appropriate.

Infrared Spectroscopy

The application of infrared spectroscopy had been limited for many years to the recognition and characterization of side chains in porphyrins and metalloporphyrins.[1,2] Today it is possible to draw structural implications from the spectra. The practical value of infrared spcetroscopy in combination with the chromatography of porphyrins, however, has remained relatively limited. The reader is therefore referred to the more extensive information contained in the book edited by Smith[1] and in the series, *The Porphyrins*, edited by Dolphin.[2]

NMR Spectroscopy[34-36]

Around 1960, NMR equipment became available to a great number of laboratories around the world, and, consequently, the method has been applied extensively for the study of a wide number of chemical compounds and also of porphyrins. This technique is especially useful for the latter class of compounds, since the large magnetic ring current (= anisotropy) spreads the proton (= ^{1}H) magnetic resonance spectrum over the unusually wide range of more than 15 ppm. Such a wealth of information has been obtained for few categories of compounds as for porphyrins, which also is expressed by the fact that in a review article written in 1975 by Scheer and Katz[36] no less than 317 references were cited.

Today, the use of NMR spectroscopy for the elucidation of structures and chemical pathways is much more widespread than a number of years ago, which is due to the highly computerized high-field (e.g., 360 MHz) equipment of our days. Even the follow-up of biochemical pathways of tetrapyrrole metabolism has become possible, especially if ^{13}C-labeled compounds are employed.

However, even though the use of NMR machines becomes more and more feasible, it only peripherally touches the subject of this book, chromatography, and very often the nonavailability of the required instrumentation poses a considerable barrier for most laboratories working with chromatographic methods. However, the amounts of compound needed for a modern NMR spectrum (*circa* 0.1 mg of porphyrin or bile pigment or even less) are sometimes small enough for the investigation of one single zone on a chromatographic plate, and the method is powerful enough to allow a wide range of investigations. For more information, the reader is referred to the literature.[1,2,34]

Mass Spectroscopy[36a]

This method, although mostly of somewhat limited practical importance, can be useful to confirm structural deductions made with other methods. Mass spectrometry, in general, is an analytical method that requires only minute amounts of material, as available, for example, from zones of TLC plates. The technique of fast atom bombardment (= FAB) mass spectrometry has become available in recent years and has proven particularly useful for compounds with very low volatility. Basically, mass spectroscopy (FAB mass spectroscopy, in particular) furnishes information on the accurate mole mass of porphyrins and metalloporphyrins. However, the fragmentation pattern of some porphyrins appears to be characteristic, too. In the absence of labile and/or "heavy functionalized" side chains, the base (= 100%) peak is usually the molecular ion. Deriving from that, fragmentation products are observed (to m/z values somewhat beneath 400). Below that, doubly charged series of ions are usually observed. In literature, major cleavage products of the side chains are discussed,[36a] considering the porphyrin ring as acting "as an inert support" allowing wide delocalization of positive charges.

A qualitative stability order for substituents has been given:[36b]

$$-H > -CH = CH_2 > -CHO > -CH_2CH_3 > -CH_2CH_2CO_2Me > -CH_2CH_2CH_3 >$$
$$-CH_2CO_2Me > -CH_2CH_2NHCOMe > -CO_2Me > -OCOMe > -CO-CH_2CO_2Me$$

The differences in mass spectra of porphyrin "type" isomers (e.g., type I and type III) are inconclusive, however.

For more detailed information, the reader is referred to the literature.[1,36a]

Melting Points

Historically, the use of melting points of porphyrin esters has been important. It has now been recognized, however, that many of the porphyrin esters crystallize with different structures,[30,31] i.e., they are polymorphic and, therefore, melt a different temperatures.

Partition Behavior (HCl Numbers)

Many porphyrins are extractable from aqueous solutions into diethyl ether or ethyl actate after the pH is adjusted to values around 3 to 4. Once in either of these organic solvents, it is possible to extract the porphyrins into basic or acidic aqueous solutions (see below); this makes possible an important clean-up step prior to chromatography. The so-called HCl number of porphyrins is mostly of historical interest. Nevertheless, it is of some practical use in the laboratory. The HCl number[32] is defined as the concentration of HCl in that percent (w/v) which extracts two thirds of the porphyrin from an equal volume of an ether solution. The HCl numbers depend jointly upon the dissociation of the porphyrin as a base and its ether-water partition coefficient. Representative values of HCl numbers are given in Table II.3 (compare Reference 33).

ESTIMATION AND SEPARATION OF PORPHYRINS

Notes on Porphyrin Stability

The term "stability" must be defined carefully. We use it to indicate that a given compound retains its structure and individual character. It need not decompose in the usual sense to be considered "unstable". Thus, any porphyrin in any solution except acidic ones can be seen as unstable because of the tendency to incorporate trace metal ions. As mentioned, metalloporphyrins have spectral characteristics quite different from the metal-free compounds. Similarly, the metalloporphyrins have considerably different chromatographic behavior in all of the systems mentioned in this section.

All solvents used for chromatographic purposes should be carefully washed with water to remove acids and alcohols. They should be dried with sodium sulfate, etc., passed over a silica gel column, and finally "glass distilled" to ensure the removal of trace metals. Those solvents, e.g., lutidine, water, etc., for which the above obviously does not apply, should have EDTA added to give a final concentration of 1 mM. As a general rule, mixtures of solvents should be prepared fresh each day. This is imperative when the components react with each other: e.g., acetic acid and methanol react to form methyl acetate.

While most porphyrins are quite stable in acidic solution, some are not stable under almost any conditions. Protoporphyrin is one of the best examples of an unstable porphyrin. In acid the vinyl groups at positions 2 and 4 tend to hydrate to produce the corresponding secondary alcohols, i.e., hydroxyethyl groups. Protoporphyrin is very unstable in the light — the photoreaction yields one of the two isomers of photoprotoporphyrin,[37-39] a green compound to be seen on most chromatograms of porphyrins. All porphyrins should, therefore, be kept in dim light.

Porphyrins in basic solution tend to be less stable. Whenever possible, they should not be stored under such conditions. It is best to store porphyrins as the dry crystalline esters or the dry dihydrochloride salts. Even the dry material should be stored in the dark.

Prechromatography Purification and Sample Preparation

A discussion of sample preparation is difficult since there exist a multitude of starting materials. Some investigators start with leaves, some with sedimentary rocks, some with microbial cultures (or supernatants), and others with urine. The first problem is to solubilize the porphyrin, when present in insoluble form. At this point it is impossible to generalize. When appropriate, i.e., with leaves and rocks, it is necessary to work with finely divided material. Once the porphyrin is in soluble form, it is necessary to perform some preliminary clean-up and concentrating. The methods vary so widely that examples of each must be considered separately (Figure 4 and Table 1).

Sample Preparation from Basic Aqueous Solutions

Basic aqueous solutions (pH 8) are prepared from leaves, rocks, liver, feces, culture supernatants, etc. Three techniques are practical for the extraction of porphyrins from these solutions:

Ethyl Acetate Extraction

Add a volume of ethyl acetate equal to $1/2$ the volume of the extract and shake it in a separatory funnel. Put the entire mixture into a beaker with a magnetic stirring bar and mix it vigorously. Adjust the pH with glacial acetic acid to about pH 3.2. Return the sample into the separatory funnel and mix the two phases thoroughly. After phase separation, remove the lower (aqueous) phase for reextraction. Place the ethyl acetate phase (containing porphyrins and metalloporphyrins) in a beaker. Extract the aqueous phase again, this time using only $1/2$ of the volume of the ethyl acetate used in the first extraction. Monitor the results of each extraction with the aid of a long-wavelength UV light in a dark room. Continue the extraction process as long as the ethyl acetate fraction continues to show a pink-red fluorescence. Take the combined ethyl acetate fractions and wash them several times with water. Next, extract the metal-free porphyrins out of the ethyl acetate with 3 M HCl using a minimum volume of the latter. The solution of the porphyrins in hydrochloric acid is now ready to be chromatographed or esterified, as the case may be (see below). If the ethyl acetate phase retains significant color, we have to suspect that stable metalloporphyrins remained. Therefore, wash the ethyl acetate with water at least twice. Then extract it in a solution of 1 M ammonia in water (minimum volume). Repeat the extraction as long as pigment continues to move from the organic to the aqueous phase. The aqueous phase now will contain metalloporphyrins. They are ready for chromatography or derivatization.

Absorption and Concentration on Talc

Add talc ($CaCO_3$, ca. 1 g/100 mℓ) to the basic solution and mix it well on a magnetic stirrer. Adjust the pH to 3.2 with glacial acetic acid. After thorough mixing and equilibration, filter off the talc on a Büchner funnel and wash with water. All the porphyrins and metalloporphyrins should be absorbed on the talc, which should fluoresce under long-wavelength UV light. The preparation is now ready for chromatography or derivatization.

Sample Preparation from Acidic Aqueous Solutions

Acidic extract of tissue etc.: extracts made 1.5 M with HCl should be protein-free and will not contain any metalloporphyrins. They should fluoresce salmon red under UV light.

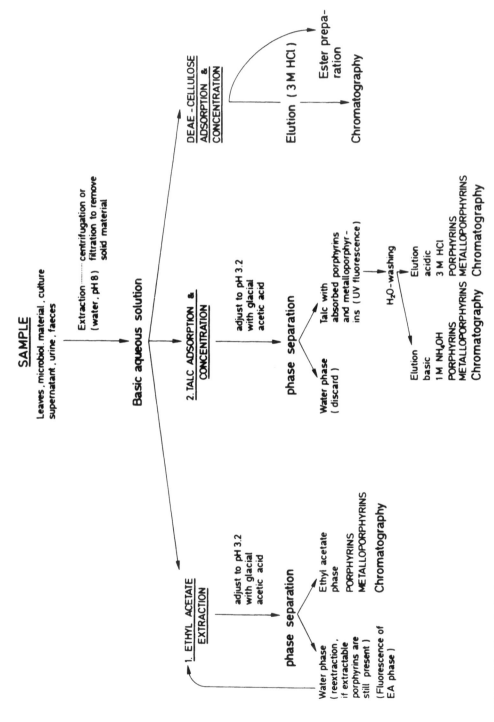

FIGURE 4. Example for the workup of porphyrin-containing samples. Further procedures are given in citations to tables dealing with "porphyrins" and in Table II HPLC 3, this Section.

Add 1 g talc per 100 mℓ and adjust the pH to 3.2 with 3 *M* NaOH (after adding some saturated sodium acetate as buffer). Filter off the talc on a Büchner funnel and wash with water. The talc should fluoresce under long wavelength UV light.

Absorption and Concentration on DEAE Cellulose

Add dry DEAE cellulose to the basic solution of the porphyrins. Mix thoroughly on a magnetic stirrer. Filter off the DEAE cellulose on a Büchner funnel and wash with water. All the porphyrins and metalloporphyrins should be on the DEAE cellulose which should fluoresce under long-wavelength UV light.

Extraction of Porphyrins
Extraction from Talc

Porphyrins may be extracted from talc with either strong mineral acids such as 3 *M* HCl or bases such as 1 *M* NH$_4$OH. They can, thus, be chromatographed directly. It is necessary, however, to pay attention to the solvent composition to be employed. For example, if one intends to use paper chromatography in lutidine water with an NH$_3$ atmosphere or reverse-phase high-performance liquid chromatography (HPLC), one must be careful about the ionic state of the applied material. A HCl extract of porphyrins will, of course, render di-cations. They will not migrate unless they are "neutralized". The same holds true for porphyrins in NH$_4$OH.

Metalloporphyrins may be extracted from talc by eluting them with 1 *M* NH$_4$OH.

Extraction from DEAE Cellulose

Porphyrins may be eluted from DEAE with 3 *M* HCl. Again, the fact that they are present in the form of a strongly acidic solution must be taken into account prior to chromatography.

Preparation and Extraction of Porphyrin Methyl Esters

The most practical method for examining almost any unknown porphyrin mixture by chromatography is to prepare the methyl ester first. Two methods are commonly used (next two paragraphs).

Esterification with methanol/mineral acid — Suspend the dry talc or DEAE cellulose to which the porphyrin is bound in methanol containing 5% concentrated sulfuric acid. After about 1 hr, filter on a Büchner funnel. Wash the talc or DEAE cellulose several times with methanol. Place the combined acidic methanol extracts in a closed container (screw-capped test tube) and place it into a boiling water bath for about 1 hr. *Caution should be observed* because of the developing internal pressure during heating. Alternatively, the acidic suspension can be incubated at room temperature in the dark overnight.

Esterification with boron trifluoride/methanol — Alternatively, porphyrins may be eluted from talc or DEAE cellulose with boron trifluoride (10 to 20% in methanol). Again, the esterification can be accelerated by heating as above, or the mixture may be allowed to stand overnight. If the sample of talc or DEAE cellulose is not thoroughly dry, it is practical to add a water-scavenging compound such as 2% *ortho*-trimethylformate.

Extraction of porphyrin esters — To the reaction mixture in acidified methanol containing the prophyrin esters, either chloroform or (better) methylene chloride is added (separatory funnel). After the contents of the funnel have been shaken, water is added to give two phases. The porphyrin ester will move quickly into the organic (lower) phase. The extraction should be repeated and monitored by observing the fluorescence under long-wavelength UV light. Further, this extraction should be carried out as quickly as possible to avoid acidic hydrolysis of the porphyrin esters. The chloroform (or methylene chloride) extract of porphyrin esters should be washed several times in the separatory funnel with distilled water. This washing is generally accompanied by a change in color from purple to a brownish red, as the porphyrin di-cation ester is neutralized. The porphyrin ester can then be crystallized by adding methanol (1:1) and allowing the chloroform to evaporate in a gentle stream of warm air. The porphyrin ester is then ready for chromatography.

Extraction from sedimentary rock — In some cases it is necessary or appropriate to extract a sample directly with organic solvents. Such is the case with sedimentary rocks. Toluene-methanol = 1:1 seems to be an efficient solvent. The filtered solution is then passed over an alumina column which removes and concentrates the porphyrins. The porphyrins are extracted from the alumina using one of the methods given above.

Chromatography

Paper Chromatography

The most practical method for preliminary studies involving porphyrins is undoubtedly chromatography. The technique is simple and does not involve complicated instrumentation. Large pickle jars are quite suitable as containers for the development of paper chromatograms. Detection of porphyrins is best done by visual observation of red fluorescent spots when the chromatogram is examined under long-wavelength UV irradiation in a dark room. Heme compounds are best visualized after spraying the chromatogram with an appropriate dye. (See, for example, Table II. PC 5.) The major limiting factor of paper chromatography is one of capacity. For clear resolution, the porphyrin should be hardly visible in the white light, but should fluoresce clearly under long-wavelength UV light.

Thin-Layer Chromatography (TLC)

TLC of porphyrins is a very useful technique which provides several advantages. First, it is possible to buy thin-layer plates with a wide variety of adsorbants. Most commonly used are silica gel, polyamide (nylon), and talc ($CaCO_3$). The latter coating is not commercially available. TLC on the analytical scale can, as a general rule, be adapted, by slight variations in the solvent mixture, to thick-layer or preparatory chromatography. Further, within limits, results obtained by TLC can be directly adapted (with modifications as they become apparent) to both large-scale liquid chromatography with the same adsorbant and/or high-performance liquid chromatography column materials. Porphyrins on developed thin-layer chromatograms tend to be considerably less stable than usual. This is, no doubt, caused by the large surface area upon which they are exposed. If such chromatograms must be stored, they should be wrapped in plastic film (Saran® wrap) or aluminum foil and stored in the freezer. TLC plates that contain a fluorescent indicator frequently cause problems when used for porphyrin chromatography. The indicator is a zinc complex and there is enough "free" zinc so that some of this metal is incorporated by the porphyrins during chromatography. This, being a continuous process, results in a streaking of each porphyrin as it is converted into the slower running zinc complex.

Liquid Chromatography (LC)

This technique is of some use in large-scale preparatory work. It is of little or no use for analytical work. The obvious advantage of LC is the potential capacity of the system, i.e., one is limited only by the size of the column. It is, however, the most difficult system with which to obtain reproducible results.

High-Performance Liquid Chromatography (HPLC)

In the last few years, the analysis of porphyrins has been revolutionized by application of HPLC. As is customary for porphyrins, medical applications have preceeded its application in plant biochemistry, which is now a rapidly developing area. HPLC is without doubt the most powerful chromatographic technique available at the present time. Separations that were impossible a few years ago are now routine laboratory procedures and take only a few minutes. The speed factor should be emphasized by the following example: in a trial run (15 min), an unknown mixture may exhibit several fast-running peaks in a solvent system such as ethyl acetate-hexane = 55:45. Greater resolution may be achieved quickly by simple adjustment of the solvent ratio, e.g., 50:50. This "fine tuning" is only practical with HPLC. In addition to the advantage of speed afforded by HPLC, there is the particular feature that quantification is implicitly automatic. It should be noted that as with all chromatographic techniques discussed, the use of standard porphyrin mixtures is imperative for definitive work.

A survey of a sample workup is given in the table (Table II. HPLC 3).

Sources and Materials for HPLC

HPLC columns are available from many commercial sources. (For sources and literature, see HPLC tables.) Even though they may bear the same description, e.g., 5 μm silica gel, 5 μm C_{18}, etc., columns are not always interchangeable. Each column, even from a single manufacturer, must be "fine tuned" with the use of standard porphyrin mixtures.

Hyperpressure Gas Chromatography (HPGC)

Porphyrins are a class of very slightly volatile compounds, although in a number of cases a purification could be achieved by sublimation.[40] It is, therefore, not surprising that although the different forms of paper-, thin-layer, and column liquid chromatography have been widely used for the separation and analysis of porphyrins, conventional gas chromatography has remained of no importance.

A few authors, however,[41-43] were successful with a special form of gas chromatography, the so-called hyperpressure gas chromatography (HPGC). Carrier gases employed were dichlorodifluoromethane (c.t. 111.5°C) or monochlorodifluoromethane (c.t. 96°C) above their critical temperatures (c.t.) at pressures of 1000 to 1400 psi. Under these conditions, the amount of solid (e.g., porphyrin or metalloporphyrin) dissolved in the gas is much larger than expected from the normal increase of vapor pressure due to external pressure. The hyperpressure gas chromatograph and the necessary modifications have been described.[41] For porphyrins, HPGC offers the only method for gas chromatography of these compounds as such. On the other hand, since metals may be separated in the form of their porphyrin metal chelates, the method offers a possibility for the separation of Cu(II), Ni(II), V(IV), and Sn(IV) in the form of their etioporphyrin metal chelates. Also, Ag(II) could be thus determined.[43]

Paper Electrophoresis

In modern clinical chemistry, electrophoretic methods are increasingly used, especially for the analysis of serum proteins. From time to time, it might happen that urine from porphyric patients has to be analyzed. We, therefore, give a brief reference on paper electrophoretic separations of porphyrins.[44,45] For analysis, free porphyrins are used. As could be expected, the speed of migration in the electric field increases with the number of carboxyl groups. To obtain a complete dissociation of the carboxylic acid groups present, an alkaline buffer has to be used (e.g., 1/20 M barbiturate, pH 8.6). Compared to modern HPLC, paper electrophoresis is somewhat slow, but within 1 to 3 hr a satisfactory separation can be achieved.[44] The method might be recommended if no other equipment than an electrophoretic setup is available and electrophoresis is routinely carried out. One also must consider that routine electrophoresis today is mostly carried out on cellulose acetate strips, whereas the literature refers to paper. For further information, the reader is referred to Table II. PEL 1.

REFERENCES

1. **Smith, K.M., Ed.**, *Porphyrins and Metalloporphyrins*, Elsevier, Amsterdam, 1975.
2. **Dolphin, D., Ed.**, *The Porphyrins*, Vol. 1 to 3 and 4 to 7, Academic Press, New York, 1978 and 1979.
3. **Cox, M. T., Jackson, A. H., and Kenner, G. W.**, *J. Chem. Soc. C.*, p. 1974, 1971.
4. **DiNello, R. K. and Chang, C. K.**, *The Porphyrins*, Vol. 1, Dolphin, D., Ed., Academic Press, New York, 1978, 290.
5. **Drabkin, D. L.**, *The Porphyrins* Vol. 1, Dolphin, D., Ed., Academic Press, New York, 1978, 31.
6. **Dolphin, D., Ed.**, *The Porphyrins*, Vol. 6 and 7, Academic Press, New York, 1979.
7. **Chlorophylls, Scheer, H.**, in *CRC Handbook of Chromatography, Plant Pigments*, Vol. 1, *Fat-Soluble Pigments*, Köst, H.-P., Ed., CRC Press, Boca Raton, 1988.
8. **Baker, E. W. and Palmer, S. E.**, *The Porphyrins*, Vol. 1, Dolphin, D., Ed., Academic Press, New York, 1978.
9. **Hodgson, G. W.**, *Ann. N.Y. Acad. Sci.*, 206, 670, 1973.
10. **Bonnet, R.**, *Ann. N.Y. Acad. Sci.*, 206, 722, 1973.
11. **Pfaltz, A., Jaun, B., Fässler, A., Eschenmoser, A., Jaenchen, R., Gilles, H. H., Diekert, G., and Thauer, R. K.**, *Helv. Chim. Acta*, 65, 828, 1982.
12. **Langhof, H., Müller, H., and Rietschel, L.**, *Arch. Klin. Exp. Dermatol.*, 212, 506, 1961.
13. **Andreoni, A. and Cubeddo, R., Eds.**, *Porphyrin in Tumor Phototherapy*, Plenum Press, New York, 1984.
14. **Sano, S.**, *The Porphyrins*, Vol. 7, Dolphin, D., Ed., Academic Press, New York, 1979, 378.
15. **Hewson, W. D. and Hager, L. P.**, *The Porphyrins*, Vol. 7, Dolphin, D., Ed., Academic Press, New York, 1979, 295.
16. **Berk, P. D. and Berlin, N. I., Eds.**, Chemistry and Physiology of Bile Pigments, Fogarty Int. Center Proc. No. 35, DHEW Publ. No. (NIH) 77-1100, National Institutes of Health, Bethesda, Md., 1977.
17. **Rüdiger, W.**, *Fortschr. Chem. Org. Naturst.*, 29, 61, 1971.
18. **Scheer, H.**, *Angew. Chem. Int. Ed. Eng.*, 20, 241, 1981.
19. **Scheer, H.**, *Light Reaction Path of Photosynthesis*, Vol. 35, Fong, F. K., Ed., Springer-Verlag, Berlin, 1982.
20. **Bennet, A. and Siegelman, H. W.**, *The Porphyrins*, Vol. 6, Dolphin, D., Ed., Academic Press, New York, 1979, 493.
21. **Brumm, P. J., Fried, J., and Friedmann, H. C.**, *Proc. Natl. Acad. Sci. U.S.A.*, 80, 3943, 1983.
22. **Benedikt, E. and Köst, H.-P.**, *Z. Naturforsch.*, 38c, 753, 1983.
23. **Lascelles, J.**, *Tetrapyrrole Biosynthesis and Its Regulation*, Benjamin, New York, 1964.
24. **Battersby, A. R. and McDonagh, A. E.**, *Porphyrins and Metalloporphyrins*, Smith, K. M., Ed., Elsevier, Amsterdam, 1975, 61.
25. **Frydman, R. B., Frydman, B., and Valasinas, A.**, *The Porphyrins*, Vol. 6, Dolphin, D., Ed., Academic Press, New York, 1979, 3.
26. **Jacobs, N. J., Jacobs, J. M., Bloomer, J. R., and Morton, K. O.**, *Enzyme*, 28, 206, 1982.
27. **White, W. I., Bachmann, R. C., and Burnham, B. F.**, *The Porphyrins*, Vol. 1, Dolphin, D., Ed., Academic Press, New York, 1978, 553.
28. **Soret, J. L.**, *Comptes Rendues*, 97, 1267, 1883.
29. **Gamgee, A.**, *Z. Biol. Munich*, 34, 505, 1897.
30. **MacDonald, S. F. and Michl, K. H.**, *Can. J. Chem.*, 34, 1768, 1956.
31. **Morsingh, F. and MacDonald, S. F.**, *J. Am. Chem. Soc.*, 82, 4377, 1960.
32. **Willstätter, R. and Mieg, W.**, *Ann. Chem.*, 350, 1, 1906.
33. **Fischer, H. and Orth, H.**, *Die Chemie des Pyrrols*, Vol. 1 and 2, Akademische Verlagsgesellschaft, Leipzig, 1934 and 1937; reprinted by Johnson Reprint, New York, 1968.
34. **Janson, T. R. and Katz, J. J.**, *The Porphyrins*, Dolphin, D., Ed., Academic Press, New York, 1979, 1.
35. **LaMar, G. N. and Walker, F. A.**, *The Porphyrins*, Dolphin, D., Ed., Academic Press, New York, 1979, 61.
36. **Scheer, H. and Katz, J. J.**, *Porphyrins and Metalloprophyrins*, Smith, K. M., Ed., Elsevier, Amsterdam, 1975, 399.
36a. **Budzikiewicz, H.**, *The Porphyrins*, Vol. 3, Dolphin, D., Ed., Academic Press, New York, 1978, 395.
36b. **Dougherty, R. C.**, *Biochemical Applications of Mass Spectrometry*, Waller, G. E., Ed., John Wiley & Sons, New York, 1972, 591.
37. **Fischer, H. and Bock, H.**, *Hoppe-Seyler's Z. Physiol. Chem.*, 255, 1, 1938.
38. **Inhoffen, H. H., Brockmann, H., Jr., and Bliesener, K.-M.**, *Ann. Chem.*, 730, 173, 1969.
39. **Barret, J.**, *Nature (London)*, 183, 1185, 1959.
40. **Smith, K. M.**, *Porphyrins and Metalloporphyrins*, Smith, K. M., Ed., Elsevier, Amsterdam, 1975, 15.

41. **Karayannis, N. M., Corwin, A. H., Baker, E. W., Klesper, E., and Walter, J. A.,** *Anal. Chem.,* 40, 1736, 1968.
42. **Klesper, E., Corwin, A. H., and Turner, D. A.,** *J. Org. Chem.,* 27, 700, 1962.
43. **Karayannis, N. M. and Corwin, A. H.,** *Anal. Biochem.,* 26, 34, 1968.
44. **With, T. K.,** *Scand. J. Clin. Lab. Invest.,* 8, 113, 1956.
45. **Lockwood, W. H. and Davis, J. L.,** *Clin. Chim. Acta,* 7, 301, 1962.

Tables for the Estimation and Separation of Porphyrins and Metalloporphyrins

GENERAL TABLES

Table II. 1
TRIVIAL NAMES AND STRUCTURES OF COMMON PORPHYRINS

Porphyrin	Substituents[a]							
	1	2	3	4	5	6	7	8
Etioporphyrin-I	Me	Et	Me	Et	Me	Et	Me	Et
Octaethylporphyrin	Et	Et	Et	Et	Et	Et	Et	Et
Deuteroporphyrin-IX	Me	H	Me	H	Me	P	P	Me
Mesoporphyrin-IX	Me	Et	Me	Et	Me	P	P	Me
Hematoporphyrin-IX	Me	–CH(OH)CH₃	Me	–CH(OH)CH₃	Me	P	P	Me
Protoporphyrin-IX	Me	V	Me	V	Me	P	P	Me
Coproporphyrin-I	Me	P	Me	P	Me	P	Me	P
Coproporphyrin-III	Me	P	Me	P	Me	P	P	Me
Uroporphyrin-I	H	P	H	P	H	P	H	P
Uroporphyrin-III	H	P	H	P	H	P	P	H
Chlorocruoroporphyrin	Me	CHO	Me	V	Me	P	P	Me
Pemptoporphyrin	Me	H	Me	V	Me	P	P	Me
Deuteroporphyrin-IX 2,4-di-acrylic acid	Me	Acr	Me	Acr	Me	P	P	Me
2,4-Diformyldeuteropor-phyrin-IX	Me	CHO	Me	CHO	Me	P	P	Me
2,4-Diacetyldeuteropor-phyrin-IX	Me	Ac	Me	Ac	Me	P	P	Me
Deuteroporphyrin-IX 2,4-disulfonic acid	Me	SO₃H	Me	SO₃H	Me	P	P	Me
Phylloporphyrin-XV[b]	Me	Et	Me	Et	Me	H	P	Me
Pyrroporphyrin-XV	Me	Et	Me	Et	Me	H	P	Me
Rhodoporphyrin-XV	Me	Et	Me	Et	Me	–CO₂H	P	Me
Phylloerythrin	Me	Et	Me	Et	Me	–CO—CH₂[c]	P	Me
Deoxophylloerythrin	Me	Et	Me	Et	Me	–CH₂—CH₂[c]	P	Me
Pheoporphyrin-a₅	Me	Et	Me	Et	Me	–CO—CH[c] | CO₂Me	P	Me

Note: Side-chain abbreviations — Me = methyl; Et = ethyl; V = vinyl; R = H or substituent; P = CH₂CH₂CO₂R; A = CH₂CO₂R; Acr = CH=CHCOOH.

[a] Substituents (1 to 8) are arranged clockwise at peripheral positions of porphyrin nucleus.

[b] Me-substituent at γ bridge.

[c] Forms ring by linkage to γ bridge.

Table II. 2
QUANTITATIVE SPECTROSCOPIC DATA:
MOLAR EXTINCTION COEFFICIENTS

UV-VIS Spectral Data of Porphyrin Esters

Compound		Soret	IV	III	II	I	Ref.
				λ_{max} (nm) — ϵ_m			
Uroporphyrin	λ	406	502	536	572	627	1
octamethyl ester[a]	ϵ	215	15.8	9.3	6.8	4.2	
Coproporphyrin	λ	400	498	532	566	621	2
tetramethyl ester[a]	ϵ	180	14.3	9.9	7.1	5.0	
Protoporphyrin IX	λ	407	505	541	575	630	2
dimethyl ester	ϵ	171	14.2	11.6	7.4	5.4	
Deuteroporphyrin IX	λ	399.5	497	530	566	621	2
dimethyl ester	ϵ	175	13.4	10.1	8.2	5.0	
Mesoporphyrin-IX	λ	400	499	533	567	621	2
dimethyl ester	ϵ	166	13.6	9.6	6.5	4.9	
Hematoporphyrin-IX	λ	402	499	534	569	622	3
dimethyl ester	ϵ	193	15.0	9.5	6.9	4.3	

[a] Spectra identical for all isomers.

REFERENCES

1. **Mauzerall, D.,** *J. Am. Chem. Soc.,* 82, 2601, 1960.
2. **Smith, K. M.,** *Porphyrins and Metalloporphyrins,* 1st ed., Elsevier, Amsterdam, 1975.
3. **Caughey, W. S., Fujimoto, W. Y., and Johnson, B. D.,** *Biochemistry,* 5, 3830, 1966.

Table II. 3
HCl-NUMBERS OF PORPHYRINS
AND PORPHYRIN ESTERS

Compound	HCl Number[a] [2,3]
Uroporphyrin III	5.0
Octamethyl ester	
Coproporphyrin III	0.9
Tetramethyl ester	1.7
Protoporphyrin IX	2.5
Dimethyl ester	5.5
Mesoporphyrin IX	0.5
Dimethyl ester	2.5
Hematoporphyrin IX	0.1
Deuteroporphyrin IX	0.3
Dimethyl ester	2.0
Chlorocruoroporphyrin	4.6
Rhodoporphyrin XV	4.0
Dimethyl ester	7.5
Pyrroporphyrin XV	1.3
Dimethyl ester	2.5
Phylloporphyrin XV	0.35
Dimethyl ester	0.9
Pheoporphyrin a_5	9.0
Phylloerythrin	7.5

[a] The HCl number is defined as the concentration of HCl in percent (w/v) which extracts two thirds of the porphyrin from an equal volume of an ether solution. The HCl numbers depend jointly upon the dissociation of the porphyrin as a base and its ether-water partition coefficient.[1]

REFERENCES

1. **Willstätter, R. and Mieg, W.,** *Ann. Chem.,* 350, 1, 1906.
2. **Fischer, H. and Orth, H.,** *Die Chemie des Pyrrols,* Vol. 1 and 2, Akademische Verlagsgesellschaft, Leipzig, 1934 and 1937.
3. **Mauzerall, D.,** *J. Am. Chem. Soc.,* 82, 2601, 1960.

PAPER CHROMATOGRAPHY OF PORPHYRINS

Table II. PC 1
FREE ACID PORPHYRINS

Paper	P1	P1	P1	P1	P1[a]	P2	P2	P3
Solvent	S1	S2	S3	S4	S5	S6	S2	S7
Technique	T1	T2[b]	T2[b]	T2	T1[c]	T2[d]	T2[e]	T3
Detection	D1	D1	D1	D1	D1	D1	D1	D1
Literature	1	1	1	2	2	1	1	3
Compound				$R_f \times 100$				
Uroporphyrin	3	6	85—95	—	—	19	26	—
Hexacarboxlic porphyrin	—	—	45—55	—	—	—	—	—
Pentacarboxylic porphyrin	—	42	—	—	—	—	—	—
Coproporphyrin	6	56	20—30	—	—	47	54	—
Tricarboxylic porphyrin	—	68	—	—	—	—	—	—
Protoporphyrin	8	83	—	10[f]	9	75	84	—
Mesoporphyrin	8	—	—	36[f]	21	81	86	—
Hematoporphyrin	8	—	5—9	100[f]	53	77	87	—
Deuteroporphyrin	8	—	—	62[f]	33	79	88	—
Pemptoporphyrin	—	—	—	36[f]	21	—	—	—
Phylloerythrin	8	—	—	—	—	—	—	—
Etioporphyrin	—	—	—	—	—	—	100	63
All porphyrin esters	100	—	—	—	—	100	100	—

[a]   Sheets 20 × 23 cm.
[b]   20°C.
[c]   0.5—7 µg per spot.
[d]   15°C.
[e]   25°C.
[f] R_f position of spot relative to hematoporphyrin.

Paper   P1 = Whatman No. 1[a]
         P2 = Whatman No. 3
         P3 = Schleicher and Schüll paper 2034 b
Solvent S1 = 2,4-2,5-lutidine-water; upper phase as saturated solution; atmosphere satu-
              rated with NH_3 vapor
         S2 = 2,6-lutidine-water = 5:3; atmosphere saturated with NH_3 vapor
         S3 = 0.1 M LiCl, 20°C; atmosphere saturated with NH_3 vapor
         S4 = ethanol-2,6-lutidine-water = 30:3:67
         S5 = pyridine-0.2 M sodium borate buffer pH 8.6 = 1:9
         S6 = 2,4-lutidine-water; upper phase as saturated solution; atmosphere saturated
              with NH_3 vapor
         S7 = CCl_4-isooctane = 7:3
Technique T1 = descending, NH_3 vapor[c]
         T2 = ascending, NH_3 vapor[b,d,e]
              Equilibration: cylindrical glass tanks (13 × 28 cm) were allowed to equili-
              brate for 30 min before use with 20 mℓ of 0.88 M ammonia placed at the
              bottom; the tanks were also lined with filter paper soaked in concentrated
              ammonia
              Developing: 20 mℓ of developing mixture was placed in a suitable-sized
              Petri dish placed at the bottom of the tanks
         T3 = horizontal in saturated atmosphere of the solvent vapor
Detection D1 = visual detection under UV light

REFERENCES

1. **Falk, J. E.**, *J. Chromatogr.*, 5, 277, 1961.
2. **Belcher, R. V., Smith, S. G., Mahler, R., and Campbell, J.**, *J. Chromatogr.*, 53, 279, 1970.
3. **Blumer, M.**, *Anal. Chem.*, 28, 1640, 1956.

Table II. PC 2
PORPHYRIN ESTERS

Paper	P1	P2	P1	P3	P3	P3	P4
Solvent	S1	S2	S3	S4	S5	S6	S7
Technique	T1$^+$	T1	T1	T1	T1^{++}	T1^{++}	T1
Detection	D1	D1	D1	D1	D1	D1	D1
Literature	1	2	3	4, 5	4, 5	4, 5	6

Compound[a]	$R_f \times 100$						
Uroporphyrin I octamethyl ester	00^{b,f}	—	—	—	—	—	—
Uroporphyrin III octamethyl ester	23^{b,f}	—	—	—	—	—	—
Heptacarboxylic porphyrin III heptamethyl ester	40^c	—	—	—	—	—	—
Hexacarboxylic porphyrin III hexamethyl ester	50^c	—	—	—	—	—	—
Pentacarboxylic porphyrin III pentamethyl ester	58^c	—	—	—	—	—	—
Coproporphyrin I tetramethyl ester	04	—	—	—	—	—	—
Coproporphyrin III tetramethyl ester	66^b	—	—	—	—	—	—
Protoporphyrin IX	—	—	83^c	—	—	—	—
Monomethyl ester	—	—	95	—	—	—	—
Dimethyl ester	74^b	80^d	100	51	23	24	7
Mesoporphyrin IX dimethyl ester	—	60^d	—	59	23	23	10
Hematoporphyrin IX							
Dimethyl ester	—	00^d	—	3	80	71	—
Dimethyl ether	—	—	—	59	63	46	—
Deuteroporphyrin IX dimethyl ester	—	40^d	—	59	45	46	—
Monovinyl monohydroxyethyl deuteroporphyrin dimethyl ester	—	—	—	20	74	61	—

[a] Methyl esters of porphyrins.
[b] Standard reference.
[c] Enzymic.
[d] Porphyrin dimethyl esters on iron-impregnated paper; R_f values estimated from areas of greatest density as spots showed long diffuse fronts.
[e] Lutidine solution of porphyrins.
[f] Caution: separation of I and III isomers does not work completely reliably.

Paper	P1	=	Whatman No. 1
	P2	=	Whatman No. 1 paper impregnated with 1.8% (w/v) ferric chloride calculated as FeCl$_3$·6 H$_2$O
	P3	=	Whatman No. 3
Solvent	S1	=	kerosene-dioxane = 4.0:1.0
	S2	=	benzene-methanol = 100:5
	S3	=	lutidine-water = 5:3.5 in ammonia atmosphere
	S4	=	kerosene-tetrahydropyran-methylbenzoate = 5:1.4:0.35
	S5	=	water-acetonitrile-n-propanol-pyridine = 3.8:1:2:0.5 (with silicone as stationary phase)
	S6	=	water-acetonitrile-dioxane = 2.3:2.8:0.8 (with silicone as stationary phase)
	S7	=	carbon tetrachloride-i-octane = 7:4

Table II. PC 2 (continued)
PORPHYRIN ESTERS

Technique   T1 = ascending

            ' = Chloroform solutions of porphyrin esters (1 μg/10 μℓ) were applied by means of a micropipette graduated in 5 μℓ along a baseline 2 cm from one edge of a 21 cm square of Whatman No. 1 paper, in such a manner that they would run with the grain of the paper. Chromatography was carried out at 22 to 26°C.

           ' ' = Second development by reversed phase paper chromatography. The partially developed paper chromatogram (Solvent Syl) was treated with a petrol ether (bp. 56 to 110°C) solution of Dow-Corning silicone No. 550 fluid (w/v 12.5/100) by dipping the paper into the solution or by pulling it through. After drying at 105 to 110°C for 3 min, the paper cylinder, made with the new basal line at the bottom, was placed in the second solvent mixture (either S5 or S6) which was used in an atmosphere saturated with water. The developing time was 1.5 hr for the solvent front (S5) resp. 50 min (S6) to ascend to about 8 cm. During the second development, the members in the unresolved spot from the first development were separated.

Detection   D1 = visual detection under UV light (366 mm)

REFERENCES

1. **Cornford, P. A. D. and Benson, A.,** *J. Chromatogr.,* 10, 141, 1963.
2. **Henderson, R. W. and Morton, T. C.,** *J. Chromatogr.,* 27, 180, 1967.
3. **Ellsworth, R. K.,** *Anal. Biochem.,* 32, 377, 1969.
4. **Falk, J. E.,** *J. Chromatogr.,* 5, 277, 1961.
5. **Chu, T. C. and Chu, E. J.,** *J. Biol. Chem.,* 208, 537, 1954.
6. **Blumer, M.,** *Anal. Chem.,* 28, 1640, 1956.

Table II. PC 3
PORPHYRIN ESTERS (TWO-SOLVENT SYSTEMS)

Paper		P1	P1	P1	P1	P1
Solvent		S1/S3	S2/S4	S2/S5	S2/S6	S2/S7
Technique		T1	T1	T1	T1	T1
Detection		D1	D1	D1	D1	D1
Literature		1	1	1	1	1

Compound[a]	$R_f \times 100$	Second run *n*-propanol with				
		Kerosene	*n*-Decane	*n*-Dodecane	*n*-Tetradecane	*n*-Hexane
Uroporphyrin I methyl ester		17	14	20	13	15
Coproporphyrin I methyl ester		47	42	52	45	47
Coproporphyrin III methyl ester		67	70	76	74	66
Protoporphyrin IX methyl ester		84	86	92	92	89
Mesoporphyrin IX methyl ester		89	92	96	95	93

[a]   Porphyrin-methylesters, Chu et al.[2]

Paper   P1 = Whatman No. 1 paper, 24°C
Solvent   S1 = chloroform-kerosene = 2.6:4.0 first run
                S2 = chloroform first run; second run propanol-alkane = 1:5
Technique   T1 = ascending; the paper was dried at 105—110°C for about 4 min after completion of the first run; the atmosphere was saturated with the same solvents as used for development
Detection   D1 = visual observation under UV light; spraying the completed chromatograms with isooctane before observation under UV light increased markedly the sensitivity of the fluorescence, 0.005 μg being observable; for spots of nonfluorescent porphyrin metal complexes, it was found that on spraying with a solution of fluoranthene in *n*-pentane, and then illuminating with light at 366 nm, the porphyrins and their metal complexes showed as dark spots against a fluorescent background, about 0.04 μg being observable

REFERENCES

1. **Falk, J. E.**, *J. Chromatogr.*, 5, 277, 1961.
2. **Chu, T. C., Green, A. A., and Chu, E. J.**, *J. Biol. Chem.*, 190, 643, 1951.

Table II. PC 4
PORPHYRIN "DERIVATIVES"

Paper	P1	P1	P1	P1
Solvent	S1	S2	S1	S2
Technique	T1	T1	T1	T1
Detection	D1	D1	D1	D1
Literature	1	1	1	1

Compound[a]	$R_f \times 100$			
Deuteroporphyrin				
Monohydroxyethyl	34	38	64[b]	68[b]
Monovinyl-mono hydroxyethyl	29	34	60[b]	66[b]
Monohydroxymethyl	19	24	54[b]	57[b]
Monohydroxymethyl-monovinyl	22	31	54[b]	56[b]
Dihydroxymethyl	1	14	56[b]	56[b]
2-Formyl-4-hydroxy-ethyl	16	26	58[b]	64[b]
2-Ethylene-glycol	30	32	64[b]	61[b]
Hematoporphyrin	3	18	56[b]	57[b]
Porphyrin a	10	26	56[b]	62[b]
Chlorin a_2	30	40	65[b]	60[b]
Mesorhodochlorin 2-α-Hydroxy	31	34	80[b]	78[b]
Mesochlorin P_6 2-α-Hydroxy	33	35	82[b]	80[b]
Mesopheophorbide a 2-α-Hydroxy	10	34	78[b]	72[b]

[a] Hydroxylated porphyrins and chlorins and their acetylated products.

[b] Acetylated compound.

Paper	P1	=	Whatman No. 1
Solvent	S1	=	chloroform-kerosene = 2.6:4[2]
	S2	=	propanol-kerosene = 1:5[2]
Technique	T1	=	ascending, 22°C
Detection	D1	=	visual

REFERENCES

1. **Barret, J.,** *Nature* 183, 1185, 1959.
2. **Chu, T.C., Green, A. G., and Chu, E. J.,** *J. Biol. Chem.,* 190, 643, 1951.

Table II. PC 5
PORPHYRINS AND METALLOPORPHYRINS

	P1	P2	P3	P3	P?	P?
Paper	P1	P2	P3	P3	P?	P?
Solvent	S1	S2	S3	S4	S5	S6
Technique	T1	T2	T3$^+$	T3^{++}	T4	T4
Detection	D$^{1/2}$	D1	D1	D1	D3	D3
Literature	1	2	2	2	2	2

Compound [a]			$R_f \times 100$			
Uroporphyrin-Fe-complex (urohemin)	—	—	—	—	96	20
Coproporphyrin-Fe-complex (coprohemin)	—	—	—	—	88	56
Protoporphyrin-dimethyl ester	7	—	—	—	—	—
Cu-complex	13	—	—	—	—	—
Fe-complex (protohemin, hemin)	—	68	63	7	34	76
Ni-complex	10	—	—	—	—	—
Mesoporphyrin-dimethyl ester	10	—	—	—	—	—
Fe-complex	—	—	—	—	45	77
Ni-complex	12	—	—	—	—	—
Hematoporphyrin-Fe-complex (hematohemin)	—	—	—	—	72	77
Deuteroporphyrin-Fe-complex (deuterohemin)	—	—	—	—	62	76
Deoxophyllerythrin-monomethyl ester	15	—	—	—	—	—
V-complex	8	—	—	—	—	—
Ni-complex	23	—	—	—	—	—
Deoxophyllerythrin-etioporphyrin	47	—	—	—	—	—
V-complex	20	—	—	—	—	—
Ni-complex	54	—	—	—	—	—
Etioporphyrin III	63	—	—	—	—	—

[a] $R_f \times 100$ values of free and esterified porphyrins and metalloporphyrins.

Paper	P1 =	Whatman No. 3
	P2 =	Whatman No. 1
	P3 =	Schleicher and Schüll paper 2043 b
Solvent	S1 =	carbon tetrachloride-i-octane = 7:3
	S2 =	2,4-2,5-lutidine-water (upper phase, saturated solution, atmosphere saturated with NH_3 vapor)
	S3 =	2,4-lutidine-water (upper phase, saturated solution, atmosphere saturated with NH_3 vapor)
	S4 =	2,6-lutidine-water = 5:3, atmosphere saturated with NH_3 vapor
	S5 =	water-*n*-propanol-pyridine = 5.5:0.1:0.4, atmosphere saturated with water and pyridine vapor
	S6 =	2,6-lutidine-water = 3.3:2.7, atmosphere saturated with water vapor
Technique	T1 =	horizontal in saturated atmosphere of the solvent vapor
	T2 =	descending, NH_3 vapor, 19°C
	T3 =	ascending, NH_3 vapor, 15° C$^+$, 25°C^{++} ($^+$, $^{++}$; see Table II. PC2)
Detection	D1 =	visual
	D2 =	porphyrin metal complexes: quenching of fluorescence in the case of metal complexes by spraying the completed and dried chromatogram with or dipping it in a saturated solution of fluoroanthene in *n*-pentane; it is then rapidly air-dried and observed under UV radiation at 366 nm
	D3 =	formation of blue spots upon spraying with benzidine reagent; by the use of this special spray, Connelly et al.[3] were able to detect as little as 3 × 10^{-4} μg of hemine.
		Important note: since benzidine is carcinogenic, it is better to use substitutes, for example, tetramethyl-benzidine (see table and text)[4,5]

Table II. PC 5 (continued)
PORPHYRINS AND METALLOPORPHYRINS

REFERENCES

1. **Blumer, M.,** *Anal. Chem.,* 28, 1640, 1956.
2. **Falk, J. E.,** *J. Chromatogr.,* 5, 277, 1961.
3. **Connelly, J. L., Morrison, M., and Stotz, E.,** *J. Biol. Chem.,* 233, 743, 1958.
4. **White, W. I.,** *J. Chromatogr.,* 138, 220, 1977.
5. For applications, see (e.g.) Köst, H.-P. and Benedikt, E., *Z. Naturforsch.,* 37c, 1057, 1982.

THIN-LAYER CHROMATOGRAPHY OF PORPHYRINS

Table II. TLC 1
FREE ACID PORPHYRINS

Layer	L1	L2	L3	L4	L5
Solvent	S1	S2	S3	S4	S5
Technique	T1	T2	T3	T4	T5
Detection	D1	D1	D1	D1	D1
Literature	1	2	3	4	5
Compound			$R_f \times 100$		
Uroporphyrin	60—80	76	—	—	—
Coproporphyrin	30—50	40	—	—	—
Protoporphyrin	10	—	19	5	33[a]
Mesoporphyrin	10	—	25	—	25[a]
Hematoporphyrin	10	—	35	—	3[a]
Deuteroporphyrin	10	2	32	—	28[a]
Deuteroporphyrin monohydroxyethyl-monovinyl	—	—	—	—	17[a]
Pemptoporphyrin	10	—	28	—	—

[a]   Approximate values.

Porphyrins were prepared from human and bovine excreta.

Layer   L1 = talc, pharmacopeia quality;[6] talc suspensions were prepared from 50 g of talc, 50 g of methanol, and 2 g of gypsum, thoroughly mixed in a shaker; the mixture was immediately poured into the applicator and placed in position on a row of clean plates of plain window glass lying on a smooth firm support; immediately after pouring the suspension into the applicator, it was moved slowly forward over the plates to spread an 0.30-mm-thick talc layer; after drying for $^1/_2$ hr at room temperature, the plates were ready for use; they can be stored for several weeks at room temperature if required

           L2 = talc on glass plates (20×20 cm²); the talc is of pharmacopeia quality and is passed through a 0.2-mm sieve prior to use; equal weights of talc and methanol (analytical) are mixed and applied on the plates; layer thickness: 0.25 mm; the plates are dried in air at room temperature

           L3 = talc plates (20×20 cm²); mixture: 40 g of talc and 70 mℓ of methanol, thickness 0.25 mm; dry at room temperature (five plates)

           L4 = 20×20 cm² silica gel Polygram® (Brinkman Instruments, Inc., Westbury, N.Y.) without fluorescent indicator, layer thickness 0.25 mm

           L5 = silica gel G, layer thickness 0.25 mm

Solvent   S1 = acetone-0.5 N HCl = 7:3 (v/v)

           S2 = acetone-0.5 N HCl = 6:4 (v/v)

           S3 = ethanol-lutidine-water = 30:3:67

           S4 = 2,6-lutidine, (practical 95%)

           S5 = benzene-methanol/formic acid = 8.5:1.5 (v/v)/0.3 M

Table II. TLC 1 (continued)
FREE ACID PORPHYRINS

Technique T1 = ascending; porphyrins were applied as solutions in HCl, ammonia, or
HCl-acetone

T2 = ascending, 1 hr; about 5—10 $\mu\ell$ solution containing 0.1—20.0 μg
of porphyrin is employed, diameter of the spots is below 5 mm

T3 = ascending; develop in the tank for at least 12 hr at a temperature range
of 20—25°C, plates can be left to develop overnight

T4 = ascending

T5 = ascending, 21—23°C, 30—45 min, solvent front 10 cm from start

Detection D1 = visual; spots of free porphyrins are detected by their red fluorescence
in UV light in a dark room

REFERENCES

1. **With, T. K.,** *J. Chromatogr.,* 42, 389, 1969.
2. **With, T. K.,** *Clin. Biochem.,* 1, 30, 1967.
3. **Belcher, R. V., Smith, S. G., Mahler, R., and Campbell, J.,** *J. Chromatogr.,*
53, 279, 1970.
4. **Ellsworth, R. K.,** *Anal. Biochem.,* 32, 377, 1969.
5. **Ellfolk, N. and Sievers, G.,** *J. Chromatogr.,* 25, 373, 1966.
6. *Pharmacopeia Nordica,* Vol. 2, Busch, Copenhagen, 1963, 594.

Table II. TLC 2
PORPHYRIN ESTERS

Compound	L1	L1	L1	L2	L2	L2	L3	L3	L3	L4	L5	L6	L7	L8
Solvent	S1	S2	S3	S1	S2	S3	S4	S5	S6	S4	S4	S7	S8	S9
Technique	T1	T1	T1	T2	T2	T2	T3	T1	T4	T3	T3	T1	T1	T5
Detection	D1	D1	D1	D1	D1	D1	D1	D1	D1	D1	D1	D1	D1	D1
Literature	1	1	1	2	2	2	3	4	4	3	3	5	6	7
							Rc × 100							
Uroporphyrin octamethyl ester	6	8	90	60[a]	80[a]	90[a]	—	15	—	—	—	—	—	14[a]
Heptacarboxylic porphyrin heptamethyl ester	—	—	—	—	—	—	—	—	—	—	—	—	—	19[a]
Hexacarboxylic porphyrin hexamethyl ester	—	—	—	—	—	—	—	—	—	—	—	—	—	25[a]
Pentacarboxylic porphyrin pentamethyl ester	—	—	—	—	—	—	—	—	—	—	—	—	—	36[a]
Coproporphyrin-tetramethyl ester	2+	30	55	20+	30+	55+	—	—	—	—	—	—	—	49[a]
Tricarboxylic porphyrin trimethyl ester	—	—	—	—	—	—	—	—	—	—	—	54	—	65[a]
Protoporphyrin IX monomethyl ester	—	—	—	—	—	—	92	—	—	55	26	84	—	70[a]
Protoporphyrin IX dimethyl ester	—	—	—	10+	10+	10+	92	—	—	24	19	—	—	—
Mesoporphyrindimethyl ester	—	—	—	—	—	—	14	—	—	3	—	—	—	—
Hematoporphyrin-dimethyl ester	—	—	—	—	—	—	—	—	—	—	—	—	20	—
Acetyl derivative	—	—	—	—	—	—	—	—	—	—	—	—	75	—
Deuteroporphyrin-dimethyl ester	—	—	—	—	—	—	88	—	—	33	21	—	—	—
Tetramethyl-tetrapropylporphyrin	—	—	—	—	—	—	—	98	—	—	—	—	—	—
Meso-tetraphenyl-porphyrin	—	—	—	—	—	—	—	—	9	—	—	—	—	—

[a] Approximate values.

Table II. TLC 2 (continued)
PORPHYRIN ESTERS

Layer   L1 = talc plates; 50 g of sieved talc and 3—5 g of burned gypsum in 50 g of methanol
           L2 = 50 g of talc, 50 g of methanol, and 2 g of gypsum, thoroughly mixed in a shaker; the mixture was immediately poured into
                the applicator; the layer was dried for 30 min at room temperature, thickness: 0.30 mm
           L3 = 0.25 mm silica gel G slurried in water
           L4 = 0.25 mm silica gel G slurried in 3% (w/v) $FeSO_4 \cdot 7 H_2O$
           L5 = 0.25 mm silica gel G slurried in 0.3% (v/v) sulfuric acid
           L2 = 20 × 20 cm silica gel Polygram® without fluorescent indicator
           L4 = silica gel H
           L8 = silica gel F_{254}

Solvent   S1 = chloroform-methanol = 1:1
           S2 = ethanol-ethyl acetate = 3:2
           S3 = acetone-glacial acetic acid-pyridine = 1:1:1
           S4 = benzene-methanol = 100:5
           S5 = acetone-*n*-hexane = 3:7
           S6 = benzene-acetone = 99:1
           S7 = 2.6 lutidine (practical 95%)
           S8 = benzene-ethylacetate-methanol-butanol = 82:14:3:1
           S9 = benzene-ethyl acetate-methanol = 85:13.5:1.5

Technique T1 = ascending
           T2 = ascending; porphyrin esters were applied as chloroform solutions
           T2 = ascending; 1—5 μg of porphyrin was applied to the plate in 10 μℓ benzene from a capillary pipette; the plates were devel-
                oped in cylindrical glass tanks 11 ø × 22 cm for about 70 min at 20°C; this allowed the solvent front to travel about 17 cm
           T4 = ascending; 2 hr in a tank that has been equilibrated with the developing solvent for at least 1 hr
           T5 = ascending, 40 min at 21°C

Detection D1 = visual; spots are detected by their red fluorescence in UV light in a dark room

REFERENCES

1. **With, T. K.,** Thin layer chromatography of porphyrins and their esters on talc plates, private communication to Burnham, B. F.
2. **With, T. K.,** *J. Chromatogr.,* 42, 389, 1969.
3. **Henderson, R. W. and Morton, T. C.,** *J. Chromatogr.,* 27, 180, 1967.
4. **Jackson, A. H.,** *Semin. Hematol.,* 14, 193, 1977.
5. **Ellsworth, R. K.,** *Anal. Biochem.,* 32, 377, 1969.
6. **Doss, M. and Bürger, H.,** *Z. Physiol. Chem.,* 348, 936, 1967.
7. **Doss, M.,** *Z. Klin. Biochem.,* 8, 197, 1970.

Table II. TLC 3
METALLOPORPHYRINS

Layer	L1	L1	L²/₃	L²/₃	L4
Solvent	S1	S2	S3	S4	S5
Technique	T1	T1	T2	T2	T3
Detection	D1	D1	D1	D1	D1
Literature	1	1	2	2	3

Compound			$R_f \times 100$		
Protoporphyrin-Fe-complex (protochemin)	5	1	47	28	31
Monomethyl ester	45	35	—	—	—
Dimethyl ester	95	55	—	—	—
Mesoporphyrin-Fe-complex (mesohemin)	—	—	51	32	49
Hematoporphyrin-Fe-complex (hematohemin)	—	—	—	13	81
Deuteroporphyrin-Fe-complex (deuterohemin)	—	—	40	20	43
Diacetyl derivative	—	—	12	8	—

Layer	L1	= Eastman chromatogram sheets (silica gel 6061)
	L²/₃	= 0.25 mm silica gel thin-layer plates/2 mm silica gel thick-layer plates
	L4	= polyamide poly-ε-caprolactam coated on ethylene terephthalate
Solvent	S1	= 2,6-lutidine-water = 20:1
	S2	= hexane-chloroform-methanol = 1:1:0.2
	S3	= *n*-butanol-water-acetic acid = 50:1.5:1.4
	S4	= hexane-*n*-propanol-acetic acid = 10:5:1.5
	S5	= methanol-acetic acid = 97.5:2.5
Technique	T1	= ascending, 20°C
	T2	= ascending; samples of the hemins = 1 μℓ of approximately 2×10^{-3} M pyridine solutions
	T3	= ascending; samples of the hemins (up to 0.25 μℓ of 1×10^{-3} M pyridine solutions were applied 1 cm from the end of 4 × 8-cm polyamide plates to give an initial spot diameter of 1—2 mm
Detection	D1	= visual

REFERENCES

1. **Asakura, T. and Lamson, D. W.,** *Anal. Biochem.*, 53, 448, 1973.
2. **Dinello, R. K. and Dolphin, D. W.,** *Anal. Biochem.*, 64, 444, 1975.
3. **Lamson, D. W., Coulson, A. F. W., and Yonetani, T.,** *Anal. Chem.*, 45, 2273, 1970.

HIGH PERFORMANCE THIN-LAYER CHROMATOGRAPHY OF PORPHYRINS

Table II. HPTLC 1
PORPHYRIN ESTERS

Layer	L1	L1	L1
Solvent	S1	S2	S3
Technique	T1	T1	T1
Detection	D1	D1	D1
Literature	1	1	1

Compound	$R_f \times 100$		
Uroporphyrin I-octamethyl ester	36	28	12
Heptacarboxylporphyrin I-hepta-methyl ester	41	30	17
Hexacarboxylporphyrin I-hexa-methyl ester	46	46	25
Pentacarboxylporphyrin I-penta-methyl ester	53	55	32
Coproporphyrin I tetramethyl ester	60	65	40
Mesoporphyrin IX dimethyl ester	75	77	59

Layer	L1	=	HPTLC-Kieselgel 60 (E. Merck) without fluorescence indicator
Solvent	S1	=	benzene-petrol ether (40—60°C b.p.)-methanol-ethyl acetate = 48.5:40.0:10.5:9.0
	S2	=	carbon tetrachloride-ethyl acetate = 1:1
	S3	=	hexane-butanone-2-acetic acid = 15:7.5:1.5
Technique	T1	=	ascending on HPTLC plates for Nano-DC (10 × 10 cm), E. Merck
Detection	D1	=	visual

REFERENCE

1. **Benedikt, E. and Köst, H.-P.**, unpublished.

LIQUID CHROMATOGRAPHY OF PORPHYRINS

Table II. LC 1
PORPHYRIN ESTERS

Packing		P1	P1	P2	P3	P3	P4	P4
Column	length (cm	10	10	10	10	10	10	10
	diameter (cm)	1.5	1.5	1.5	1.5	1.5	1.5	1.5
	material	G	G	G	G	G	G	G
Solvent[a]$_1$		S1-1	S2-1	S3-1	S4-1	S5-1	S6-1	S7-1
Solvent$_2$		S1-2	S2-2	S3-2	S4-2	S5-2	S6-2	S7-2
Solvent$_3$		S1-3	S2-3	S3-3	S4-3	S5-3	S6-3	S7-3
Temperature		Ambient	Ambient	Ambient	Ambient	Ambient	Ambient	Ambient
Detection		D1	D1	D1	D1	D1	D1	D1
Literature		1	1	1	1	1	1	1

Order of elution of components (1 = first, 2 = second, 3 = third)

Compound

Uroporphyrin-octamethyl esters	3	3	3	3	3	1	1
Coproporphyrin-tetramethyl esters	2	2	2	2	2	2	2
Dicarboxylic porphyrin-dimethyl esters	1	1	1	1	1	3	3

[a] Stepwise elution with three consecutive solvents. Each solvent system elutes one component of the system of porphyrins given above in the order of elution given. Intermediate compounds, e.g., heptacarboxylic porphyrin heptamethyl ester, etc., will elute in-between uroporphyrin octamethyl ester and coproporphyrin tetramethyl ester.

Packing	P1	= aluminum oxide grade IV
	P2	= aluminum oxide grade II
	P3	= calcium carbonate grade V
	P4	= magnesium oxide grade III
Solvent	S1-1	= benzene-chloroform = 10:1
	S1-2	= benzene-chloroform = 1:1
	S1-3	= chloroform-methanol = 100:1
	S2-1	= petroleum ether (b.p. 40—60°C)-chloroform = 1:1
	S2-2	= petroleum ether (b.p. 40—60°C)-chloroform = 1:6
	S2-3	= chloroform-methanol = 100:1
	S3-1	= petroleum ether (b.p. 40—60 °C)-chloroform = 6:1
	S3-2	= petroleum ether (b.p. 40—60°C)-chloroform = 4:1
	S3-3	= petroleum ether (b.p. 40—60°C)-chlorodorm = 1:1
	S4-1	= benzene
	S4-2	= benzene-chloroform = 10:6
	S4-3	= chloroform-methanol = 100:1
	S5-1	= petroleum ether (b.p. 40—60°C)-chloroform = 3:1
	S5-2	= petroleum ether (b.p. 40—60°C)-chloroform = 1:1
	S5-3	= petroleum ether (b.p. 40—60°C)-chloroform = 1:7
	S6-1	= chloroform-methanol = 100:0.5
	S6-2	= chloroform-methanol = 100:1
	S6-2	= chloroform-methanol = 100:2
	S7-1	= benzene-methanol = 100:4
	S7-2	= benzene-methanol = 100:8
	S7-3	= benzene-methanol = 100:10
Detection	D1	= visual

REFERENCE

1. **Nicholas, R. E.**, *Biochem. J.*, 48, 309, 1951.

Table II. LC 2
PORPHYRINS, HEMINS, AND ESTERS
(SILICA GEL)

Packing		P1	P1	P1
Column	length (cm)	60	60	15
	diameter (cm)	2.5	2.5	n.a.
	material	G	G	G
Solvent		S1	S2	S3
Temperature		Ambient	Ambient	Ambient
Detection		D1	D1	D1
Literature		1	1	2

Compound		V_e (mℓ)		
Protoporphyrin IX		—	—	+ + +
Fe-complex		—	575	—
Monomethyl ester		—	—	+ +
Monomethyl ester-Fe-complex		300	—	—
Dimethyl ester		—	—	+
Dimethyl ester-Fe-complex		230	—	—

Note: + elutes first, + + elutes second, + + + elutes last

Packing	P1	=	silica gel
Solvent	S1	=	hexane-chloroform-methanol = 1:1:0.3
	S1	=	chloroform-methanol = 1:1
	S3	=	2,6-lutidine-water = 12:1
Detection	D1	=	visual

REFERENCES

1. **Asakura, T. and Lamson, D. W.**, *Anal. Biochem.*, 53, 448, 1973.
2. **Ellsworth, R. K.**, *Anal. Biochem.*, 32, 377, 1969.

Table II. LC 3
PORPHYRINS AND PORPHYRIN ESTERS (SEPHADEX)

Packing		P1	P1	P1	P1	P2
Column	length (cm)	60	60	60	60	87
	diameter (mm)	8	8	8	8	25
	material	G	G	G	G	G
Solvent		S1	S2	S3	S4	S5
Temperature		Ambient	Ambient	Ambient	Ambient	Ambient
Detection		D1	D1	D1	D1	D2
Technique		T1	T1	T1	T1	T2
Literature		1	1	1	1	2

Compound			V_e (mℓ)		
Uroporphyrin	18	—	23	—	—
Octamethyl ester.	—	—	—	—	203
Octa-*n*-butyl ester.	—	—	—	—	163
Coproporphyrin	78	25	34	32	—
Tetramethyl ester	—	—	—	—	225
Tetra-*n*-butyl ester	—	—	—	—	176
Mesoporphyrin dimethyl ester	—	—	—	—	246
Di-*n*-butyl ester	—	—	—	—	215
Hematoporphyrin	—	—	—	109	—
Deuteroporphyrin	570	38	—	—	—
Dimethyl ester	—	—	—	—	271
Di-*n*-butyl ester	—	—	—	—	238
Porphyrin C	—	—	—	65	—

Packing   P1   =   Sephadex G-25 (Dextran gel)
Packing   P2   =   Sephadex LH-20
Solvent   S1   =   0.2 *M* borate buffer pH 8.6
         S1   =   0.002 *M* borate buffer pH 8.6
         S3   =   0.01 *M* borate buffer pH 8.6
         S4   =   0.05 *M* borate buffer pH 8.6
         S5   =   chloroform-methanol = 1:1, containing 1 g Tris base per liter
Detection   D1   =   optical absorption at the Soret peak (cells of 1-cm light path)
           D2   =   visual
Technique   T1   =   for application to the column, porphyrins were dissolved in a minimum quantity (0.5—1 mℓ) of 0.2 *M* sodium borate buffer pH 8.6; to each 2-mℓ fraction, 2 mℓ of 3 *N* HCl were added and the mixture was further diluted with 1.5 *M* HCl, if necessary
           T2   =   The elution profile was taken from porphyrin ester run separately on a 2.5 × 87-cm column of Sephadex LH-20. Solvent, chloroform-methanol containing 1 g Tris base per liter.

REFERENCES

1. **Rimington, C. and Belcher, R. V.**, *J. Chromatogr.*, 28, 112, 1967.
2. **Bachmann, R. C. and Burnham, B. F.**, *J. Chromatogr.*, 41, 344, 1969.

HIGH PERFORMANCE LIQUID CHROMATOGRAPHY OF PORPHYRINS

Table II. HPLC 1
PORPHYRINS AND PORPHYRIN ESTERS

Packing	P1	P2	P3	P3	P3	P4	P5	P5	P5
Temperature	Ambient	Ambient	Ambient	Ambient	Ambient	Ambient	Ambient	Ambient	Ambient
Solvent	S1	S2	S3	S4	S5	S6	S7	S8	S9
Flow rate (ml $\times$ min^{-1})	0.5	0.8—1.3	1	1	1	1.5	2—3[a]	2	2
Column length (cm)	25	25	183	183	183	20	n.a.	30	30
diameter(mm) (ID)	26	20	n.a.	n.a.	n.a.	4	n.a.	3.9	3.9
form	Straight	Straight	n.a.	n.a.	n.a.	Straight	Straight	Straight	Straight
material	SS	SS	SS 316	SS 316	SS 316	SS	SS	SS	SS
Detector	D1	D2	D3	D3	D3	D3	D4	D5	D5
Literature	1	2	3	3	3	4	5	6	6
Compound					t_R (min)				
Uroporphyrin	1.8[b,c]	—	—	—	—	—	—	—	—
Octamethyl ester	—	8.6	5.8	—	—	7	6.4	6.2[d]	8.6[c]
Heptacarboxylic porphyrin	2.2[b,c]	—	—	—	—	—	—	—	—
Heptamethyl ester	—	5.2	—	—	—	4.5	4.1	4.6[d]	6.2[c]
Hexacarboxylic porphyrin	3.4[b,c]	—	—	—	—	—	—	—	—
Hexamethyl ester	—	3.2	—	—	—	3.5	2.8	3.5[d]	4.5[c]
Pentacarboxylic porphyrin	5.0[b,c]	—	—	—	—	—	—	—	—
Pentamethyl ester	—	2.0	—	—	—	2.1	2.0	2.8[d]	3.4[c]
Coproporphyrin	7.4[b,c]	—	—	—	—	—	—	—	—
Tetramethyl ester	—	1.2	—	—	—	1.8	1.4	2.2[d]	2.7[c]
Protoporphyrin IX-dimethyl ester	—	—	—	—	7.1	—	—	—	—
Mesoporphyrin IX	12.2[b,c]	—	—	—	—	—	—	—	—
Dimethyl ester	—	—	—	—	6.3	—	—	1.8[d]	2.0[c]
Harderioporphyrin-trimethyl ester	—	—	—	13.0	—	—	—	—	—
Isoharderioporphyrin-trimethyl ester	—	—	—	11.2	—	—	—	—	—

a   Flow rate at 2 mℓ × min^{-1} until the pentacarboxyl porphyrin has passed, then at 3 mℓ × min^{-1}.

b   Standard free acid porphyrins (Porphyrin Products, Logan, Utah) in acetone-0.1 N HCl = 10:1, v/v; 0.04 nmol of each compound.

c   Approximate values.

d   k' = 0.62 t_R-0.96.

e   k' = 0.66 t_R-0.92.

d,e   The capacity factor k' is defined as:
k$_i'$ = (V$_i$-V$_o$) × V$_o$$^{-1}$
V$_i$ = elution volume of the ith component of mixture
V$_o$ = void volume of column

Packing		
P1	=	Perkin Elmer Silica A 10 μm
P2	=	Micro Pak CN, 10 μm (Varian)
P3	=	Corasil II
P4	=	5 μm Partisil
P5	=	5 μm Porasil

Solvent		
S1	=	elution was performed for 25 min with a linear gradient of acetone-dilute acetic acid (2—90%A) as follows: acetone-0.23 M acetic acid = 70:30, v/v delivered from pump B in the reverse-pump exchange mode and 10% acetic acid delivered from pump A in the reversed-pump exchange mode
S2	=	ethyl acetate-n-heptane-isopropanol = 40:60:0.5
S3	=	ethyl acetate-petrol ether (b.p. 60—80°C) = 2.5:97.5
S4	=	ethyl acetate-petrol ether (b.p. 60—80°C) = 25:75
S5	=	ethyl acetate-petrol ether (b.p. 60—80°C) = 30:70
S6	=	ethyl acetate-cyclohexane = 60:40
S7	=	benzene-ethyl acetate-chloroform = 70:10:20
S8	=	benzene-ethyl acetate-methanol = 85:13.5:0.75
S9	=	ethyl acetate-heptane = 55:45

Table II. HPLC 1 (continued)
PORPHYRINS AND PORPHYRIN ESTERS

Detector	D1	= Perkin Elmer series 3 liquid chromatograph equipped with an LC-55 UV-Vis digital spectrophotometer; porphyrins were detected by their absorbance at 403 nm and by fluorescence in a Perkin Elmer fluorescence spectrophotometer model 240 A
	D2	= spectrophotometric at 400—402 nm (Variscan)
	D3	= Cecil variable wavelength detector set at 400 nm and fitted with a 10-μℓ flow cell
	D4	= Beckman model 25 spectrophotometer set at 400 nm and fitted with a microflow cell
	D5	= Waters 44 monitor set at 403 nm

REFERENCES

1. **Longas, M. O. and Pols-Fitzpatrick, M. B.,** High-pressure liquid chromatography of plasma free acid porphyrins, *Anal. Biochem.,* 104, 268, 1980.
2. **Miller, V. and Malina, L.,** High-performance liquid chromatographic analysis of biologically important porphyrins, *J. Chromatogr.,* 145, 290, 1978.
3. **Evans, N., Games, D. E., Jackson, A. H., and Matlin, S. A.,** Applications of high pressure liquid chromatography and field desorption mass spectrometry in studies of natural porphyrins and chlorophyll derivatives, *J. Chromatogr.,* 115, 325, 1975.
4. **Evans, N., Jackson, A. H., Matlin, S. A., and Towill, R.,** High performance liquid chromatographic analysis of porphyrins in clinical materials, *J. Chromatogr.,* 125, 345, 1976.
5. **Petryka, Z. J. and Watson, C. J.,** A new rapid method for isolation of naturally occurring porphyrins and their quantitation after high performance liquid chromatography, *Anal. Biochem.,* 84, 173, 1978.
6. **Straka, J. G., Kushner, J. P., and Burnham, B. F.,** High-performance liquid chromatography of porphyrin esters. Identification of mixed esters generated in sample preparation, *Anal. Biochem.,* 111, 269, 1981.

Table II. HPLC 2
PORPHYRINS AND PORPHYRIN ESTERS: SEPARATION OF ISOMERS

Column packing	P1	P1	P2	P2	P3
Temperature	Ambient	Ambient	Ambient	Ambient	Ambient
Solvent	S1	S2	S3	S4	S5$^-$
Flow rate (mℓ $\times$ min^{-1})	1.5^{a,b}	1.5^b	0.9	1.0	1.5
Column length (cm)	30	30	30	30	30
diameter (mm, (I.D.)	3.9	3.9	4.0	3.9	3.9
form	st	st	st	st	st
material	SS	SS	SS	SS	SS
Detector	D1	D1	D2	D2	D1
Literature	1	1	2	2	3

Compound			t_R **(min)**		
Uroporphyrin					
I-isomer	—	—	22	11	9.5
II-isomer	—	—	38	—	—
III-isomer	—	—	29	14	12
IV-isomer	—	—	31.5	—	—
Uroporphyrin-octamethyl ester					
I-isomer	98	—	—	—	—
III-isomer	102	—	—	—	—
Heptacarboxylic porphyrin					
I-isomer	—	—	—	—	3.8
I-isomer heptamethyl ester	—	5	—	—	—
III-isomer	—	—	—	—	5.3
Hexacarboxylic porphyrin					
I-isomer	—	—	—	—	3.2
I-isomer hexamethyl ester	—	6	—	—	—
III-isomer	—	—	—	—	5.3
Pentacarboxylic porphyrin					
I-isomer	—	—	—	—	2.8
I-isomer pentamethyl ester	—	7	—	—	—
III-isomer	—	—	—	—	4.7
Coproporphyrin					
I-isomer	—	—	—	—	3.6
I-isomer tetramethyl ester	—	10	—	—	—
III-isomer	—	—	—	—	9.5

[a] Two columns in series.

[b] Recycling mode.

Packing	P1	=	μ-Porasil (Waters column model A, LC 202)
	P2	=	μ-Bondapak C$_{18}$ (Waters); column equipped with a Whatman 50 $\times$ 4.6-mm CO:PEL (ODS) 37- to 50-μm precolumn
	P3	=	μ-Bondapak (Waters)
Solvent	S1	=	*n*-heptane-glacial acetic acid-acetone-water = 90:60:30:0.5
	S2	=	*n*-heptane-glacial acetic acid-acetone-water = 90:60:90:0.5
	S3	=	acetonitrile-10^{-2} *M* phosphate buffer pH 6.95 = 4:96
	S4	=	acetonitrile-10^{-2} *M* phosphate buffer pH 6.95 = 5:95
	S5$^+$	=	separation of *n*-carboxylic porphyrins, n = 8...4

Table II. HPLC 2 (continued)
PORPHYRINS AND PORPHYRIN ESTERS: SEPARATION OF ISOMERS

No. of carboxyl groups	Percent acetonitrile in phosphate buffer (10^{-2} M, pH 6.85; 5 × 10^{-4} M EDTA
8	2.5
7	5.0
6	10.0
5	12.5
4	15.0

Detector   D1 = Waters 440 detector, compounds monitored at 405 nm
            D2 = Schoeffel FS 970 fluorometer

REFERENCES

1. **Bommer, J. C., Burnham, B. F., Carlson, R. E., and Dolphin, D.,** The chromatographic separation of uroporphyrin I and III octamethyl esters, *Anal. Biochem.*, 95, 444, 1979.
2. **Wayne, A. W., Straight, R. C., Wales, E. E., and Englert, E., Jr.,** Isomers of uroporphyrin free acids separated by HPLC, *J. HRC and CC*, 2, 621, 1979.
3. **Prior, M. et al.,** personal communication.

Table II. HPLC 3
SURVEY OF SAMPLE WORKUP FOR HPLC OF PORPHYRINS
(CONDENSED TABLE)

Procedure	Ref.
Preadsorption on silica gel	1
Separation of uroporphyrin isomers	2
Direct injection after addition of mesoporphyrin standard	3
Solvent extraction method	4
Direct injection of urine preserved with sodium carbonate/tetrasodium EDTA; external standard solutions	5
Isolation of uroporphyrins from urine and feces; conversion into methyl esters (methanol/sulfuric acid)	6
Coprecipitation of uroporphyrin and coproporphyrin with calcium hydroxide; redissolution in hydrochloric acid; final pH<2	7
Sample preparation using purchased porphyrins	7
Adsorption of porphyrins onto talc at pH 3.5; (urine) feces are extracted with HCl/diethyl ether/water, then use of talc as described for urine sample preparation	8
Urine is injected directly without sample pretreatment; feces are pretreated with HCl/diethyl ether/water and the water injected directly	9
Direct injection onto HPLC	10
Urines low on porphyrin are treated with talc. The porphyrins are eluted with acetone/1 M/ℓ HCl (1:9 v/v); after removal of acetone in a stream of nitrogen at 45°C the remaining solution is used for HPLC analysis	
Feces are freeze-dried and an aliquot is extracted with 60 mM t-butyl ammonium phosphate in methanol/water (80:20); aliquots of the solution are injected	11
Standards are acidified with 1 N or 6 N HCl or conc H$_2$SO$_4$; analyte is first adsorbed on a precolumn and then eluted with 5% acetonitrile in 0.01 M phosphate buffer, pH 6.95, at a flow rate of 1 mℓ/min	12
Separation of coproporphyrin isomers I to IV	13
Urine is acidified (pH 1.0) with 6M/ℓ HCl and adsorbed at a 12-cm XAD-2 precolumn; the porphyrins are eluted with 3 × 5 mℓ of 2% conc HCl in acetone; esterification is with 10% H$_2$SO$_4$ in methanol; workup with chloroform/water; feces are freeze-dried, treated with 10% H$_2$SO$_4$ in methanol, and the methyl esters extracted into chloroform	14

REFERENCES

1. **Barwise, A. J. G., Whitehead, E. V.,** Manuscript presented at meeting of Division of Petroleum Chemistry, American Chemical Society, Houston, Texas, March 1980.
2. **Bommer, J.,** personal communication, 1984.
3. **Ford, R. E., Ou, C.-N., and Ellefson, R. D.,** *Clin. Chem.,* 27, 397, 1981.
4. **Gaetani, E., Laureri, C. F., Vitto, M., Rocci, E., Gibertini, P., Farina, F., Cassanelli, M., and Ventura, E.,** *J. Chromat.,* 231, 425, 1982.
5. **Hill, R. H., Jr., Bailey, S. L., and Needham, L. L.,** *J. Chromat.,* 232, 251, 1982.
6. **Jackson, A. H., Rao, K. R. N., Smith, S. G.,** *Biochem. J.,* 203, 515, 1982.
7. **Johansson, B. and Nilsson, B.,** *J. Chromat.,* 229, 49, 1982.
8. **Lim, C. K. and Chan, J. Y. Y.,** *J. Chromat.,* 228, 305, 1982.
9. **Lim, C. K., Rideout, J. R., and Wright, D. J.,** *Biochem. J.* 211, 435-438 (1983).
10. **Lim, C. K. and Peters, T. J.,** *Clin. Chim. Acta,* 139, 55, 1984.
11. **Meyer, H. D., Jacob, K., Vogt, W., and Knedel, M.,** *J. Chromatogr.,* 217, 473, 1981.
12. **Wayne, A. W., Straight, R. C., and Englet, E., Jr.,** J. HRC and CC 2, 621, 1979.
13. **Wright, D. J., Rideout, J. M., Lim, C. K.,** *Biochem. J.,* 209, 553, 1983.
14. **Wilson, J. H. P., Van den Berg, J. W. O., Edixhoven-Bosdijk, A., and van Gastel-Quist, L. H. M.,** *Clin. Chim. Acta,* 89, 165, 1978.

HYPERPRESSURE GAS CHROMATOGRAPHY (HPGC) OF PORPHYRINS AND METALLOPORPHYRINS

Table II. HPCG 1
PORPHYRINS AND METALLOPORPHYRINS

Column packing	P1	P1	P2	P3	P3
Temperature (°C)	145	140	153	145	140
Gas	G1	G1	G1	G1	G1
Flow rate (mℓ × min^{-1})	162	366	522	136	395
Pressure (psi g)	1500	1475	1950	1350	1550
Column length (in.)	64	31	25	46	43
diameter (in., I.D.)	0.125	0.125	0.125	0.125	0.125
form	n.a.	n.a.	n.a.	n.a.	n.a.
material	M_1	M_1	M1	M1	M1
Detector	D1	D2	D2	D1	D2
Literature	1	2	2	1	2

Compound			t_R (min)		
Mesoporphyrin IX-dimethyl ester	—	16.0	11.0	—	16.1
Mesoporphyrin IX-diethyl ester	—	24.8	8.2	—	11.4
Deuteroporphyrin IX-dimethyl ester	—	36.7	16.6	—	19.9
Etioporphyrin II	29.3	5.0	7.2	21.4	9.1
Mg (II) complex	26.4	4.7	6.8	20.5	7.4
Cu (II) complex	29.3	5.6	7.0	18.6	6.8
Ag (II) complex	35.2	—	—	20.5	—
Zn (II) complex	61.5	—	—	19.2	—
TiO complex	67.4	—	—	23.3	—
VO complex	70.3	12.0	13.4	43.9	19.5
Mn (II) complex	61.5	—	—	20.5	—
Co(II) complex	32.2	—	—	16.0	—
Co (III) complex	41.0	—	—	20.5	—
Ni (II) complex	38.0	6.4	8.2	20.5	9.0
Pd (II) complex	85.0	—	—	40.9	—
Pt (II) complex	67.4	—	—	31.6	—
Deoxophyllerythrin	—	21.6	17.8	—	18.6
Deoxophyllerythrin-etioporphyrin	—	—	14.4	—	—

Column packing	P1	= 10% Epon 1001 on Chromosorb W
	P2	= 12% Versamid 900 on Chromosorb W
	P3	= 20% silicone gum rubber XE-60 on Chromosorb W
Gas	G1	= freon or Genetron 12 (dichlorodifluoromethane)
Column material	M1	= titanium or stainless steel
Detector	D1	= Beckman DK 2 recording spectrophotometer

REFERENCES

1. **Karayannis, N. M. and Corwin, A. H.**, *J. Chromatogr.*, 47, 247, 1970.
2. **Karayannis, N. M. and Corwin, A. H.**, *Anal. Biochem.*, 26, 34, 1968.

PAPER ELECTROPHORESIS (PEL) OF PORPHYRINS

Table II. PEL 1
PORPHYRINS

Paper	P1	P2
Buffer	B1	B2
Voltage	V1	V2
Current (mA)	5—10	?
Time (hr)	1—3	3
Technique	T1	T2
Literature	1	2

Compound[a]	Relative mobilities	
Uroporphyrin	1.00[b,c]	1.00[b]
Heptacarboxylic porphyrin	—	0.93[b]
Hexacarboxylic porphyrin	—	0.75[b]
Pentacarboxylic porphyrin	—	0.42[b]
Coproporphyrin	0.10[b]	0.15[b]
Protoporphyrin	0.02[b]	--

[a] Porphyrin solutions or urines.
[b] Calculated from the distance of the respective spot from the origin (near the cathode) relative to uroporphyrin.
[c] Migrating distance of uroporphyrin ca. 12.5 cm.

Paper   P1 = Whatman No. 1, sheets 20—25 × 25 cm
        P2 = Whatman 3 MM
Buffer   B1 = 0.05 M barbiturate buffer pH 8.6; 200—250 mℓ were placed in each of the electrode vessels
        B2 = 0.04 M Na_2CO_3/10^{-4} EDTA
Voltage   V1 = 7.5—8.0 V × cm^{-1}, corresponding to ca. 200 V for the dimensions of the paper employed, but up to 500 V could be applied with corresponding increased speed of movement
        V2 = 10 V × cm^{-1}
Technique   T1 = no cooling is used, the paper is supported by two glass rods running parallel to the long side of the tank from one of its short sides to the other side; it is placed ca. 3 cm below the lid of the tank; the ends of the paper dip 0.5—1 cm below the surface of the buffer solution without touching the sides of the glass vessel
        T2 = the paper is freely suspended horizontally between glass supports at the ends, 20 cm apart; the porphyrins are applied from solution in conc. NH_4OH at the cathode end of the wet paper in streaks containing 0.5—50 μg of porphyrin per centimeter; equilibration time prior to switching on the power supply is 1 hr

REFERENCES

1. **With, T. K.,** *Scand. J. Clin. Lab. Invest.,* 8, 113, 1956.
2. **Lockwood, W. H. and Davies, J. L.,** *Clin. Chim. Acta,* 7, 301, 1962.

Part III: Chlorophylls

Hugo Scheer

Chromatographic Methods for the Separation of Chlorophylls

> . . . the pigments are separated from the top to the bottom (of the calcium carbonate column) in differently colored zones . . . I call such a preparation a *chromatogram,* and the corresponding method the *chromatographic method.* **M. Tswett, 1906**

INTRODUCTORY, FUNCTIONAL, AND BIOSYNTHETIC CONSIDERATIONS

Introduction

Tswett's separation of a plant pigment extract into colored bands of individual chlorophylls and carotenoids, due to their differential velocity on a stationary solid or liquid phase in a mobile liquid phase, marked the invention of chromatography.[1] Of the more than 100 tested adsorbents listed already in the first publication, $CaCO_3$ and powdered sugar were recommended. The prevalence of the latter for preparative chromatography of chlorophylls, until today, marks the excellence of this choice. Sugar columns are still used for preparative work, and their analytical importance has been diminished only recently with the adaptation of high-performance liquid chromatography (HPLC).

The literature on chlorophyll chromatography prior to 1974 has been excellently and critically reviewed by Svec.[2a] The three survey tables, III. PC 1, III. TLC 1, and III. LC 1, adapted from his work and completed with some more recent results, give a survey of preparative and analytical chromatographic procedures. All chlorophylls are rather unstable pigments which require certain precautions (see Table III. 6). The article of Svec is recommended, too, for the discussion of many practical aspects involving their extraction and safe handling. The analysis of bacteriochlorophylls has recently been reviewed by Oelze[2b] and a short critical review on chlorophyll chromatography is available.[2c] This review covers pertinent earlier work, but is mainly focused on major developments of the past 15 years. It is separated into five parts. In the first two parts, the structure and spectroscopic characterization of chlorophylls are briefly surveyed. It is followed by a general discussion of the separation and estimation of chlorophylls. The detailed information on chlorophyll chromatography, focusing on more recent developments, is incorporated into the tables and an appendix on some special experimental techniques.

The review deals entirely with free chlorophylls, with the discussions focused on the chlorophylls and their derivatives isolated from natural sources. The procedures can generally be extended to synthetic or more extremely modified pigments.[3]* The free chlorophylls are usually isolated from natural sources by solvent extraction, preferentially with acetone and/ or methanol being used (see Appendix A). However, most, if not all, chlorophylls occur naturally complexed to proteins, and often in a single species a single pigment (e.g., chlorophyll *a*) is associated with a variety of different polypeptides serving different functions. The solvent extraction dissociates these chlorophyll-protein complexes and, thus, results in a net pigment analysis. A better insight into the structures and functions of the chlorophylls in vivo has recently been obtained with the isolation of the intact pigment-protein complexes. These isolations are based on protein separation techniques which are beyond the scope of this review. For leading references on this important aspect of chlorophylls, see References 4a to 4i.

Structure, Function, Occurrence, Biosynthesis

Chlorophylls are the main photosynthetic pigments of plants, most algae, and photosynthetic bacteria (see Figures 1 and 2). They comprise a comparably small group of structurally closely related, but ubiquitous, magnesium porphyrins, all of which carry an extra five-

---

* The extensive work published on the chemistry and chromatography of chlorophyll-related pigments precludes a comprehensive survey. For leading references, examples from only a few laboratories are given (References 3a to 3g).

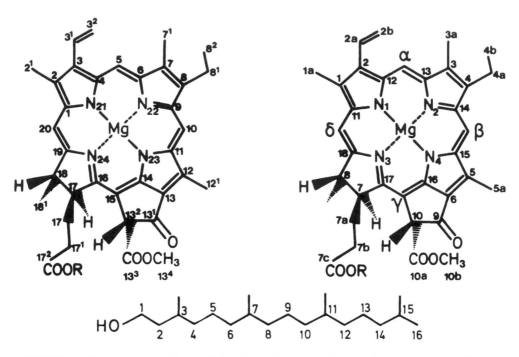

FIGURE 1.   Nomenclature of chlorophylls based on a "numbers-only system" (left), compared with the older nomenclature (right). For further information, see text. The structure and numbering of phytanol is shown below. Most chlorophylls contain phytol (= Δ 2 - phytaenol) as esterifying alcohol (R).

membered isocyclic ring derived from the 13-propionic acid side chain* of protoporphyrin (Table III. 1, Figure 3). They include members of the porphyrin, the 17,18-dihydroporphyrin and the 7,8,17,18-tetrahydroporphyrin conjugation system. The magnesium-free derivatives, the pheophytins and pheophorbides, occur mainly as products of chlorophyll breakdown in feces[7] and sediments.[8] However, demetalation has also been detected as one of the first steps in the biodegradation of chlorophylls within the parent organisms,[9] and small but important amounts of pheophytins serve as primary electron acceptors in the reaction centers of bacterial photosynthesis[10] and photosystem II of oxygenic photosynthesis.[11] Pigments bearing the 17,18-dihydroporphyrin (= "chlorin") conjugation system typical of chlorophyll *a,* and the 2,3,17,18-tetrahydroporphyrin (= "isobacteriochlorin") conjugation system isomeric to that of bacteriochlorophyll *a* have also been isolated from several organisms where they serve nonphotosynthetic functions. They are intermediates in vitamin B_{12} synthesis,[12] but are also important on their own as coenzymes in sulfate and nitrate metabolism[13] and as sexual pigments in some organisms.[14] Related structures (mainly as V and Ni complexes) are found as minor components in geological deposits and are probably mainly derived from chlorophylls.[8] Several books[7,15] and reviews[2,16] deal with different aspects of chlorophylls and related pigments.

With the exception of chlorophyll *c,* all chlorophylls carry a long-chain esterifying alcohol at the 17^3-carboxyl group. It is commonly, but not always, Δ2-phytaenol(phytol) (see Table III. 1). The resulting amphiphilic nature of most chlorophylls is an important, but still largely unexplored, facet of chlorophyll structure and function.

The Mg-tetrapyrroles, e.g., the chlorophylls and the Fe-tetrapyrroles, are synthesized

---

*   The nomenclature of chlorophylls has changed several times in the past. A semisystematic nomenclature has been recommended, but is still far from being generally accepted. Here, the IUPAC numbering scheme[5] proposed has been used throughout in combination with a substitution nomenclature for the chlorophyll derivatives. The most common numbering schemes are compared in Figure 1. For further comments, see Reference 6.

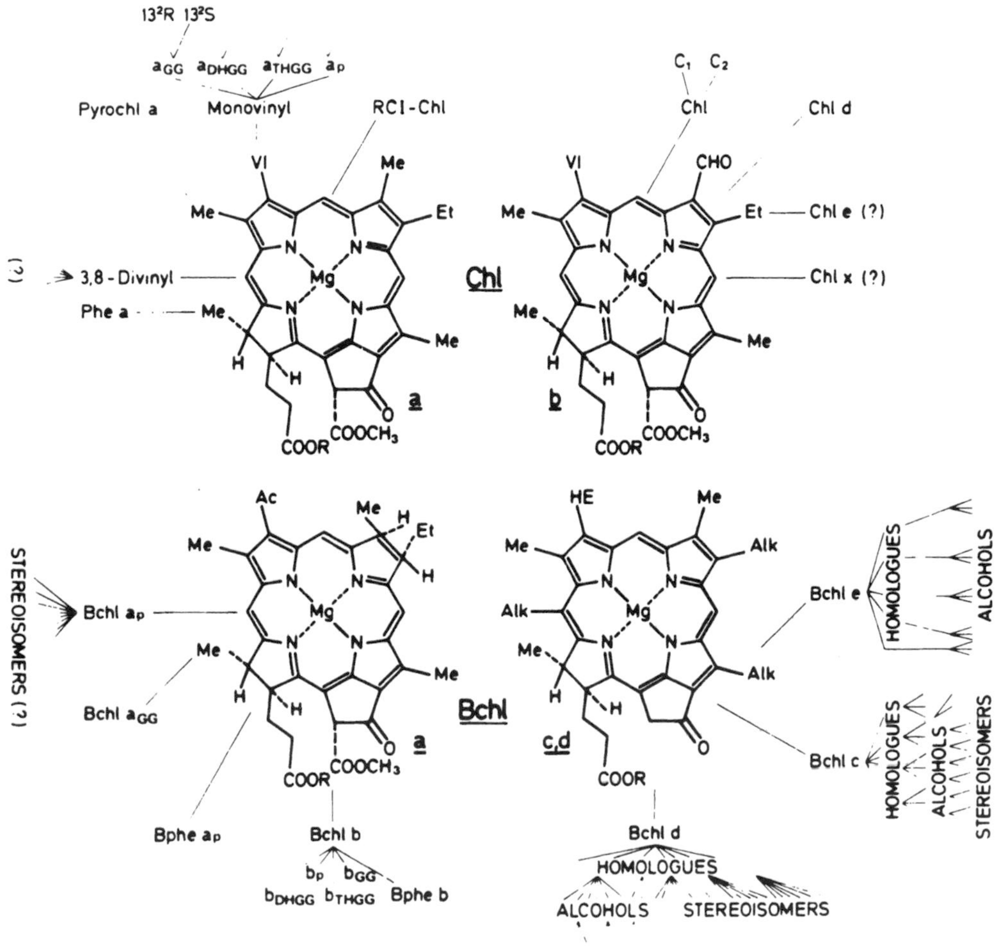

FIGURE 2. Our knowledge of chlorophylls is growing "logarithmically". Today, more than 100 different chlorophylls are known. Also, compare tables.

from the common precursor, δ-aminolevulinate. Different pathways to this key intermediate exist in different organizms. The C_4 or Shemin pathway, starting with succinate and glycin, is prevalent in purple bacteria and vertebrates, whereas other bacteria and plants use the C_5 pathway, starting with glutamate. In plants, the chlorophylls are probably formed by the latter, the hemes by the former (see References 16, 16p, and 16q). Tetrapyrrole formation presently seems to involve the same biosynthetic route irrespective of the pathway leading to δ-aminolevulinate. The biosynthesis of the tetrapyrrole moiety of the chlorophylls ("magnesium-branch") (Figure 3) separates from that of the hemes ("iron-branch") at protoporphyrin (see References 16a and 16b for leading references). Most of the steps leading to chlorophyll *a* are now known in some detail. After insertion of Mg* and methylation of the C-13^3 carboxyl group, the isocyclic ring is formed and the 8-vinyl group reduced. The subsequent reduction of the peripheral Δ17 double bond requires light in most higher plants and is an important regulation point. At or after this reduction, the tetrapyrrol and isoprenoid pathways meet. The C-17^3 carboxyl group is first esterified with geranylgeraniol, and the

---

* Zinc porphyrins related to Mg-porphyrins of the chlorophyll biosynthetic pathway have been identified and may be derived from unspecific metalation by the Fe-chelatase,[17] but have also been invoked as another branch leading to biliproteins.[18] However, chromophores of the latter seem to be formed generally from hemes.

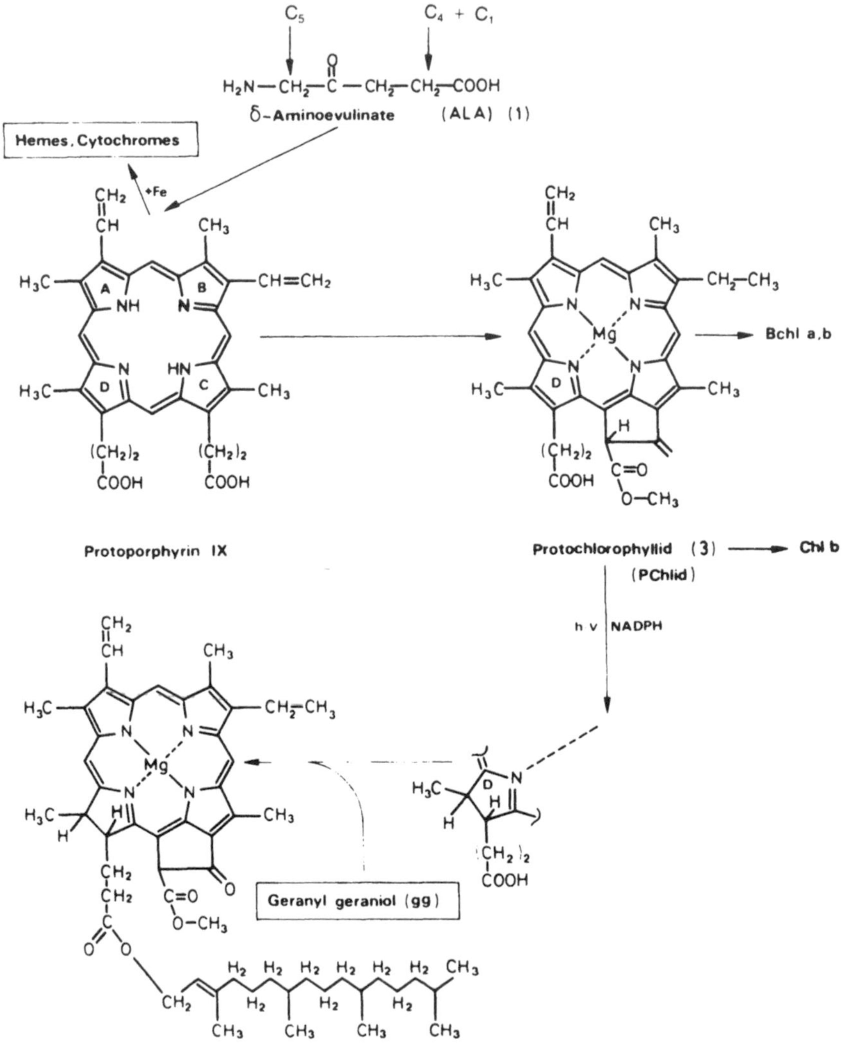

FIGURE 3.   Biosynthesis of chlorophyll. Protoporphyrin IX serves as common precursor of hemes and cytochromes, on the one hand, and protochlorophyllide, on the other. Esterification of hydrogenated protochlorophyllide with geranyl geraniol and stepwise further hydrogenation leads to chlorophylls (e.g., Chl *a*, Chl *b*). (From Scheer, H. and Inhoffen, H. H., in *The Porphyrins*, Vol. 2, Dolphin, D., Ed., Academic Press, New York, 1978, chap. 2. With permission.)

reduction of the $\Delta14'$, $\Delta10'$, and $\Delta6'$ double bonds of the terpenoid moiety occurs in that order, probably after it has become part of the chlorophyll structure.[19]

Bacteriochlorophyll *a* formation largely follows this scheme and may actually proceed via chlorophyll *a* or a closely related pigment, followed by reduction of the $\Delta7$ double bond and transformation of the 3-vinyl into an acetyl group (see Reference 12 for references). Bacteriochlorophyll *b* has been suggested as an intermediate in the former process, but for chemical reasons only.[20] Chlorophyll *b* has been believed to be formed via chlorophyll *a*,[21a] but evidence to the contrary has been published.[21b] The biosynthesis of the other chlorophylls is largely unknown.

The scheme is complicated (1) by indication of parallel pathways at several stages[16] and (2) by the discovery of a multitude of often only partially characterized pigments of unknown function, which are structurally closely related to the common chlorophylls.[8,19,22-27] The finding of these "new" chlorophylls is mainly due to the advances in chlorophyll chromatography.

Esterifying alcohols other than phytol have been identified in plants only as biosynthetic precursors.[16j,19] The bacteriochlorophylls are much more variable. The finding of $\Delta 2,6$[28] and $\Delta 2,10$ phytadienol[29] in species containing bacteriochlorophyll c and b, respectively, indicates different sequences of hydrogenation of geranyl geraniol ($\Delta 2,6,10,14$-phytatetraenol) in photosynthetic bacteria. The latter is the major esterifying alcohol of bacteriochlorophyll a in some species,[30] and the bacteriochlorophylls c, d, and e contain its C_{15} analog, farnesol, as the main alcohol.[26-28,31] The biosynthetic routes leading to bacteriochlorophylls with more unusual alcohols[28] are not known.

The pheophytins also have a photosynthetic function, but little is known of the details of their biosynthesis. The finding of a different esterifying alcohol (geranyl geraniol and phytol, respectively) in bacteriochlorophyll a and bacteriopheophytin a in reaction centers from *Rhodospirillum rubrum* indicates different biosynthetic pathways and not only a simple demetalation step.[32]

The biodegradation of chlorophylls in their parent organisms is largely unknown. Pheophorbides,[9,33] as well as a red pigment of unknown type,[34] have been characterized spectroscopically. Pyropheophytin a has been identified in dark-adapting *Euglena*, where it seems to be a primary product of chlorophyll breakdown.[9]

CHARACTERIZATION OF CHLOROPHYLLS: SPECTROSCOPIC METHODS

UV-Vis Spectra

All chlorophylls have comparatively narrow-banded and characteristic spectra which set them apart from most plant pigments, except the porphyrins of which they are a subclass. For structural correlation, chlorophylls have to be extracted from the natural sources, because the protein-chromophore interactions may profoundly influence their spectra in the native environment.

Chlorophylls belong to three different types of conjugation systems and can be classified accordingly by UV-vis spectroscopy. Due mainly to aggregation,[35] the UV-vis spectra are markedly solvent dependent. The moderately polar disaggregating solvent diethyl ether has been widely used for the characterization of chlorophylls. It can be replaced by the less volatile dioxan without serious changes. Methanol or acetone, which are often used for extraction and (mixed with a small percentage of water) for HPLC, may give rise to distinct spectral changes.* As a pronounced example, the Q_x-band of bacteriochlorophylls absorbing around 580 nm is red-shifted in these solvents by up to 50 nm, as compared to ethereal solutions.[35b] Care has to be taken if chlorophylls have to be identified in the presence of detergents. Depending on type and concentration of the detergent, extraordinary spectral variations occur due to the formation of aggregates with structures very similar to the ones found or believed to be present in chlorophyll proteins. In the author's experience,[35c] all such aggregates can be dissociated with the cationic detergent cetyl-trimethyl ammonium bromide (CTAB) at a concentration of 1%. Under these conditions, the spectra of the (monomeric) solutions are similar to the solutions in methanol.

The spectra of the pheophytins are distinct from the respective chlorphylls (Figures 4 to 7). Since most** chlorophylls can be demetalated without other structural changes (see

---

* Considerable artifact formation may also take place during extraction, which is discussed below.

** The acid-labile $\Delta 8,8^1$ double bond poses problems in the demetalation of bacteriochlorophyll b and g. It yields considerable amounts of by-products of the pheophytin a spectral type,[20,37] even under the best conditions presently known (methanolic HCl, N_2 flushed[38]).

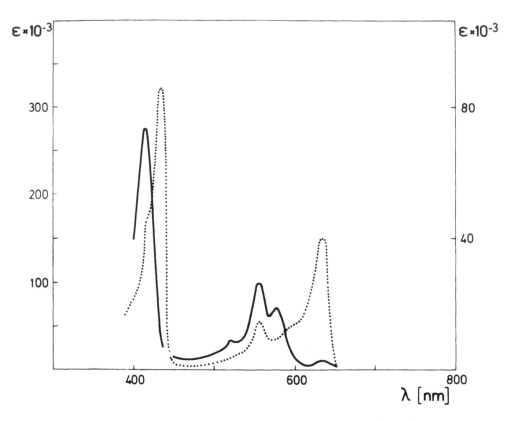

FIGURE 4.   UV-vis spectra typical for metalloporphyrins (- - -). The spectrum is characterized by two bands in the visible range and an intense Soret band. The oxorhodo-type spectrum (—) is found in pheophytins as long as the isocyclic ring is still present. (From Scheer, H. and Inhoffen, H. H., in *The Porphyrins*, Vol. 2, Dolphin, D., Ed., Academic Press, New York, 1978, chap. 2. With permission.)

Appendix), the spectra of the resulting pheophytins can be used for further identification. As an additional bonus, the spectra of pheophytins are less solvent dependent.

The *c*-type chlorophylls,[39,60] protochlorophyll(ide), and many other chlorophyll precursors are true metalloporphyrins. They are characterized by two bands in the visible range ($\epsilon \approx$ 15,000), and an intense near-UV or Soret band ($\epsilon \approx$ 150,000) (Figure 4, dashed line). The pheophytins have oxorhodo-type spectra[16f,41] so long as the isocyclic ring is still present (Figure 4, solid line). The chlorophylls *a, b,* and *d* and the bacteriochlorophylls *c, d,* and *e* are metal complexes of 17,18-dihydroporphyrins (chlorins). They have a single, intense, visible band ($\epsilon \approx$ 100,000) and a structured Soret band of comparable maximum absorption intensity (Figure 5, dashed lines). In their pheophytins, the visible band drops in intensity and remains at roughly the same position, while the Soret band is blue-shifted and retains its intensity. An additional band around 530 nm and a second, often structured one around 500 nm are characteristic for these pheophytins (Figure 5, solid line). The bacteriochlorophylls *a, b* and *q* are metal complexes of 7,8,17,18-tetrahydroporphyrins (''bacteriochlorins''). They have an intense band in the near IR ($\epsilon \approx$ 100,000), a split Soret band ($\epsilon \approx$ 100,000), and a third band around 580 nm (Figure 6, dashed line). The most indicative change upon demetalation is a shift of this third band to about 525 nm (Figure 6, solid line). For completion, the typical spectra of isobacteriochlorins and their metal complexes are shown in Figure 7.

The absorptions of some of the chlorophylls are listed in Table III. 2. A vast amount of data has accumulated on the effect of chemical modification of chlorophylls and pheophytins on their absorption spectra. Some of these effects have been sampled in Table III. 3. They

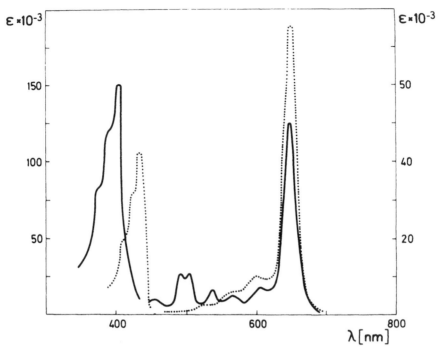

FIGURE 5. Typical UV-vis spectra of Mg-chlorins (- - -) and free base chlorins (—). To the former belong the chlorophylls *a, b,* and *d,* as well as the bacteriochlorophylls *c, d, e.* Their pheophytins belong to the latter type. (From Scheer, H. and Inhoffen, H. H., in *The Porphyrins,* Vol. 2, Dolphin, D., Ed., Academic Press, New York, 1978, chap. 2. With permission.)

can be helpful in the first classification of unknown chlorophylls. It is apparent from Table III. 3, however, that many structural changes produce only insignificant spectral changes. Most notable for analytical purposes is probably the fact that many changes at the isocyclic ring, bearing the chemically very reactive enolizable β-ketoester function, belong to this group. It should be noted that many of the alterations which may arise during extraction, work-up, and purification of chlorophylls involve reactions at this part of the molecule yielding products having spectra which are almost indistinguishable from those of the parent pigments.

Fluorescence Spectra

The natural chlorophylls and many of their derivatives are highly fluorescent in monomeric solution. The pheophorbides have lower, but still appreciable, quantum yields (e.g. o.1 for pheophytin *a*). Accordingly, fluorescence detection is the most sensitive technique for routine analysis of chlorophylls and their derivatives, including pheophorbides. In combination with HPLC or HPTLC, the femtomole range is accessible with standard equipment. A further advantage of fluorescence spectroscopy is the possibility of mixture analysis without or with only partial separation of the components. A systematic variation of excitation and emission wavelengths yields a minimum number of components present in the mixture and also important structural information (see References 23 and 42 for examples). Unfortunately, the quantitative analysis not only requires specialized equipment, but it is also prone to systematic difficulties (aggregation, concentration quenching, solvent effects, energy transfer, etc.). The fluorescence excitation spectra of chlorophylls are very similar to their absorption spectra (Section 1) and the Stokes shifts of the emission spectra are in the range of 20 nm. Detailed fluorescence data are available for most of the "common" chlorophylls listed in Table III. 1,[23,42] but not for the majority of derivatives of known structure which are otherwise fully characterized. Structure assignments by comparative fluorescence spec-

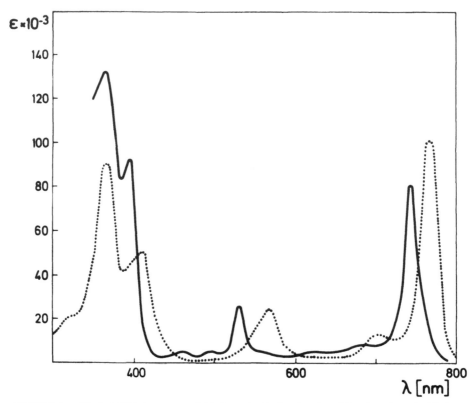

FIGURE 6.   Typical UV-vis spectra of Mg-bacteriochlorins (- - -) and free base bacteriochlorins. Bacteriochlorophylls *a* and *b* are examples of the former, their pheophytins of the latter.

troscopy alone are, thus, severely limited, if not impossible, but may be feasible in combination with high-resolution chromatographic techniques.

Circular Dichroism

Spectropolarimetric techniques and especially circular dichroism are important for stereochemical investigations of chlorophylls and for the analysis of pigment-pigment interactions. Spectropolarimetric detection has, to the author's knowledge, not yet been applied to the chromatography of chlorophylls, but may become useful for special problems. One example is the analysis of C-13^2 epimers, the so-called "prime" pigments which accompany the chlorophylls of natural stereochemistry due to epimerization in solution and which have recently been implicated as components of photosystem I reaction centers. Other stereoisomers have also been suggested as minor components of bacteriochlorophyll extracts.[24] The spectropolarimetry of pheophorbides derived from chlorophylls *a* and *b* has been reviewed.[43] A detailed discussion of this and the following two spectroscopic methods is inappropriate in this context, although they are indispensible for the structure elucidation of any new or suspectedly new chlorophylls.

NMR Spectroscopy

Proton magnetic resonance is probably the most powerful single tool for chlorophyll structure analysis. This is due to the ring-current of the aromatic tetrapyrrole skeleton, which acts as a built-in shift reagent.[44] ^{1}HMR is also a valuable aid in the identification of the esterifying alcohol,[30] and for stereochemical analysis up to the determination of optical purity.[44a,45] ^{1}HMR has successfully been applied to the investigation of mixtures,[39] but pure compounds are desirable. Measurements down to 100 μg can routinely be performed on FT

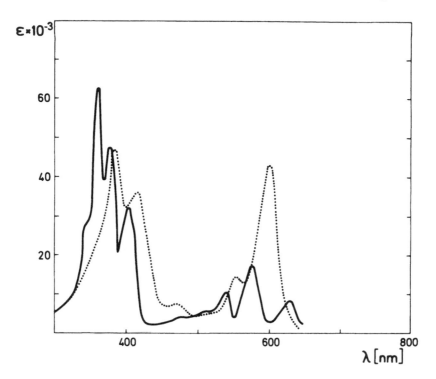

FIGURE 7. Typical UV-vis spectra of metallo-isobacteriochlorins (- - -) and their free bases (—). Pigments of this type are not considered as chlorophylls, but are found as cofactors of reductases and as biosynthetic intermediates of vitamin B_{12}.

instruments. The limit for special problems is as low as a few micrograms, provided there has been meticulous sample preparation. Like other solution studies, care has to be taken with solvent selection. The common solvent is C_2HCl_3, which is available in high isotopic purity but is generally contaminated with acid, leading easily to demetalation. Chloroform may also lead to aggregation. For comparison with authentic samples, the concentration then has to be similar in both measurements. Alternatively, addition of small amounts of deuterated methanol will lead to disaggregation, or the spectra can be taken in the strongly dissociating solvent, pyridine. Again, most problems can be circumvented if the pheophytins are studied instead. [13]CMR has evolved as an additional important technique during the past decade.[44] It is less affected by aggregation, but still restricted in its comparatively low sensitivity, generally requiring ≥ 1 mg of the pure compound. Other magnetic resonance techniques ([15]NMR, ESR, ENDOR) are only referred to in this brief survey. They have yielded valuable information on chlorophylls, but are still restricted to special problems.[44,46,47]

Mass Spectroscopy

Mass spectroscopy is routinely used in structure analysis of pheophytins and their derivatives.[48] Molecular ions are generally observed in the EI mode, but may be accompanied by M + 2 as well as M − 2 ions. The fragmentation pattern is analogous to that of porphyrins; the most prominent benzylic cleavage leads to characteristic fragments for the substituents of the isocyclic ring and the reduced ring(s) including the esterifying alcohols. The analysis of magnesium complexes became possible by the use of other ionization techniques.[49] With [257]Cf-plasma ionization, even chlorophyll aggregates have been investigated.[50] Gas chromatography coupled with mass spectroscopy is severely restricted by the instability of chlorophylls and their derivatives at the required temperatures. It has been used, however, for the analysis of chlorophyll-derived porphyrins in geochemical studies.[8]

ESTIMATION AND SEPARATION OF CHLOROPHYLLS

Nonchromatographic Analytical Techniques

Chlorophylls are geographically widely distributed and are common to all photosynthetic organisms with the exception of halobacteria. Their electronic absorption and fluorescence in spectral ranges, comparably free from overlap with other pigments, have attracted researchers from various disciplines to use them analytically as well as for reference purposes. This includes their use as reference standards in plant and photosynthetic bacterial physiology, in chemotaxonomy, and in certain aspects of geology. A number of nonchromatographic techniques have been developed which allow a more or less accurate estimate of the major components for the different purposes. Most of these methods rely on absorption or fluorescence measurements at two or three defined wavelengths to solve a system of two or three linear equations yielding the concentrations of, e.g., the chlorophylls *a* and *b* in plants. Several of these methods are listed in Table III. 4. The basic assumption of all these methods is that only a limited number of pigments of known absorption or fluorescence under the conditions of measurement contribute to the spectrum at the points of absorption of fluorescence measurement. This requirement is met to a very varying degree depending on the type of sample, in particular if chlorophyll degradation problems can be present (see Table III. 6). All such techniques require well-defined extraction methods for partial purification and definition of the solution state of the desired chlorophylls. The extraction can, in favorable cases, also be used to separate part of the chlorophyll pigments, e.g., the free acids, from the C-17[3] esters by partition between aqueous ammonia and organic solvents. The limitations of nonchromatographic techniques are exemplified in two recent publications dealing with complex samples, viz., sediments[69] and phytoplankton.[70] In both cases a chromatographic (HPLC) step proved necessary to give reliable results. For further discussions of specific problems and limitations, the reader is referred to the references in Table III. 3.

Prechromatography Purification

Chlorophylls are generally extracted with prechilled acetone, methanol, or mixtures thereof, if possible, under inert gas or sometimes under reducing conditions. Methanol is the more powerful extraction agent, but may lead to oxidation at C-13[2] ("allomerization")[65] and hydrolysis or transesterification by the action of chlorophyllase.[66] (See Table III. 6 for some alteration products of chlorophylls.) On the other hand, acetone has been shown to promote the ready dehydrogenation of bacteriochlorophyll *a*,[72] and this solvent also promotes degradation in bacteriochlorophyll *b* and related pigments bearing the 8-ethylidene group (see Table III. 6). A suitable prepurification of chlorophylls — especially for preparative work — can greatly reduce chromatographic needs, both due to the removal of, e.g., lipids, and also to the inhibition of side reactions during repeated chromatography. One common method is the solvent exchange from methanol/acetone to petroleum ether, from which many chlorophylls precipitate upon cooling and can be readily collected by centrifugation. This method is inappropriate, however, for more polar chlorophylls and derivatives, e.g., pigments containing a free propionic acid chain, since they are not (completely) transferred into petroleum ether. Extraction and solvent exchange can be combined in a two-phase extraction.[67] Another established prepurification[51] involves the precipitation of the chlorophylls from the crude extracts with dioxane or similar reagents which cross-link the chlorophylls by bidentate binding to the central Mg.

Accordingly, the Mg-free pheophytins cannot be purified this way. They can, on the other hand, often be enriched by extraction procedures involving acids, due to the different basicity of the pigments with different substituents or conjugation. Partition between ether and hydrochloric acid of different concentrations[7] was one of the major separation techniques used routinely before the widespread application of chromatography. Modification of the original system is still superior to chromatography, e.g., for large-scale separation of chlorins and porphyrins[52] or of the pheophorbides of the *a* and *b* series.[53]

The above procedures are extremely useful in the large-scale preparation of some chlorophylls and derivatives. They can be supplemented by a careful choice of the parent organisms. As an example, the separation of chlorophylls *a* and *b* which is necessary after the extraction from plant material, can be circumvented by using blue-green algae if chlorophyll *a* is the desired product. Similarly, the use of carotenoidless mutants of photosynthetic bacteria facilitates the purification of bacteriochlorophylls.

Prepurifications have to be met with much greater scepticism in analytical work. They have generally been developed for the "typical" chlorophylls, but may, for various reasons, lead to a loss of "less-typical" pigments, e.g., free acids, pheophytins, etc. In these cases potentially important pigments present in smaller amounts may be lost even if the total balance of pigments is barely affected. It is thus desirable to use the total methanol extracts, which probably remove most of the pigments, for all analytical problems. Fortunately, this has become possible and practical with reverse-phase HPLC.

Chromatography

Introductory Remarks

Free chlorophylls are unstable to light, acid, base, oxygen, and many other reagents. A major problem in chlorophyll chromatography is then to avoid irreversible reactions and/or adsorption during the entire separation procedure, and especially on the column. A very important factor contributing to this lability is the central magnesium atom. Since it can be removed from most chlorophylls without additional modifications of the rest of the molecule, demetalation (viz., pheophytinization) has often been used prior to the separation proper. A rapid demetalation procedure for samples in the submilligram range (which can be scaled up easily) is given in the Appendix. After pheophytinization the choice of suitable chromatographic systems is greatly increased, and the formation of artifacts during handling and chromatography is greatly reduced. This strategy should be considered for any given problem. It is also highly recommended as a control of chromatographies carried out with genuine chlorophylls.

There are two major limitations. One arises with chlorophylls for which acid-promoted side reactions are anticipated. An example is bacteriochlorophyll *b* and its derivatives still carrying the acid-labile exocyclic $\Delta 8,8^1$-double bond at ring B.[20] The other arises if the chlorophylls proper are the desired products, since the remetalation process of pheophorbides is less straightforward than demetalation. It is impractical on an analytical scale, but has successfully been used for preparative purposes after modifications of chlorophylls. In the absence of the enolizable β-ketoester system at the isocyclic ring, Mg-insertion is possible by a modification[54] of Eisner's method[55] ($Mg[ClO_4]_2$ in pyridine) developed for porphyrins. More sophisticated procedures have been developed for pigments bearing this important structure.[56-58] In any case, a final purification of the remetalation products is obligatory.

According to these introductory remarks, the chromatography of the chlorophylls proper and the metal-free pheophorbides is dealt with separately. A further division has been made between analytical and micropreparative separations, and preparative chromatography, respectively. A survey of various chromatographic systems is given in Tables III. PC 1, III. TLC 1, and III. LC 1, updated from Svec's review.[2]

Chlorophylls

Preparative Chromatography

Pertinent examples of preparative chromatography of chlorophylls are given in Table III. LC 3. Powdered sugar is still a powerful method. Advantages are its mild and nonreactive adsorbing properties, its homogeneity, easy availability, and, last but not least, price. An example of the resolving power under optimum conditions is given in Table III. LC 4. The commercial material contains up to 3% starch to prevent caking; 2 to 3% more may be added under humid conditions. The mixture should be dried overnight at 60 to 80°C and sieved immediately before use. Very sensitive chlorophylls (e.g., bacteriochlorophyll *b*) can be protected by the addition of 0.1% powdered sodium ascorbate to the adsorbent after drying, but before sieving. The columns are packed either dry[2] in successive small portions (1 to 2 cm thick), or in a slurry.[59] As a rule of thumb, they can be loaded with up to 0.4 mg/g adsorbent. Sugar can be substituted with powdered cellulose with very similar results.

Mixtures of petroleum ether with traces of *n*-propanol are generally used as eluents. These mixtures do not readily dissolve higher concentrations of most chlorophylls, however. It can then be helpful to first dissolve the sample in a *minimum* amount of diethyl ether, load this solution on the column, and subsequently start elution with pure petroleum ether prior to adding the necessary amount of propanol. Larger amounts of the polar diethyl ether can lead to spreading and, hence, decreased resolution. An alternative way of sample application is by adding an amount of adsorbent corresponding to 0.5 to 1 cm column-height to an ethanol or ethereal solution of the pigment mixture, removing the solvent in a rotary evaporator in vacuum, packing the dried residue carrying the pigment onto the column, and adding another 1-cm layer of adsorbent before starting the elution. In the latter procedure, care should be taken to avoid reactions of the chlorophylls adsorbed in a finely dispersed state on the dried adsorbent (light, temperature during evaporation of solvent).

The columns are generally developed by applying a slight negative pressure (approximately 400 torr delivered by an aspirator and controlled by a needle valve). It is useful to increase successively the propanol concentration in increments of 0.25 to 0.5%. This allows a better control of the band movement, which is markedly affected, e.g., by the varying residual water content of adsorbents and eluents, under varying environmental conditions.

Since the original completion of this article (1983), the author has appreciated polysaccharide materials as potentially very useful in methods involving separation on cross-linked Sephadex[61] and on ion exchangers.[62] Their main advantage is a comparatively high capacity, although the latter tend to react with labile chlorophylls. Comparably cheap, noncrystalline DEAE-cellulose seems to be a very good alternative to the sometimes tricky sugar columns. If the somewhat tedious conditioning given by Sato and Murata[62] is followed, reliable and quick separations of many chlorophylls and derivatives can be achieved with slightly more polar eluents than used with sugar (e.g., chloroform instead of petrol ether as major component). This allows at the same time the application of more concentrated solutions. The conditioned material can be stored for several weeks.

The separated pigments can be removed in two different ways from the column. One is by elution from the bottom of the column. The other is to gently push the wet adsorbent column out of the tube with pressurized nitrogen, cutting out the desired fractions, and removing the pigments from the adsorbent with an eluent containing an increased amount of propanol (or even with pure ethanol or the like) in a Büchner funnel or a small column.

After elution, a thorough washing is necessary to remove any components dissolved from the solid phase. In general, the chlorophyll is eluted from the adsorbent in a solvent mixture rich in petroleum ether. It is washed first with 80% aqueous methanol, and subsequently with 50 and 25% methanol. After a final washing with water, it is dried quickly over NaCl. The chlorophyll generally precipitates from this solution upon cooling in the deep freezer and can be collected by centrifugation. It can be stored for years in sealed vials under vacuum or purified argon or nitrogen.

The separation on these polar phases depends mainly on differential polar interactions with the chlorophylls. Where the differences between the pigments exist mainly within nonpolar parts of the molecules, hydrophobic phases have been successfully employed (see Table III. LC 1 for leading references). Polyethylene powder is a cheap adsorbent, although of apparently varying quality. The separation of the chlorophylls c_1 and c_2, porphyrin-free acids which differ by a double bond in the C-17 side chain, has been achieved on a material of special sorptive properties.[40,60] The bacteriochlorophylls c and d (formerly chlorobium chlorophylls 660 and 650), comprising a complex mixture of homologs with different alkyl side chains, different esterifying alcohols, and stereoisomers,[26] have been partially separated by preparative HPLC on polyethylene powder.[28] The event of highly reproducible alkylated silica phases has greatly facilitated reverse-phase chromatography of chlorophylls, although the costs are still very high for large-scale separations. Examples for this widely applicable adsorbent are the micropreparative separations of plant[19] and bacterial chlorophylls[28,29,32] differing only in the esterifying alcohols.

Analytical Chromatography on TLC Plates

Thin-layer and the currently less important paper chromatographic systems applied to the analysis of chlorophylls have been listed in Tables III. PC 1 and III. TLC 1. Examples for the analysis of crude plant extracts are given in Table III. TLC 2. A complete separation is not possible by this means. Cellulose has long been the preferred stationary phase, but silica and reverse phases have been used as well. Due to the large potential contact areas to air and oxygen, care has to be taken to avoid artifacts during chromatography. An inert atmosphere and dim light are recommended. Extensive drying of the plates after application of the samples and between subsequent developments should be avoided. If plates are used for micropreparative purposes, the zones should be scraped off under a hood before the plates become completely dry. The material should be desorbed with solvents of only moderately increased polarity to leave (the generally more polar) artifactual products behind.

Silica has the advantages of high capacity and high resolving power, but artifactual problems are multiplied. It is not only strongly adsorbing, but also more or less acidic. We found that sodium ascorbate added in concentrations ≈1% w/w before making the slurry *may* be useful as a mild protectant (reductant and buffer) with only minor losses in resolution. Because even these protective measures are not completely safe, the results obtained on silica should be carefully checked using less reactive adsorbents whenever possible. Low temperatures do considerably increase the resolution of stability of chlorophylls on silica, at least in column chromatography.[73]

Conventional TLC plates can be coated quite inexpensively with the desired adsorbent in every lab. Since it is possible to use, e.g., silica which is completely free of additives like fluorophor, binder, etc., they are useful for micropreparative work. Precoated plates are more expensive and always contain a binder (which may, however, affect the separation). High-resolution precoated plates have proved extremely useful in the author's laboratory. They can be used repeatedly and give excellent resolution in very short (5 to 10 min) times. With respect to silica, the same considerations apply as are discussed above, but side reactions are less prominent due to separation speed.

The chemically bonded reverse-phase microplates (e.g., Merck HPTLC) avoid most of the problems and have still very high resolution. For general applications in qualitative analysis, with the eye as a very sensitive and spectrally responsive detector, they appear to be excellent, and with the proper scanning equipment they may prove superior to columnar HPLC with less costs. A simple but effective system applicable to most chlorophylls, pheophytins, and methylpheophorbides is given in Table III. LC 3.

Analytical Column Chromatography

Several applications of conventional column chromatography to total plant extracts are listed in Table III. LC 2. Powdered sugar has been used only occasionally for analytical and micropreparative work due to the problems involved with narrow columns. HPLC has opened a wide field of applications. The contact with air and light can easily be kept at a minimum, and a wide variety of adsorbents are available. For general purposes, C-8 or C-18 chemically bonded reversed phases with methanol (or acetonitrile)/water and methanol/acetone mixtures, respectively, give excellent resolution, and adaptations to the different brands are simple. An incomplete list of chlorophylls (esters) separated by this method if given in Tables III. LC 5 through III. HPLC 9. HPLC appears, to date, the method of choice for the analysis of total extracts (Table III. HPLC 1). The system of Braumann and Grimme[63] separates most of the relevant pigments in a single run. A ternary solvent system has recently been studied systematically.[68] Commercial material occasionally contains residual free OH groups. They can be covered by on-column silylation (Appendix A). Chlorophylls as sensitive as bacteriochlorophyll *b* can be separated without artifact formation if traces of sodium ascorbate are added to the eluent.[29]

The separation is obviously most sensitive to changes in the hydrophobic parts of the molecule. Potential uses are for (1) the separation of chlorophyll and pyrochlorophyll,[9] (2) the separation of chlorophylls esterified with different alcohols differing only in the number of double bonds,[19] and (3) the isolation of trace quantities of what appears to be naturally occurring stereoisomers.[35b]

In the latter cases, even structural information may be obtained. As an example, the elution order of chlorophylls esterified with different alcohols is similar to that expected for the free alcohols upon GLC on a reverse-phase capillary column. Also, the ratio of retention times for pairs of pigments differing only in the alcohol has been shown to be constant for a series of pigments derived from chlorophyll *b*.[36] An increased resolution for a series of chlorophylls of yet undetermined structure has been reported by the use of silica gel.[23] The authors have, thus, been able to isolate a surprisingly large number of fractions from greening etioplasts and tissues, although most of them still appeared to be pigment mixtures by fluorescence analysis. A corn mutant has recently been shown to contain large amounts of one of these newly characterized chlorophylls.[25] HPLC on silica at low temperatures gives excellent resolution for (e.g.) chlorophyll *a* and its epimer *a'*.[73]

A particular problem is the separation of pigments bearing a free propionic acid side chain. Hydrophobic phases have been used for preparative separations (see above); recently, HPLC on reverse-phase materials has successfully been used to separate free and esterified pigments in a single run with gradient elution (Table III. HPLC 3; see also Table III. HPLC 10). It should be noted, however, that ethyl acetate is often used as one of the solvents, which then gives good resolution, even if no ion-suppression or ion-pair techniques are applied.

Continuous Liquid-Liquid Partition

Liquid-liquid partition techniques have only occasionally been used for chlorophyll separations. A recent example to the analysis of chlorophyll *a*-degradation products demonstrates its resolving power (Table III. LC 6). The advantage of extremely mild conditons has been outweighed by the complexity of the conventional equipment. Technological progress is likely to influence the separation of chlorophylls as well. For single-step partition as a pre-purification step, see Analytical Chromatography on TLC Plates.

Pheophytins
Preparative Separations

All chromatography systems which have been applied to chlorophylls can be used as well for the pheophytins. Due to increased stability, however, silica and (the only rarely explored) alumina can be used with less restrictions. However, their stability is only comparative and the precautions mentioned above apply to the pheophorbides as well. CCl_4/acetone mixtures, sometimes with third components added, are of almost universal applicability as eluents for esterified pheophorbides. The group of Inhoffen has developed a method for the isolation of pheophytin *a* and *b* (as a starting material to many derivatives) by which up to 20 g of a crude pheophytin *a/b* mixture, containing carotenoids and degradation products as well, can be separated on a single 8 ø × 20 cm column (see Appendix A). In cases where chlorinated systems must be avoided, carbon tetrachloride can be replaced by toluene.

Unfortunately, the solvent system is (see Appendix A) highly corrosive to stainless steel and, thus, not suitable for HPLC. Difficult separations have successfully been dealt with by preparative TLC and two to three development cycles. In a sandwich chamber containing eight 1-m × 20-cm plates coated with 12 mm silica H (Merck), up to 500 mg of material can be separated at a time. The only disadvantage of this cheap procedure is that it can be rather messy (dust, solvent vapors). High-quality medium-pressure columns have been used by Risch et al.[64] for the (partial) separation of the bacteriochlorophylls *c, d,* and *e,* and packing problems have been discussed there (Table III. LC 5). Preparative high-pressure (performance) chromatography has so far only been used occasionally.[28] A drawback of this powerful method is the time and amount of solvent necessary for optimization.

Analytical Chromatography

Most of the conventional methods listed in Tables III. PC 1 through III. LC 2 have become outdated by high-performance stationary phases. Pheophytins are usually less mobile than the corresponding chlorphylls on RP-phases and more mobile on polar phases, but chromatography systems employed for the latter are usually suited for the former as well (Tables III. HPLC 1 through III. HPLC 10). Some special considerations for HPLC of pheophytins have been summarized in Appendix A. HPLC is easy to quantify with moderate precision. Again, the use of advance TLC plates is highly recommended for qualitative analysis, and also for first steps in the development of a HPLC separation.

Special Considerations for Chlorophyll High-Performance Liquid Chromatography (Appendix A)

Pumps

Pumps, injector system, and system software do not require special considerations for chlorophyll-HPLC. The only special consideration for the pumps is recycling ability, which is standard on many reciprocating pumps. It allows a very straightforward check of on-column stability.

Columns

Columns can be chosen according to the general criteria discussed above; a C-8 or C-18 reverse phase is, in the author's experience, sufficient for 90% of the problems. Silylation of residual free SiOH surface may be necessary with some commercial materials (Appendix B). Increased temperatures (40° C) are recommended for reverse phases, whereas low temperatures greatly enhance resolution on silica.

Silylation of Reverse-Phase HPLC Columns

Reverse-phase HPLC columns on covalently modified silica bases may contain varying amounts of free silica surface. This may lead both to unwanted adsorption and to sample alterations (see Appendix B). We have found the following simple procedure useful to minimize free silica surface by on-stream silylation: the column if flushed with dry, peroxide-free tetrahydrofurane (3 mℓ/min). Within 1 hr, 4 injections, each of 25 µℓ bis-trimethylsilyl-trifluoroacetamide (e.g., Pierce Chemical), are made.

New material can be silylated before filling a column by treating a slurry in dry, peroxide-free tetrahydrofurane with the above reagent (20 µℓ/g adsorbent). After heating for 30 min to 70°C in a stoppered vessel, the absorbent is washed, subsequently, with tetrahydrofurane, methanol, and (mandatory) toluene.

Eluents

Eluents: Methanol-water mixtures are most widely used with reversed-phase columns. Acetonitrile can replace methanol, but presents a greater health hazard. It is considered superior to methanol because the latter may promote allomerization. With very sensitive chlorophylls, water can be replaced by a freshly prepared 1% (w/v) aqueous solution of sodium ascorbate, which inhibits oxidation and at the same time slightly buffers the eluent and adsorbent. Unused solvent should be discarded, in this case, at the end of the day. On silica, prime eluents for conventional chromatography are mixtures of chlorinated hydrocarbons, especially carbon tetrachloride, with acetone. These mixtures should *not* be used in HPLC, because they are *highly corrosive* to the stainless steel used for pumps, tubing, and the column material. They can be replaced by toluene/acetone or toluene/methanol mixtures with some loss in resolution.

Detectors

Absorption and fluorescence detectors are suitable for most problems; the former can detect picomoles, and the latter is even up to three orders of magnitude more sensitive with the highly fluorescent chlorophylls. It should be kept in mind, however, that many chlorophyll derivatives, e.g., pheophytins, are only weakly fluorescent. For both types of detectors, a high sensitivity up to 750 nm or better, even 850 nm for the bacteriochlorophylls, is critical for a selective detection of chlorophylls. Only few other natural dyestuffs absorb in this spectral region and they generally have much broader absorption bands than the chlorophylls (Table III. 5). A filter photometer is, therefore, sufficient for many routine applications, but a variable-wavelength machine is superior for discrimination among chlorophyll derivatives with different absorption maxima. The latter may be shifted by the eluent and are best determined *in situ*. A scanning detector is an even better device to get maximum information in analytical HPLC. It is now commercially available from several suppliers, although with rather different maximum sensitivity. With electronic experience and/or help, it is also possible to convert a good conventional variable wavelength detector into a highly sensitive scanning detector.[71] As an example, spectra have been taken of peaks containing down to 10 pmol in such a home-built scanning detector. Most modern, scanning detectors require catching the peak in the detector, e.g., by means of a four-way valve to avoid drifts after stopping the flow, or to allow a by-pass with the pumps still running. Detectors which allow on-stream scanning have been introduced and may be the future prime choice, in particular diode-array detectors which are capable of scanning within a few tenths of a second a full spectrum. A large memory and/or fast storage capabilities of the computer interfaced (or integrated) with the detector are required to take full advantage of the scanning speed.

Derivatization Reactions and Formation of Artifacts
(Appendix B)

Pheophytinization of Chlorophylls

The demetalation of the different chlorophylls requires different amounts of acid. This procedure will produce pheophytins from all known chlorophylls. The amount of acid can be reduced if necessary. Since demetalation results in a color change (bright bluish-green to brown in "chlorin-type" chlorophylls, blue to pink in "bacteriochlorin-type" chlorophylls), it can be followed by visual observation. See Table III. 1 for the UV-vis absorption spectra of chlorophylls and pheophytins. For very labile chlorophylls, the procedure can be carried out under argon or nitrogen.

The pigment is dissolved in peroxide-free diethylether. If no micro-preparative equipment is used, the total volume should be a minimum of 5 mℓ, even for very small amounts of pigments. The solution is transferred into a separation funnel and the same amount of 1% hydrochloric acid is added. The mixture is gently swirled and changes color if sufficient acid has been added. After phase separation the acid (lower phase) is removed. The etheral solution is then washed once with water, twice with a saturated aqueous solution of $NaHCO_3$, and finally twice again with water. After drying with solid NaCl, the ether is evaporated in a stream of nitrogen (analytical) or in vacuum (preparative samples). With suitable equipment, the procedure can be scaled down by a factor of 50, if less volatile ethers (e.g., tetrahydrofuran) are used. This method fails with pigments like bacteriochlorophyll *b* bearing an ethylidene group. See Reference 38 for an improved technique.

Common Artifacts

Several distinct sites of the chlorophyll molecule are susceptible to ready alteration. Svec has listed the most common ones in his review,[2] and Table III. 3 on the UV-vis spectra of chlorophyll derivatives may serve as a preliminary guide for any suspicious compound. Some studies on the subject are listed in Tables III.6 and III.HPLC 5 to 8, to which the reader is referred for references. Chromatography on polar phases has been one of the most common points to produce a variety of alteration products. With the advent of reverse-phase HPLC, the focus has somewhat shifted to alterations during the extraction and handling of the biological material. It should be kept in mind, however, that the reverse stationary phases now in HPLC are generally bonded to silica and may have residual free, reactive surface. With the increasing sensitivity and importance of trace analysis, alteration then still remains a problem of concern.

Alterations are generally thought of as unwanted chemical processes. There is, however, also the possibility of enzymatic reactions involved in chlorophyll breakdown (e.g., of chlorophyllase). The ease of some of the chemical modifications of chlorophylls is finally also potentially beneficial in allowing derivatization as an aid for structure elucidation. Below is a summary of the alterations which should always be critically evaluated if a new chlorophyll is encountered.

Artifacts Involving the Central Magnesium Atom

The central Mg is susceptible to acid-catalyzed exchange against hydrogen (pheophytinization), but also against other metals. Pheophytinization is further induced by light. The different chlorophylls show a rather broad range of stabilities towards acid (e.g., bacteriochlorophyll *a* > chlorphyll *b* > chlorophyll *a*), which can lead (and also be useful) to selective demetalation. There is recent evidence that demetalation is also an enzymatic process in chlorophyll degradation.

A second effect of Mg is the change in the reduction potential of the macrocycle towards a more ready oxidation. Many chlorophyll oxidations proceed via π-cation radicals and have recently been shown to be also dependent on the coordination state of the central Mg. This metal ion is coordinatively unsaturated and bears either one or two extra ligands, depending mainly on the solvent system. The macrocycle of bacteriochlorophyll *a* is more readily oxidized with the Mg-bearing one (e.g., in acetone solution) rather than two "extra" ligands (e.g., in methanolic solution).

Last but not least are the spectroscopic properties of chlorophylls dependent on the solvation — and by consequence aggregation — of chlorophylls. Any identification by spectroscopic comparison with authentic material must, therefore, be done under identical conditions.

Artifacts Involving the Isocyclic Ring

The isocyclic ring E bears the enolizable β-ketoester system, which is responsible for most of the observed reactions. The susceptible C-13 is further activated by being in a quasi-benzylic position and by the strain of the ring. The ready enolization of C-13^2 is responsible for its epimerization to the so-called "prime" pigments, and to the production of "allomerization'' products (13-alkoxylated, -hydroxylated, or -acetoxylated pigments). In all allomerization products, C-13^2 is no longer enolizable. This is the basis of the negative ''phase test'' for these compounds. Important with respect to artifacts is the inevitable formation of *two* quite stable epimers, thus increasing the number of separable fractions. Another well-characterized allomerization product is the 13^1 α-oxa-13^2-methoxy-chlorophyll (''10-methoxylactone''). Heat treatment leads to the loss of the 13^2-COOCH$_3$-substituent to yield the pyrochlorophylls. The situation has become more complicated by the implication of chlorophyll *a'*, e.g., the 13^2-epimer of chlorophyll *a* as a constituent of photosystem I and by the finding of 13^2-hydroxychlorophylls, 13^2-hydroxypheophorbides, and pyropheophorbides in degreening *Euglena* and in aged bacterial and plant cell cultures. Here enzymatic reactions may occur besides the ready chemical reactions.

Artifacts Involving Carbon C-20

The 20-methine bridge of the chlorin macrocycle is susceptible to electrophilic attack. Although no common alteration product, the 20-Cl derivative has recently been reported to be produced during the washing of chlorophyll solutions with tap water (which is chlorinated in most parts of the world). Again, these compounds are also implicated as natural products of important function. The occurrence of 20-Cl chlorophyll(s) has been demonstrated in photosystem I in stoichiometric amounts relative to P-700.

Artifacts Involving Oxidation of the Macrocycle

This problem is important in bacteriochlorophylls *a* and *b*, which are readily oxidized to chlorophyll-type pigments. 3-Devinyl-3-acetylchlorophyll *a* and its derivatives are common contaminants of bacteriochlorphyll *a* and especially *b* preparations. Acetone is notorious as solvent, which has been related to the ready oxidation of solvent. Since methanol is, on the other hand, prone to induce allomerization, mixed systems have been found most safe. Chlorophylls of the chlorin-type (plant chlorophylls and bacteriochlorophylls *c, d,* and *e*) are stable towards oxidation under the common extraction conditions, but care should be taken in the presence of oxidants. Quinones of high redox potential (e.g., tetrachloroquinones) have been used as selective oxidants during chemical structure correlation studies for both the conversion of bacteriochlorins to chlorins and of chlorins to porphyrins.

Artifacts Involving the 3-Vinyl Group

Although generally quite reactive, alterations of this substituent are generally much slower than at the aforementioned sites. Pigments of this type have been found as by-products during quinone oxidation. Some of the less common chlorophylls (e.g., chlorophyll *d*) are derived from (probably biosynthetic?) modifications of the 3-vinyl group, and the bacteriochlorophylls *c,d,* and *e* contain the 3-hydroxyethyl substituent.

Artifacts Involving Propionic Ester Side Chains

The 17^2-ester group is attacked by chlorophyllase to produce transesterified pigments (e.g., methyl esters in methanolic solution) and/or the free acid. Chlorophyllase is active in the common extraction media, and its activity varies greatly with the biological material. The 13^2-carbomethoxy group is stable to transesterification and hydrolysis under all common extraction conditions.

REFERENCES

1. **Tswett, M.,** Adsorptionsanalyse und Chromatographische Methode. Anwendung auf die Chemie des Chlorophylls, *Ber. Dtsch. Bot. Ges.*, 24, 384, 1906.

2a. **Svec, W. A.,** The isolation, preparation, characterization, and estimation of the chlorophylls and the bacteriochlorophylls, in *The Porphyrins*, Vol. 5, Dolphin, D., Ed., Academic Press, New York, 1978, 341.

2b. **Oelze, J.,** Analysis of bacteriochlorophylls, in *Methods in Microbiology*, Vol. 18, Academic Press, London, 1985, 252.

2c. **Cavaleiro, J. A. S. and Smith, K. M.,** Chromatography of chlorophylls and bacteriochlorophylls, *Talanta*, 33, 963, 1986.

3a. **Falk, H., Moornaert, G., Isenring, H. P., and Eschenmoser, A.,** Über Enolderivate der Chlorophyllreihe. Darstellung von 13^2, 17^3-Cyclophäophorbide Enolen, *Helv. Chim. Acta*, 58, 2347, 1975.

3b. **Hynninen, P.,** Application of elution analysis to the study of chlorophyll transformations by column chromatography on sucrose, *J. Chromatogr.*, 175, 75, 1979.

3c. **Inhoffen, H. H., Jäger, P., and Mählhop, R.,** Partialsynthese von Rhoidin-g$_7$-trimethylester aus Chlorine$_6$-trimethylester, zugleich Vollendung der Harvard-Synthese des Chlorophylls a zum Chlorophyll b, *Justus Liebigs Ann. Chem.*, 749, 109, 1971.

3d. **Risch, N., Brockmann, H., Jr., and Gloe, A.,** Strukturaufklärung von neuartigen Bakteriochlorophyllen aus Chloroflexus aurianticus, *Justus Liebigs Ann. Chem.*, p. 408, 1979.

3e. **Smith, K. M.,** Partial synthesis of chlorophyll-A from rhodochlorin, *Tetrahedron*, 37, 399, 1981.

3f. **Wasielewski, M. R. and Thompson, J. F.,** 9-Desoxo-9, 10-dehydrochlorophyll a, *Tetrahedron Lett.*, p. 1043, 1978.

3g. **Wolf, H. and Scheer, H.,** Photochemische Hydrierung von Phäophyrinen: 7,8-cis Phäophorbide, *Justus Liebigs Ann. Chem.*, p. 1710, 1973.

4a. **Thornber, J. P., Markwell, J. P., and Reinman, S.,** Plant chlorophyll protein complexes: recent advances, *Photochem. Photobiol.*, 29, 1205, 1979.

4b. **Cogdell, R. J. and Thornber, J. P.,** Light-harvesting pigment-protein complexes of purple photosynthetic bacteria, *FEBS Lett.*, 122, 1, 1980.

4c. **Gingras, G.,** Comparative review of photochemical reaction center preparations from photosynthetic bacteria, in *The Photosynthetic Bacteria*, Clayton, R. K. and Sistrom, W. R., Eds., Plenum Press, New York, 1978, chap. 6.

4d. **Thornber, J. P., Trosper, T. L., and Strouse, C. E.,** Bacteriochlorophyll *in vivo*: relationship of spectral forms to specific membrane components, in *The Photosynthetic Bacteria*, Clayton, R. K. and Sistrom, W. R., Eds., Plenum Press, New York, 1978, chap. 7.

4e. **Olson, J. M.,** Bacteriochlorophyll a-proteins from green bacteria, in *The Photosynthetic Bacteria*, Clayton, R. K. and Sistrom, W. R., Eds., Plenum Press, New York, 1978, chap. 8.

4f. **Anderson, J. M. and Barrett, J.,** Light-harvesting pigment-protein complexes of algae, in *Encyclopedia of Plant Physiology*, n.s., Vol. 19, *Photosynthesis III*, Staehelin, L. A. and Arntzen, C. J., Eds., Springer Verlag, Berlin, 1986, 269.

4g. **Anderson, B. and Anderson, J. M.,** The chloroplast thylakoid membrane — isolation, subfractionation, and purification of its supramolecular complexes, in *Modern Methods of Plant Analysis*, n.s., Vol. 1, *Cell Components*, Linskens, H. F. and Jackson, J. F., Eds., Springer Verlag, Berlin, 1985, 231.

4h. **Thornber, J. P.,** in *Encyclopedia of Plant Physiology*, n.s., Vol. 19, *Photosynthesis III*, Staehelin, L. A. and Arntzen, C. J., Eds., Springer Verlag, Berlin, 1986, chap.3.

5. IUPAC-IUB Joint Commission on Biochemical Nomenclature (JCBN), Nomenclature of tetrapyrroles, *Pure Appl. Chem.*, 51, 2251, 1979.

6. **Bonnett, R.,** Nomenclature, in *The Porphyrins*, Vol. 1, Dolphin, D., Ed., Academic Press, New York, 1978, 1.

7. **Fischer, H. and Orth, H.,** *Die Chemie des Pyrrols*, Vol. 2, 2nd half, Akademische Verlagsgesellschaft, Leipzig, 1940; reprinted by Johnson Reprint Corp., New York, 1968.

8. **Baker, E. W. and Palmer, S. E.,** Geochemistry of porphyrins, in *The Porphyrins*, Vol. 1, Dolphin, D., Ed., Academic Press, New York, 1978, 485.

9. **Schoch, S., Scheer, H., Schiff, J. A., Siegelman, H. W., and Rüdiger, W.,** Pyropheophytin accompanies pheophytin in darkened light grown cells of *Euglena*, *Z. Naturforsch.*, 36c, 827, 1981.

10. **Feher, G. and Okamura, M.Y.,** Chemical composition and properties of reaction centers, in *The Photosynthetic Bacteria*, Clayton, R. K. and Sistrom, W. R., Eds., Plenum Press, New York, 1978, chap. 19.

11. **Klimov, V. V., Dolan, E., Shaw, E. R., and Ke, B.,** Interaction between the intermediary electron acceptor (pheophytin) and a possible Plastoquinone-iron complex, *Photosystem II*, 77, 7227, 1980.

12a. **Scott, A. I., Irwin, A. J., Siegel, L., and Shoolery, J. S.**, Sirohydrochlorin. Prosthetic group of sulfite and nitrite reductase in its role in the biosynthesis of vitamin B_{12}, *J. Am. Chem. Soc.*, 100, 7987, 1978.

12b. **Deeg, R., Kriemler, H. P., Bergmann, K.-H., and Müller, G.**, Neuartige, methylierte Hydroporphyvine und deren Bedeutung bei der Cobyrinsäure-Bildung, *Z. Physiol. Chem.*, 358, 339, 1977.

12c. **Imfeld, M., Arigoni, D., Deeg, R., and Mueller, G.**, Factor I ex *Clostridium tetanomorphum:* proof of structure and relationship to vitamin B12 synthesis, in *Vitamin B12 and Intrinsic Factor, 3rd. Eur. Symp.*, de Gruyter, Berlin, 1979, 315.

12d. **Battersby, A. R. and McDonald, E.**, Origin of the pigments of life: the type-III problem in porphyrin biosynthesis, *Acc. Chem. Res.*, 12, 14, 1979.

13a. **Siegel, L. M., Murphy, M. J., and Kamin, H.**, Reduced nicotinamide adenine dinucleotide phosphate-sulfite reductase of enterobacteria. I, *J. Biol. Chem.*, 248, 251, 1973.

13b. **Vega, J. M. and Kamin, H.**, Spinach nitrite reductase, *J. Biol. Chem.*, 252, 896, 1977.

14a. **Agins, L., Ballantine, J. A., Ferrito, V., Jaccarini, J., Murray-Rust, P., Pelter, A., Psaila, A. F., and Schembri, P. J.**, Bonellin, *Pure Appl. Chem.*, 51, 1847, 1979.

14b. **Matthews, J. I., Braslavsky, S. E., and Camilleri, P.**, The photophysics of bonellin: a chlorin found in marine animals, *Photochem. Photobiol.*, 32, 733, 1980.

15. **Vernon, L. P. and Seely, G. R., Eds.**, *The Chlorophylls*, Academic Press, New York, 1966.

16a. **Katz, J. J., Norris, J. R., Shipmann, L. S., Thurnauer, M. C., and Wasielewski, M. R.**, Chlorophyll functions in the photosynthetic reaction center, *Annu. Rev. Biophys. Bioeng.*, 7, 393, 1978.

16b. **Battersby, A. R. and McDonald, E.**, Biosynthesis of porphyrins, chlorins and corrins, in *Porphyrins and Metalloporphyrins*, Smith, K. M., Ed., Elsevier, Amsterdam, 1975, chap. 3.

16c. **Katz, J. J.**, Chlorophyll, in *Inorganic Biochemistry*, Eichhorn, G., Ed., Elsevier, Amsterdam, 1973, 1022.

16d. **Scheer, H. and Inhoffen, H. H.**, Hydroporphyrins: reactivity, spectroscopy, and hydroporphyrin analogues, in *The Porphyrins*, Dolphin, D., Ed., Vol. 2 (Part B), Academic Press, New York, 1978.

16e. **Jones, O. T. G.**, Chlorophyll biosynthesis, in *The Porphyrins*, Vol. 3, Dolphin, D., Ed., Academic Press, New York, 1978, chap. 3.

16f. **Weiss, C.**, Optical spectra of chlorophylls, in *The Porphyrins*, Vol. 3, Dolphin, D., Ed., Academic Press, New York, 1978, chap. 3.

16g. **Brockmann, H., Jr.**, Stereochemistry and absolute configuration of chlorophylls and linear tetrapytrroles, in *The Porphyrins*, Vol. 2, Dolphin, D., Ed., Academic Press, New York, 1978, chap. 9.

16h. **Sauer, K.**, Primary events and the trapping of energy, in *Bioenergetics of Photosynthesis*, Govindjee, Ed., Academic Press, New York, 1978, chap. 3.

16i. **Papageorgiou**, Chlorophyll fluorescence: an intrinsic probe of photosynthesis, in *Bioenergetics of Photosynthesis*, Govindjee, Ed., Academic Press, New York, 1975, chap. 6.

16j. **Rüdiger, W. and Schoch, S.**, Chlorophylls, in *Chemistry and Biochemistry of Plant Pigments*, Vol. 2, Goodwin, Ed., in press.

16k. **Bogorad, L.**, in *Chemistry and Biochemistry of Plant Pigments*, Vol. 1, Goodwin, T. W., Ed., Academic Press, London, 1976, 64.

16l. **Holden, M.**, in *Chemistry and Biochemistry of Plant Pigments*, Vol. 2, Goodwin, T. W., Ed., in press.

16m. **Jackson, A. H.**, in *Chemistry and Biochemistry of Plant Pigments*, Vol. 1, Goodwin, T. W., Ed., Academic Press, London, 1976, 1.

16n. **Schneider, H. A. W.**, in *Pigments in Plants*, Czygan, F. C., Ed., Gustav Fischer Verlag, Stuttgart, 1980, 237.

16o. **Castelfranco, P. A. and Beale, S. I.**, in *The Biochemistry of Plants*, Vol. 8, Stumpf, P. K. and Conn, E. E., Eds., Academic Press, New York, 1981.

16p. **Castelfranco, P. A. and Beale, S. I.**, *Annu. Rev. Plant Physiol.*, 34, 241, 1983.

16q. **Porra, R. J. and Meisch, H.-U.**, *TIBS*, 9, 99, 1983.

16r. **Leeper, F. J.**, *Nat. Prod. Rep.*, 2, 19 and 561, 1985.

16s. **Larkum, A. W. D. and Barrett, J.**, *Adv. Bot. Res.*, 10, 1, 1983.

17. **Jones, M. S. and Jones, O. T. G.**, Ferrochelatase of *Rhodopseudomonas spheroides*, *Biochem. J.*, 119, 453, 1970.

18. **Csatorday, K., MacColl, R., and Berns, D. S.**, Accumulation of protoporphyrin IX and zinc protoporphyrin IX in *Cyanidium caldarium*, *Proc. Natl. Acad. Sci. U.S.A.*, 78, 1700, 1981.

19. **Schoch, S., Lempert, U., and Rüdiger, W.**, Über die letzten Stufen der Chlorophyllbiosynthese: Zwischenprodukte zwischen Chlorophyllid und phytohaltigem Chlorophyll, *Z. Pflanzenphysiol.*, 83, 427, 1977.

20. **Scheer, H., Svec, W. A., Cope, B. T., Studier, M. H., Scott, R. G., and Katz, J. J.**, Structure of bacteriochlorophyll *b*, *J. Am. Chem. Soc.*, 96, 3714, 1974.

21a. **Shlyk, A. A., Fradkin, L. I., Rudoi, A. B., Prudnikova, I. V., and Savchenko, G. E.**, Group mechanism of pigment assembly in centers of chlorophyll biosynthesis, in *Chloroplast Development, Developments in Plant Biology*, Vol. 2, Akoyonoglu, G. and Akoyonoglu, J. H., Eds., Elsevier, New York, 1978, 119.

21b. **Bednarik, D. P. and Hoober, K. J.**, *Science*, 230, 450, 1985.

22. **Dörnemann, D. and Senger, H.,** Isolation and partial characterization of a new chlorophyll associated with the reaction centre of photosystem I of scenedesmus, *FEBS Lett.,* 126, 323, 1981.
23. **Rebeiz, C. A., Balanger, F. C., Freyssinet, G., and Saab, D. S.,** Chloroplast biogenesis. XXIX. The occurrence of several novel chlorophyll *a* and *b* chromophores in higher plants, *Biochim. Biophys. Acta,* 50, 234, 1980.
24. **Scholz, B. and Ballschmiter, K.,** Do all 8 diastereomeric bacteriochlorophylls exist in nature, *Angew. Chem.,* 20, 956, 1981.
25. **Bazzaz, M. B.,** New chlorophyll chromophores isolated from a chlorophyll-deficient mutant of maize, *Photobiochemistry,* 2, 199, 1981.
26. **Smith, K. M., Bisset, G. M. F., and Bushell, M. J.,** Partial synthesis of optically pure methyl bacteriopheophorbides *c* and *d* from methyl pheophorbide *a, J. Org. Chem.,* 45, 2218, 1980.
27. **Brockmann, H., Jr.,** Bacteriochlorophyll e: structure and stereochemistry of a new type of chlorophyll from Chlorobiaceae, *Philos. Trans. R. Soc. London B,* 273, 277, 1976.
28. **Caple, M. B., Chow, H. C., and Strouse, C. E.,** Photosynthetic pigments of green sulfur bacteria (the esterifying alcohols of bacteriochlorophylls c from *Chlorobium limicola, J. Biol. Chem.,* 253, 6730, 1978.
29. **Steiner, R., Schäfer, W., Blos, I., Wieschhoff, H., and Scheer, H.,** Δ2,10-Phytadienol as esterifying alcohol of bacteriochlorophyll *b* from *Ectothiorhodospira halochloris, Z. Naturforsch.,* 36C, 417, 1981.
30. **Katz, J. J., Strain, H. H., Harkness, A. L., Studier, M. H., Svec, W. A., Janson, T. R., and Cope, B. T.,** Esterifying alcohols in the chlorophylls of purple photosynthetic bacteria. A new chlorophyll, bacteriochlorophyll (gg), all-trans geranylgeranyl bacteriochlorophyllide a, *J. Am. Soc.,* 94, 7938, 1972.
31a. **Holt, A. S.,** Recently characterized chlorophylls, in *The Chlorophylls,* Vernon, L. P. and Seely, G. R., Eds., Academic Press, New York, 1966, 111.
31b. **Gloe, A., Pfennig, N., Brockmann, H., Jr., and Trowitzsch, W.,** A new bacteriochlorophyll from brown-colored Chlorobiaceae, *Arch. Mikrobiol.,* 102, 103, 1975.
32. **Walter, E., Schreiber, J., Zass, E., and Eschenmoser, A.,** Bchl a$_{GG}$ und Bphe a$_p$ in den photosynthetischen Reaktionszentren von *R. rubrum* G 9, *Helv. Chim. Acta,* 62, 899, 1979.
33a. **Egle, K.,** Biologischer Chlorophyllabbau, in *Handbuch der Pflanzenphysiologie,* Vol. 5, Part 1, Ruhland, W., Ed., Springer-Verlag, Berlin, 1960, 354.
33b. **Yentsch, CH. S.,** The relationship between chlorophyll and photosynthetic carbon production with reference to the measurements of decomposition products of chloroplastic pigments, *Mem. 1st Ital. Idrobiol.,* 18 (Suppl.), 322, 1965.
34. **Morris, M. M., Park, K., and Mackinney, G.,** On the photodecomposition of chlorophyll in vitro, *J. Agric. Food Chem.,* 21, 277, 1973.
35a. **Katz, J. J. and Janson, T. R.,** Chlorophyll-chlorophyll interactions from ^{1}H and ^{13}C nuclear magnetic resonance spectroscopy, *Ann. N.Y. Acad. Sci.,* 206, 579, 1973.
35b. **Steiner, R., Wieschhoff, H., and Scheer, H.,** HPLC of bacteriochlorophyll b and its derivatives as an aid for structure analysis, *J. Chromatogr.,* 242, 127, 1982.
35c. **Gottstein, J. and Scheer, H.,** unpublished.
36a. **Gottstein, J. and Scheer, H.,** Long-wavelength absorbing forms of bacteriochlorophyll a in solutions of Triton-X 100, *Proc. Natl. Acad. Sci. U.S.A.,* 80, 2231, 1981.
36b. **Scherz, A. and Parson, W. W.,** Oligomers of bacteriochlorophyll and bacteriopheophytin with spectroscopic properties resembling those found in photosynthetic bacteria, *Biochim. Biophys. Acta,* 766, 653, 1984; Exciton interactions in dimers of bacteriochlorophyll and related molecules, *Biochim. Biophys. Acta,* 766, 666, 1984.
36c. **Scheer, H., Paulke, B., and Gottstein, J.,** Long-wavelength absorbing forms of bacteriochlorophylls, in *Optical Properties and Structure of Tetrapyrroles,* Blaur, G. and Sund, H., Eds., de Gruyter, Berlin, 1985, 507.
36d. **Scherz, A., Rosenbach, V., and Malkin, S.,** *Biochim. Biophys. Acta,* in press.
37. **Brockmann, H., Jr., and Kleber, I.,** Bacteriochlorophyll *b, Tetrahedron Lett.,* p. 2195, 1970.
38. **Davis, M. S., Forman, A., Hanson, L. K., Thornber, J. P., and Fajer, J.,** Anion and cation radicals of bacteriochlorophyll and bacteriopheophytin b. Their role in the primary charge separation of *Rhodopseudomonas viridis, J. Phys. Chem.,* 83, 3325, 1979.
39. **Dougherty, R. C., Strain, H. H., Svec, W. A., Uphans, R. A., and Katz, J. J.,** The structures, properties and distribution of Chlorophyll *c, J. Am. Chem. Soc.,* 92, 2826, 1970.
40. **Budzikiewicz, H. and Taraz, K.,** Chlorophyll c, *Tetrahedron,* 27, 1447, 1971.
41. **Smith, K. M.,** General features of the structure and chemistry of porphyrin compounds, in *Porphyrins and Metalloporphyrins,* Smith, K. M., Ed., Elsevier, New York, 1975, chap. 1.
42. **Goedheer, J. C.,** Visible absorption and fluorescence of chlorophyll and its aggregates in solution, in *The Chlorophylls,* Vernon, L. P. and Seely, G. R., Eds., Academic Press, New York.
43. **Wolf, H. and Scheer, H.,** Stereochemistry and chiroptic properties of pheophorbides and related compounds, *Ann. N.Y. Acad. Sci.,* 206, 549, 1973.

44a. **Scheer, H. and Katz, J. J.,** Nuclear magnetic resonance spectroscopy of porphyrins and metalloporphyrins, in *Porphyrins and Metalloporphyrins,* 2nd ed., Smith, K. M., Ed., Elsevier, New York, 1975.

44b. **Janson, T. R. and Katz, J. J.,** NMR spectra of diamagnetic porphyrins, in *The Porphyrins,* Vol. 4, Dolphin, D., Ed., Academic Press, New York, 1978, chap. 1.

45. **Scheer, H.,** Darstellung und absolute Konfiguration von 7,8-cis Phäophorbiden und 9-Hydroxy-phäphorbiden, Ph.D. thesis, Technical Braunschweig, University, West Germany, 1971.

46. **Norris, J. R., Scheer, H., and Katz, J. J.,** ENDOR spectroscopy of chlorophylls and the photosynthetic light conversion apparatus, in *The Porphyrins,* Vol. 4, Dolphin, D., Ed., Academic Press, New York, 1978, chap. 3.

47. **Boxer, S. G., Closs, G. L., and Katz, J. J.,** The effect of magnesium coordination on the ^{13}C and ^{15}N magnetic resonance spectra of chlorophyll *a*. The relative energies of nitrogen Nπ^* states as deduced from a complete assignment of chemical shifts, *J. Am. Chem. Soc.,* 96, 7058, 1974.

48. **Budzikiewicz, H.,** Mass spectra of porphyrins and related compounds, in *The Porphyrins,* Vol. 3, Dolphin, D., Ed., Academic Press, New York, 1978, chap. 9.

49a. **Constantin, E., Nakatani, Y., Teller, G., Hueber, R., and Ourisson, G.,** Electron-impact and chemical ionization mass-spectrometry of chlorophylls, phaeophytins and phaeophorbides by fast desorption on a gold support, *Bull. Soc. Chim. Fr.,* p. 303, 1981.

49b. **Grotemeyer, J., Bosel, U., Walter, K., and Schlag, E. W.,** Multiphoton-ionization mass spectroscopy of native chlorophylls, *J. Am. Chem. Soc.,* 108, 4233, 1986.

49c. **Dougherty, R. C., Dreifuss, P. A., Sphon, J., and Katz, J. J.,** Hydration behavior of chlorophyll a: a field desorption mass spectral study, *J. Am. Chem. Soc.,* 102, 416, 1980.

49d. **Tabet, J. C., Jablonski, M., Cotter, R. J., and Hunt, J. E.,** Time resolved laser desorption. III. The metastable decomposition of chlorophyll a and some derivatives, *Int. J. Mass Spectrom. Ion Phys.,* 65, 105, 1985.

50. **Hunt, J. E., MacFarlane, R. D., Katz, J. J., and Dougherty, R. C.,** High-energy fragmentation of chlorophyll *a* and its fully deuterated analogue by ^{252}Cf plasma desorption mass spectrometry, *J. Am. Chem. Soc.,* 103, 6775, 1981.

51a. **Scholz, B. and Ballschmitter, K.,** Preparation and reversed-phase high-performance liquid-chromatography of chlorophylls (technical note), *J. Chromatogr.,* 208, 148, 1981.

51b. **Iriyama, K., Shiraki, M., and Yoshiura, M.,** An improved method for extraction, partial purification, separation and isolation of chlorophyll from spinach, *J. Liq. Chromatogr.,* 2, 255, 1979.

51c. **Gleixner, G., Karg, V., and Kis, P.,** Rapid preparation of pure chlorophyll *a, Experientia,* 38, 303, 1982.

52. **Whitlock, H. W., Jr., Hanauer, R., Oester, M. Y., and Bower, B. K.,** Diimide reduction of porphyrins, *J. Am. Chem. Soc.,* 91, 7485, 1969.

53. **Risch, N., Reich, H., Schormann, A., and Brockmann, H., Jr.,** Note on a simple method for the separation of chlorophyll derivatives of the A-series and B-series, *Justus Liebigs Ann. Chem.,* p. 1519, 1981.

54. **Scheer, H., Katz, J. J., and Norris, J. R.,** Proton-electron hyperfine coupling constants of the chlorophyll a cation radical by ENDOR spectroscopy, *J. Am. Chem. Soc.,* 99, 1372, 1977.

55. **Baum, S. J., Burnham, B. F., and Plane, R. A.,** Studies on the biosynthesis of chlorophyll: chemical incorporation of magnesium into porphyrins, *Proc. Natl. Acad. Sci. U.S.A.,* 52, 1439, 1964.

56. **Isenring, H. P., Zass, E., Smith, K., Falk, H., Le Luisier, J., and Eschenmoser, A.,** Enolisierte Derivate der Chlorophyllreihe: 132-Desmethoxycarbonyl-17³-desoxy-cyclochlorophyllid *a*-enol und eine Methode zur Einführung von Mg unter milden Bedingungen, *Helv. Chim. Acta,* 58, 2357, 1975.

57. **Wasielewski, M. R.,** A mild method for the introduction of magnesium into bacteriopheophytin a, *Tetrahedron Lett.,* p. 1373, 1977.

58. **Bucks, R. R. and Boxer, S. G.,** Synthesis and spectroscopic properties of a novel cofacial chlorophyll-based dimer, *J. Am. Chem. Soc.,* 104, 340, 1982.

59. **Hynninen, P. H.,** Application of elution analysis to the study of chlorophyll transformations by column chromatography on sucrose, *J. Chromatogr.,* 175, 75, 1979.

60. **Jeffrey, S. W.,** Properties of two spectrally different components in chlorophyll *c* preparations, *Biochim. Biophys. Acta,* 177, 456, 1969.

61. **Iriyama, K. and Yoshiura, M.,** Separation of chlorophyll *a* and chlorophyll *b* by column chromatography with sephadex LH-20 or powdered sugar, *J. Chromatogr.,* 177, 154, 1979.

62. **Sato, N. and Murata, N.,** Preparations of chlorophyll *a,* chlorophyll *b* and bacteriochlorophyll *a* by means of column chromatography with diethylaminoethylcellulose, *Biochim. Biophys. Acta,* 501, 103, 1981.

63. **Braumann, I. and Grimme, L. H.,** Reversed-phase high-performance liquid-chromatography of chlorophylls and carotenoids, *Biochim. Biophys. Acta,* 637, 8, 1981.

64. **Risch, N., Kemmer, T., and Brockmann, H., Jr.,** Chromatographische Trennung von Bchl *e, Justus Liebigs Ann. Chem.,* 1978, 585, 1978.

65. **Brereton, R. G., Rajananda, V., Blake, T. J., Sanders, J. K., and Williams, D. H.,** "In beam" electron impact mass spectrometry: the structure of a bacteriochlorophyll allomer, *Tetrahedron Lett.,* p. 1671, 1980.

66. **Ellsworth, R. K., Tsuk, R. M., and St. Pierre, L. A.,** Attribution of hydrolytic and esterifying "chlorophylase" activities observed in vitro to two enzymes, *Photosynthetica,* 10, 312, 1970 **Aiga, I. and Sasa, T.,** Formation of atypical chlorophyllide *a, Plant Cell. Physiol.,* 11, 161, 1970.

67. **Hynninen, P.,** Isolation of chlorophylls *a* and *b* using an improved two-phase extraction method followed by precipitation and a separation on a sucrose column, *Acta Chem. Scand. B,* 31, 829, 1977.

68. **Goyens, L., Post, E., Dehairs, F., Vandenhout, A., and Bayens, W.,** The use of HPLC with fluorimetric detection for chlorophyll *a* determination in natural extracts of chloropigments and their degradation products, *Int. J. Environ. Anal. Chem.,* 12, 51, 1982.

69. **Brown, L. M., Hargrave, B. T., and MacKinnon, M. D.,** Analysis of chlorophyll *a* in sediments by HPLC, *Can. J. Fish. Aquat. Sci.,* 38, 205, 1981.

70. **Bessiere, J. and Montel, A.,** Methode rapide de dosage selectif des chlorophylls *a* et *b:* utilisation de la separation par HPLC, *Water Res.,* 16, 987, 1982.

71. **Scheer, H. and Rauscher, G.,** Empfindliche und flexible Kopplung von HPLC und AS, *Labor Praxis,* 4—7, 24, 1980.

72. **Brereton, and Sanders,** private communication.

73. **Watanabe, T., Nakazato, M., Mazaki, H., Hongu, A., Konno, M., Saitoh, S., and Honda, K.,** *Biochim. Biophys. Acta,* 807, 110, 1985.

Tables for the Estimation and Separation of Chlorophylls

GENERAL TABLES

Table III.1

NAME LIST: STRUCTURES, FUNCTIONS, OCCURRENCE, AND SPECTRA OF CHLOROPHYLLS

Structure	Pigment	R[a]	Occurrence	Function[b]	Chlorophyll	Pheophytin
	Chlorophyll *a* $R_1 = H$, $R_2 = C_2H_5$, $R_3 = COOCH_3$, $R_4 = H$ $C_{55}H_{72}N_4O_5Mg$ MW = 892	$\Delta 2^c$	All oxygenic photosynthetic organisms	A + RC[d]	662, 430[i]	667, 535, 505,[1] 408
	Chlorophyll *b* $C_{55}H_{70}N_4O_6Mg$ MW = 906	$\Delta 2$	Green plants Algae[e] Prochloro	A	644, 430	655, 525, 412

Table III.1 (continued)
NAME LIST: STRUCTURES, FUNCTIONS, OCCURRENCE, AND SPECTRA OF CHLOROPHYLLS

Structure	Pigment	R^a	Occurrence	Function[b]	Chlorophyll	Pheophytin
	Chlorophylls c_1, c_2 $C_{35}H_{30}N_4O_5Mg$ MW = 610 (Chl c_2 has 2H less)	H	Pheophyta Cryptophyta Pyrrophyta Chrysophyta Bacillariophyta Prasynophyta	A[f]	626, 576, 444[g] (627, 578, 448)	650, 592, 579, 532, 433[h]
	Chlorophyll d $C_{54}H_{70}N_4O_6Mg$ MW = 894	Δ2	Rhodophyta Chlorella (?)	A[i]	688, 447	692, 547, 516, 421

Structure		Organism		Absorption maxima	Absorption maxima
Protochlorophyllide ($R' = C_2H_5$ or C_2H_3) $C_{35}H_{32}N_4O_5Mg$ MW = 612	H^j	Oxygenic photosynthetic organisms	P	623, 432[k]	638, 586, 564, 525, 417
Bacterioprotochlorophyllide ($R' = C_2H_3$)	H^j	Photosynthetic bacteria	P		
Bacteriochlorophyll a $C_{55}H_{74}N_4O_6Mg$ MW = 910 ($R = \Delta2$)	$\Delta2;\ \Delta2,6,10,14$	Photosynthetic bacteria	A + RC	773, 577, 358	749, 525, 385, 357
Bacteriochlorophyll b ($R_1 = COCH_3$) $C_{55}H_{72}N_4O_6Mg$ MW = 908 ($R = \Delta2$)	$\Delta2;\ \Delta2,10$	Few species of photosynthetic bacteria[m]	A + RC	794, 580, 368	776, 528, 398, 368
Bacteriochlorophyll g ($R_1 = C_2H_3$) $C_{55}H_{72}N_4O_6Mg$ MW = 886	$\Delta2,6,10,14$	*Heliobacterium chlorum*	A + RC	763, 575, 470, 418, 408	753, 518, 396, 388

Table III.1 (continued)
NAME LIST: STRUCTURES, FUNCTIONS, OCCURRENCE, AND SPECTRA OF CHLOROPHYLLS

Structure	Pigment	R[a]	Occurrence	Function[b]	Chlorophyll	Pheophytin
	Bacteriochlorophylls c[n]	Mainly farnesol, many others	Chlorobiaceae, Chloroflexaceae	A[o]	660, 432	664, 547, 515, 408
	Bacteriochlorophylls d[h]		Chlorobiaceae, Chloroflexaceae	A	646, 458	658, 548, 505, 406
	Bacteriochlorophylls e[n] (isomer mixture)		Chlorobiaceae, Chloroflexaceae	A	654, 424	654, 534, 439

a   See formulas at the end of the table for the alcohol.

b   A = antenna or light-harvesting pigments; RC = reaction center pigments; P = biosynthetic precursor.

c   Accompanied by $\Delta2,6,10,14$; $\Delta2,10,14$; and $\Delta2,14$ as biosynthetic precursors. Further precursors contain a second vinyl group at C-8 ($R_2 = C_2H_3$).

d   The reaction center of photosystem II probably contains pheophytin a (no central Mg) as intermediary electron acceptor. Chlorophyll-RCI (13^2-hydroxy-20-chloro-chlorophyll a) ($R_1 = Cl$, $R_2 = C_2H_5$, $R_3 = COOCH_3$, $R_4 = OH$) and the 13^2-epimer of chlorophyll a, e.g., chlorophyll a' ($R_1 = H$, $R_2 = C_2H_5$, $R_3 = H$, $R_4 = COOCH_3$) have been found in photosystem I reaction centers.

e   Not in xanthophytes, rhodophytes, cryptophytes, or cyanobacteria.

f   Marine algae contain up to 50% of the chlorophylls c_1 and c_2.

g C_1 values for c_2 in brackets.

h   Mixture of c_1 and c_2 in CH_2Cl_2.

i   Possibly an artifact. However, some species are reported to contain up to 33% of the chlorophylls as chlorophyll d.

j   Protochlorophyll occurs, in part, in the esterified form, e.g., with $\Delta2$-phytaenol (= phytol) and its precursors.

k   The spectra are solvent dependent. See Table III.2 for UV-vis spectra of chlorophylls in different solvents.

l   Monovinyl, the divinyl derivatives has a pronounced red shift (7 nm) in the Soret, and only a small red shift in the long-wavelength region.

m   *Rhodopseudomonas viridis*, *Rp. sulfoviridis*, *Thiocapsa pfennigii*, *Ectothiorhodospira halochloris*, *Et. abdelmalekii* contain bacteriochlorophyll *b*, and *Heliobacterium chlorum*, bacteriochlorophyll *g*.

Bchl	c	d	e
R_1	CH$_3$	CH$_3$	CHO
R_2	Ethyl, *n*-propyl, *i*-butyl or neopentyl		
R_3	Ethyl or Methyl		
R_4	Phtyl, farnesyl, cetyl, others		
R_5	CH$_3$	H	CH$_3$

n   Very variable structure; also variable stereochemistry at C-31

o   Bacteriopheophytin *c* (or a similar pigment) has recently been reported to occur in reaction centers of green bacteria.

Table III.2
QUANTITATIVE SPECTROSCOPIC DATA: MOLAR EXTINCTION COEFFICIENTS (cm^{-1} mol^{-1}) OF CHLOROPHYLLS. COMPLEMENTARY SPECTROSCOPIC DATA[a]

(ϵ) = Molar Extinction Coefficient × 10^{-3}

Compound[b]	Solvent	λ max(nm) (ε)	λ max(nm) (ε)	λ max(nm) (ε)	λ max(nm) (ε)	λ max(nm) (ε)	λ max(nm) (ε)	Ref.
Chlorophyll a	Diethyl ether	410 (76.0)	430 (117.3)	533 (3.76)	578 (8.27)	615 (14.5)	662 (90.0)	1
	Acetone	—	429 (111.6)	—	—	—	661 (86.2)	15
	Acetone, 80% aqueous	—	430 (94.6)	—	582 (10.3)	618 (17.5)	663 (82.6)	16
	Ethanol, 96%	—	433 (90.5)	536 (4.26)	—	—	665 (81.0)	3
		—	432 (74.2)	—	—	—	665 (74.4)	17
Pheophytin a	Diethyl ether	408 (114.8)	471 (4.44)	505 (12.7)	534 (11.0)	609.5 (8.53)	667 (55.4)	2
	Dioxane	—	—	506 (10.1)	535 (8.65)	609 (6.98)	667 (43.0)	5
	Acetone, 80% aqueous	409 (113.8)	472 (5.48)	505 (13.1)	536 (11.4)	610 (10.4)	667 (49.2)	3
Methylpheophorbide a 8-Deethyl 8-vinylchlorophyll a	Dioxane	—	—	506 (11.4)	535 (9.27)	610 (7.77)	666 (52.8)	6
Chlorophyll b	Ether	430 (56.8)	455 (158.4)	549 (6.40)	595 (11.5)	—	644 (56.2)	2
	Acetone	—	453 (158.8)	—	—	—	642 (56.0)	15
	Acetone, 80% aqueous	—	455 (133.1)	536 (5.77)	558 (7.17)	600 (13.0)	645 (46.9)	16
	Ethanol, 96%	—	460 (184.1)	—	—	—	649 (47.6)	3
		—	464 (107.2)	—	—	—	649 (40.0)	1
Pheophytin b	Diethylether	412 (73.4)	434 (190.9)	525 (12.6)	555 (7.69)	599 (8.40)	655 (37.2)	2
	Dioxane	—	—	—	—	—	655 (81.6)	3
	Acetone, 80% aqueous	—	436 (160.0)	527 (13.2)	558 (8.38)	600 (9.46)	652 (30.6)	7
Methylpheophorbide b	Dioxane	—	—	525 (11.8)	552 (7.72)	600 (7.8)	—	2
Chlorophyll c	Diethylether[c]	447 (138.9)	—	—	579 (12.6)	628 (13.5)	627 (18.1)	18
	Diethyl ether	447 (159.9)	—	—	—	—	—	—
	Acetone	442 (70.7)	—	—	—	—	628 (9.6)	19
Chlorophyll c₁	Acetone	446 (212.3)	—	—	—	—	629 (23.9)	20
	Pyridine	461 (211.1)	—	—	—	—	640 (21.4)	20
Chlorophyll c₂	Acetone	445 (195.8)	—	—	—	—	630 (22.7)	20
	Pyride	466 (280)	—	—	—	—	642 (19.4)	20

Pigment	Solvent							Ref.
Chlorophyll *d*	Diethyl ether[c]	392 (52.7), 445 (87.4)	447 (87.4)	548 (3.6)	595 (8.47)	643 (12.8)	688 (98.7)	21
Bacteriochlorophyll a_p	Diethyl ether	358 (73.3)	391 (48.0)	530 (2.73)	577 (20.8)	697 (9.11)	686 (105)	21
	Diethyl ether						773 (91.0)	2
	Acetone	358 (85.3)					770 (96.0)	9
	Dioxane	358 (40.2)					772 (95.6)	18
	Carbon tetrachloride						775 (20.1)	22
	Acetone/methanol = 7:2						772 (115)	9
	Methanol	365 (53.9)				608 (15.4)	781 (88.0)	9
							767 (76.0)	2
							772 (42.0)	9
Bacteriopheophytin *a*	Diethyl ether	357 (113.6)	384 (62.7)	525 (28.3)	625 (3.64)	680 (10.7)	749 (67.5)	2
	Chloroform	363 (99.4)	390 (59.4)	533 (26.2)	630 (3.91)	687 (11.5)	757 (63.4)	2
a_p	Diethyl ether	357 (105.9)	385 (61.0)	525 (28.2)	623 (3.10)	678 (9.51)	749 (71.4)	10
a_{CG}	Diethyl ether	358 (109.6)	386 (58.9)	526 (27.5)	622 (3.24)	680 (8.32)	751 (67.9)	10
a'_p	Diethyl ether	357 (113.0)	384 (61.0)	525 (28.3)	622 (3.0)	678 (9.51)	749 (71.6)	10
a'_{CG}	Diethyl ether	358 (109.6)	385 (58.9)	526 (27.5)	624 (3.09)	680 (9.55)	751 (67.6)	10
Bacteriochlorophyll b_p	Diethyl ether[c]	368 (94)	407 (82)		580 (27)		792 (100)	9
	Dioxane[a]	368 (81)	408 (78)		578 (25)		795 (100)	9
Bacteriopheophytin $b_{\Delta 2.10}$	Acetone	368 (226)	398 (237)		528 (50)		794 (100)	12
Bacteriopheophytin *bp* or $b_{\Delta 2.10}$	Diethyl ether[c]					676 (18)[d]	794 (100)	13
	Diethyl ether[c]					678 (25)[u]	776 (100)	13
Bacteriochlorophyll *c*	Acetone[b]	356 (56)	384 (65)	431 (100)	574 (11)	624 (14)	668 (64)	11
	Diethyl ether[c]		419 (55.4)	429 (112)		624 (13)	659 (73)	14
				431 (113.2)			660 (75.6)	18
Bacteriopheophytin *c*	Acetone[b]	379 (Sh)	408 (100)	515 (17)	547 (22)	604 (15)	664 (65)	11
	Diethyl ether		406 (86.2)	512 (8.5)	544 (13)	607 (5)	663 (46)	14
Bacteriochlorophyll *d*	Acetone[b]	330 (45)	408 (87)	424 (100)		608 (17)	654 (61)	11
	Diethyl ether[f]		405 (68.7)	448 (117)	602 (10)	618 (11)	651 (88.3)	14
				425 (113.6)			650 (88.3)	18
Bacteriopheophytin *d*	Acetone[b]	390 (Sh)	406 (100)	505 (14)	533 (14)	548 (16)	658 (55)	11
	Diethyl ether[f]		403 (84.7)	501 (8.8)	531 (9.4)	602 (65)	658 (44.1)	14

Table III.2 (continued)
QUANTITATIVE SPECTROSCOPIC DATA: MOLAR EXTINCTION COEFFICIENTS (cm⁻¹ mol⁻¹) OF CHLOROPHYLLS. COMPLEMENTARY SPECTROSCOPIC DATA[a]

$$(\epsilon) = \text{Molar Extinction Coefficient} \times 10^{-3}$$

Compound[b]	Solvent	λ max(nm) (ε)	λ max(nm) (ε)	λ max(nm) (ε)	λ max(nm) (ε)	λ max(nm) (ε)	λ max(nm) (ε)	Ref.
Bacteriochlorophyll e	Acetone[b,g]	337 (48.5)	458 (100)	—		592 (19)	647 (34)	11
Bacteriopheophytin e	Acetone[b,g]	378 (19.5)	439 (100)	534 (11)	571 (8)	598 (9)	654 (24)	11
Protochlorophyll	Diethyl ether		432 (187)				623 (22.6)	2
	Acetone		432 (165.6)				623 (21.4)	2

a   All extinctions have been given in molar units. For the often-used weight units (cm⁻¹·g⁻¹·ℓ), these values have to be multiplied by the molecular weight. The latter is given in Table III.1.

b   The subscripts GG, P, and Δ 2,10 refer to the esterifying alcohols geranylgeraniol, phytol, and 2,10-phytadienol, respectively.

c   Arbitrary units.

d   In part due to the absorption of oxidation/isomerization product(s).

e   Mixture of several homologues; ϵ calculated from the data of Reference 14, assuming the substituents $R_1 = R_3 = R_5 - CH_3$, $R_2 = C_2H_5$, $R_4 = $ farnesyl.

f   Mixture of several homologues; ϵ calculated from the data of Reference 14, assuming the substituents $R_1 = R_3 = CH_3$, $R_2 = C_2H_5$, $R_4 = $ farnesyl, $R_5 = H$ (Table 1).

g   Mean of bchl e-fractions isolated from six different species; the peak position varies by ± 2 nm, the relative intensities up to 50%.

REFERENCES

1. **Anderson, A. F. H. and Calvin, M.,** *Nature,* 194, 285, 1962.
2. **Smith, J. H. C. and Benitez, A.,** *Modern Methods of Plant Analysis,* Vol. 4, Paech, K. and Tracey, M., Eds., Springer, Berlin, 1955, 142.
3. **Vernon, L. P.,** *Anal. Chem.,* 32, 1144, 1960.
4. **Jeffrey, S. W.,** *Nature,* 194, 600, 1962.
5. **Stern, A. and Wenderlein, H.,** *Hoppe-Seyler's Z. Physiol. Chem.,* 175, 405, 1936.
6. **Stern, A. and Wenderlein, H.,** *Hoppe-Seyler's Z. Physiol. Chem.,* 174, 81, 1935.
7. **Stern, A. and Wenderlein, H.,** *Hoppe-Seyler's Z. Physiol. Chem.,* 174, 32, 1935.
8. **Weigl, J. W.,** *J. Am. Chem. Soc.,* 75, 999, 1953.
9. **Clayton, R. K.,** *Photochem. Photobiol.,* 5, 669, 1966.
10. **Walter, E., Schreiber, J., Zass, E., and Eschemoser, A.,** *Helv. Chim. Acta,* 62, 899, 1979.

11. **Gloe, A.,** Dissertation, Universität Göttingen, Göttingen, 1977.

12. **Baumgarten, D. (with Sauer, K.),** M.S. thesis, University of California, Berkeley, 1970.

13. **Steiner, R.,** Zulassungsarbeit, University of Munich, Munich, 1981.

14. **Holt, A. S.,** *Chemistry and Biochemistry of Plant Pigments,* Goodwin, T. W., Ed., Academic Press, London, 1965, 3.

15. **Strain, H. H., Thomas, M. R., and Katz, J. J.,** *Biochim. Biophys. Acta,* 75, 306, 1963.

16. **Hoffmann, P. and Werner, D.,** *Jena Rev.,* 11, 1114, 1960.

17. **Wintermans, J.F.M.C.,** *Photosynthetica,* 3, 11, 199.

18. **Strain, H. H. and Svec, W.,** *The Chlorophylls,* Vernon, L. P. and Seely, G. R., Eds., Academic Press, New York, 1966.

19. **Jeffrey, S. W.,** *Biochim. Biophys. Acta,* 177, 456, 1969.

20. **Jeffrey, S. W.,** *Biochim. Biophys. Acta,* 279, 15, 1972.

21. **Holt, A. S. and Morley, H. V.,** *Can. J. Chem.,* 37, 507, 1959.

22. **Kim, W. S.,** *Biochim. Biophys. Acta,* 112, 392, 1966.

Table III.3
EFFECTS OF CHEMICAL MODIFICATIONS ON CHLOROPHYLL ABSORPTION

Reference compound	Change	Red	Soret	Remarks	Ref.
Changes at C-3					
Isochlorin e_4-dme	H	−12	−16	a	1
Chlorin e_6-tme	COCH₃	+3	+1		1
	H	−14	−7	a	1
	C₂H₅	−16	−5	a	1
Methylpheophorbide *a*	C₂H₅	−8	−6	d	2
Pheoporphyrin a_5	CH(OCOCH₃)CH₃	+1	+1	c	2
	CHOH-CH₃	+2	+1	c	2
Methylpheophorbide *a*	CHOH-CH₃	−6	−6	d	2
	CH(OCOCH₃)CH₃	−4	−5	d	2
	COCH₃	+10	—	b	2
Pyromethylpheophorbide *a*	CHOHCH₃	−11	−8	a, d	3
	COCH₃	+11	−1	b	3
Chlorophyll *a*	COCH₃	−10	−4	a	4
Pheophytin *a*	C₂H₅	−10	—	d	2
Chlorophyll *a*	CHO(= Chl *d*)	+20	+17	b	5, 6
Pheophytin *a*	CHO(= Chl *d*)	+23	+13	b	5, 6
Bacteriopheophytin *c*	3-5-Ethylene	−9	−4		
Changes at C-7					
Chlorophyll *a*	CH₂NH₂	−6	+5	d	7
	CH₂NH₃	−10	+5	f	7
	CHO (= Chl *b*)	−19	+25	a (twofold)	7
Pheophytin *a*	CHO (= Phe *b*)	−12	+26	a	8

Changes at C-8

Compound	Substituent				Ref.
Pyromethylpheophorbide a	CH(OCH₃)CH₃, CHOH-CH₃	+5	+5	d	3
	C₂H₅	+8	+13	a	3
	COCH₃	−2	+30	a, e	3
Bacteriochlorophyll a	Ethylidene (= Bchl b)	+20	+10	d	9
Bacteriopheophytin a	Ethylidene (= Bphe b)	+27	+33	d	10
Bacteriochlorophytin c,d,e. / Bacteriopheophytin c,d,e	Alkyl other than ethyl	0	0	c	6, 11—14
Pheophytin a	CHOH-CH₃	+5	+5	d	10

Changes at C-12

Compound	Substituent				Ref.
Bacteriochlorophyll a	Alkyl	±0	±0	c	6, 11—14

Changes at the Isocyclic Ring

Compound	Substituent				Ref.
Chlorophyll a	13²-Epimer	±0	±0	c	15
	13²-Silylated enol	+2	+6	d, h	15
Pheophytin a	13¹-Silylated enol	−23	−66	a (sevenfold). n, j, k	15
Chlorophyll a	Cyclochlorophyll enol	+28	+12	c	16
Methylpheophorbide b	Pyro (= 13²-H₂)	+2	+1	a	17
	13²-Methoxy	0	0	c, g	17
	13²-Acetoxy	0	0	c, g	17
Methylpheophorbide a	13²-Methoxy-pyro / Peripheral Mg-complex	+15	+20 / −50	a (threefold). h. j	18
Methylpheophorbide b	Peripheral Mg-complex	+3	+75 / +8	a (threefold). h. j	18
Bacteriomethylpheophorbide a	Peripheral Mg-complex	+20 / −40¹	+25	a. h. m	18
Bacteriopheophytin b	Peripheral Mg-complex	+30 / −70¹	+30 / 0	a. h. j. m	18

Table III.3 (continued)
EFFECTS OF CHEMICAL MODIFICATIONS ON CHLOROPHYLL ABSORPTION

Reference compound	Change	Red	Soret	Remarks	Ref.
Protopheophytin *a*	Peripheral Mg-complex	+80	+55	n	18
Methylpheophorbide *a*	13^2-Alkoxy	±0	±0	d, g	19
	Isochlorine-e_4-dme (ring open, no 13-substituent)	+13	±0	a	1
	Chlorine-e_6-dme (ring opened between 13^1 and 13^2)	+6	±0	a	1
	13-Acetyl-isochlorine-e_4 dme (ring open, $COCH_3$ at 13^1C)	−8	−6		1
Protochlorophyll *a*	13^2-H_2 (= Pyro)	−4	+1	c	20
Protopheophytin *a*	13^2-Methoxy	−2	+1	c	20
Methylpheophorbide *a*	13^1-H,OH	−10	−16	a, o	21, 22
Chlorophyll *a*	13^1-H,OH	−28	±0	a (twofold), o	23
Pheoporphyrin a_5 Zn-complex	13^2-H_2	−2	−3	b	2
Pyromethylpheophorbide *a*	13^1-H_2	−18	−22	a	3, 24
10-Alkoxy-methylpheophorbide *a*	13^1-H_2	−16	−20	a	10
Cu-pyromethylchlorophyllide *a*	13^1-Cl,Δ-13^1,13^2-CHO	−16	+17 / −65	a, h	25
	13^1-Cl,Δ-13^1,13^2-C≡NOH	−6	−5 / −65	a, h	25
	13^1-Cl,Δ-13^1,13^2-CN	−24	+17 / −65	a, h	25
Ring D					
Meso-pyromethylpheophorbide *a*	17,18-*cis*				26—28
Meso-10-methoxymethylpheophorbide *a*	17,18-*cis*	+5	+3		
Chlorophyll *a*, pheophytin *a*, bacteriochlorophyll *a,b,c*, bacteriopheophytin *a,b,c*	Other esterifying alcohols	±0	±0	c	29—32

Changes at C-20

Meso-chlorin e₆-dme	Cl	+13	+6	g, p	1
Meso-isochlorin e₄-dme	Cl	+10	+7	a, p	1
Bacteriochlorophyll *d*	CH₃ (= Bchl *c*)	+10	+8	a, p	6, 11
Bacteriopheophytin *d*	CH₃ (= Bphe *c*)	+8	+6	a, p	4
Pyromethylpheophorbide *a* (3-CHONCH₃)	Br	+18	+8	p	25
Chlorophyll *a*	Cl	+5	+1	Q = 1.54	45
13²-OH-Chlorophyll *a*	Cl	+5	+1	Q = 1.54	43
Pheophytin *a*	Cl	+7	+1	Q = 2.04	43
Methylpheophorbide *a*	Cl	+7	+1	Q = 2.04	44

Central Metals

Methylchlorophyllide *a*	Ni	−12	−14	b, h	3, 33
			−39		
	Pd	−22	−14	b, h	33
			−42		
Methylchlorophyllide *a* (Q = 1.35)	Ni²⁺	−13	−13	h, Q = 1.48	34
	Ag²⁺	−18	−2	h, Q = 1.54	34
	Mn³⁺	0	−63	Q = 0.37	34
	Cd²⁺	−5	−12	Q = 1.76	34
	Co²⁺	−18	−13	Q = 2.08	34
	Cu²⁺	−13	−21	Q = 1.54	34
	Fe³⁺	−52	−38	Q = 2.64	34
	Sn⁴⁺	−9	−12	Q = 1.84	34
	Zn²⁺	−9	−2	Q = 4.08	34
Chlorine₆-tme (Mg-complex³⁴) (Q = 3.13)	Ag²⁺	−16	−1	Q = 2.26	35
	Pb²⁺	+4	+1	Q = 2.62	34
	Ni²⁺	+21	−17	Q = 1.56	34
		−4	−6	Q = 1.88	35
	Sn⁴⁺	0	−4	Q = 1.88	34
		0	−4	Q = 2.78	34
	Zn²⁺	−4	−1	Q = 2.54	35
		+4	−2	Q = 2.60	34

Table III.3 (continued)
EFFECTS OF CHEMICAL MODIFICATIONS ON CHLOROPHYLL ABSORPTION

Reference compound	Change	Red	Soret	Remarks	Ref.
	Co^{2+}	+2	−5	Q = 1.82	35
		−4	−14	Q = 1.94	34
	Pd^{2+}	−19	−15	Q = 1.29 (h)	35
			−25		
	Pt	−27	−15	Q = 2.27	34
		−27	−13	Q = 1.76	35
	Cd	−3	+1	Q = 2.89	35
		+4	+3	Q = 2.92	34
	Hg	−6	−1	Q = 4.3	35
	Cu^{2+}	0	−4	Q = 2.14	34
	V^{4+}	−46	+6	Q = 10	34
	Ni^{2+}	0	−4	Q = 1.88	34
	Fe^{3+}-Cl	−20	−28	Q = 3.44	34
	Fe^{3+}-OH	+44	−25	Q = 4.01	34
	Mn^{3+}-I	−35	−45	Q = 2.81	34
Chlorophyll *a*	Zn^{2+}	−6	n.g.		34
	Cu^{2+}	−10	n.g.		34
Chlorophyll *b*	Zn^{2+}	−6	n.g.		34
	Cu^{2+}	−14	n.g.		34
Pyromethylchlorophyllide *a*	Zn^{2+}	−9	−7	Q = 1.4	25
	Cu^{2+}	−9	−8	Q = 1.3	25
	Ni^{2+}	−21	−14	Q = 1.05	
	Pd^{2+}	−24	−12	Q = 0.84	25
		Other Changes			
Bacteriochlorophyll *a*	From 5 to 6 coordinate Mg	—	—	Red shift of 585-nm band by ~25 nm	36, 37

Chlorophyll *a*, bacteriochlorophyll *a*	Dimerization	Pronounced dependence on the mode of linkage	38, 39
	Aggregation	Small red or blue shifts, very large red shifts of red band in "hydrated aggregates" ("crystalline chlorophyll", "hydrates", and micellar complexes)	40, 41, 46, 47
Chlorophyll *a*, bacteriochlorophyll *a*, isobacteriochlorin	Dehydrogenation at "reduced" rings (D and B)	Change from "bacteriochlorin", "isobacteriochlorin" to "chlorin" and "porphyrin"-type spectra	Figures 2—5
Pheophytin *a,b*, bacteriopheophytin *c,d,e*	Metalation	See section Central metals	Figure 2
Bacteriopheophytin *a,b*	Metalation	Intermediate band strongly red-shifted	Figure 3
Chlorophyll *a,b,c*, bacteriochlorophyll *a-e*	Reduction at methine-bridges	Several products with widely varying spectra	42 (for leading references)

a   Q increased.
b   Q decreased.
c   Very similar spectra.
d   Very similar spectra, except for shift of all bands in the same direction.
e   Intermediate bands (λ_{max} = 500—600 nm) red-shifted.
f   Q similar.
g   All stereoisomers at C-13^1 and/or C-13^2 have essentially identical spectra.
h   Split Soret band.
j   Broadened bands.
k   Additional band around 750 nm.
l   Possibly artifact due to oxidation of a derivative of 3-acetylmethyl-pheophorbide *a*.
m   No band around 530 nm.
n   Only difference spectrum reported.
o   Small differences among the stereoisomers at C-13^1 and/or C-13^2.
p   Band around 535 nm increased.
q   Q similar.
r   Reference = 3-divinyl-3-acetyl-pyromethylpheophorbide *a*.

REFERENCES

1. **Jeckel, G.**, Dissertation, Technische Hochschule, Braunschweig, 1967.
2. **Scheer, H.**, Dissertation, Technische Hochschule, Braunschweig, 1971.
3. **Mengler, C.-D.**, Dissertation, Technische Hochschule, Braunschweig, 1966.
4. **Smith, J. R. L. and Calvin, M.**, *J. Am. Chem. Soc.*, 88, 4500, 1966.
5. **Holt, A. S. and Morley, H. V.**, *Can. J. Chem.*, 37, 507, 1959.
6. **Holt, A. S.**, *The Chlorophylls*, Vernon, L. P. and Seely, G. R., Eds., Academic Press, New York, 1966, 111.
7. **Davis, R. C., Ditson, S. L., Fentiman, A. F., and Pearlstein, R. M.**, *J. Am. Chem. Soc.*, 103, 6823, 1981.
8. **Goedheer, J. C.**, *The Chlorophylls*, Vernon, L. P. and Seely, G. R., Eds., Academic Press, New York, 1966.
9. **Scheer, H., Svec, W. A., Cope, B. T., Studier, M. H., Scott, R. G., and Katz, J. J.**, *J. Am. Chem. Soc.*, 96, 3714, 1974.
10. **Scheer, H. and Steiner, R.**, unpublished.
11. **Brockmann, H., Jr. and Tacke-Karimdadian, R.**, *Justus Liebigs Ann. Chem.*, p. 419, 1979.
12. **Brockmann, H., Jr.**, *Philos. Trans. R. Soc. London*, 273, 277, 1976.
13. **Smith, K. M., Bisset, G. M. F., and Bushell, M. J.**, *J. Org. Chem.*, 45, 2218, 1980.
14. **Smith, K. M., Kehres, L. A., and Tabba, H. D.**, *J. Am. Chem. Soc.*, 102, 7149, 1980.
15. **Hynninen, P. H., Wasielewski, M., and Katz, J. J.**, *Acta Chem. Scand.*, 33, 637, 1979.
16. **Falk, H., Hoornaert, G., Isenring, H. P., and Eschenmoser, A.**, *Helv. Chim. Acta*, 58, 2347, 1975.
17. **Wolf, H., Richter, I., and Inhoffen, H. H.**, *Justus Liebigs Ann. Chem.*, 725, 177, 1969.
18. **Scheer, H. and Katz, J. J.**, *J. Am. Chem. Soc.*, 100, 561, 1978.
19. **Wolf, H., Brockmann, H., Jr., Richter, I., Mengler, C.-D., and Inhoffen, H. H.**, *Justus Liebigs Ann. Chem.*, 718, 162, 1968.
20. **Biere, H.**, Dissertation, Technische Hochschule, Braunschweig, 1966.
21. **Wolf, H. and Scheer, H.**, *Justus Liebigs Ann. Chem.*, 745, 87, 1971.
22. **Wolf, H. and Scheer, H.**, *Tetrahedron*, 28, 5839, 1972.
23. **Hynninen, P.**, *J. Chromatogr.*, 175, 89, 1979.
24. **Pennington, F. C., Strain, H. H., Svec, W. A., and Katz, J. J.**, *J. Am. Chem. Soc.*, 86, 1418, 1964.
25. **Trowitzsch, W.**, Dissertation, Technische Universität, Braunschweig, 1974.
26. **Wolf, H. and Scheer, H.**, *Justus Liebigs Ann. Chem.*, p. 1710, 1973.
27. **Wolf, H. and Scheer, H.**, *Tetrahedron Lett.*, p. 1115, 1972.
28. **Suboch, V. P., Losev, A. P., and Gurinovich, G. P.**, *Photochem. Photobiol.*, 20, 183, 1974.
29. **Steiner, R., Schäfer, W., Blos, I., Wieschoff, H., and Scheer, H.**, *Z. Naturforsch.*, 36c, 417, 1981.
30. **Caple, M. B., Chow, H.-C., and Strouse, C. E.**, *J. Biol. Chem.*, 253, 6730, 1978.
31. **Walter, E., Schreiber, J., Zass, E., and Eschenmoser, A.**, *Helv. Chim. Acta*, 62, 899, 1979.
32. **Schoch, S., Lempert, U., and Rüdiger, W.**, *Z. Pflanzenphysiol.*, 83, 427, 1977.
33. **Richter, I.**, Dissertation, Technische Hochschule, Braunschweig, 1969.
34. **Urumov, T.**, Dissertation, Technische Universität, Munich, 1979.
35. **Somaya, O.**, Dissertation, Technische Hochschule, Braunschweig, 1969.
36. **Evans, T. A. and Katz, J. J.**, *Biochim. Biophys. Acta*, 396, 414, 1975.
37. **Sanders, J. K. M.**, private communication.
38. **Wasielewski, M.**, *Light Reaction Path of Photosynthesis*, Fong, F. K., Ed., Springer-Verlag, Heidelberg, 1982, chap. 7.
39. **Bucks, R. R. and Boxer, S. G.**, *J. Am. Chem. Soc.*, 104, 340, 1982.
40. **Katz, J. J., Shipman, L. L., Cotton, T. M., and Janson, T. R.**, *The Porphyrins*, Vol. 5, Dolphin, D., Ed., Academic Press, New York, 1978, chap. 9.
41. **Sauer, K.**, *Bioenergetics of Photosynthesis*, Govindjee, Ed., Academic Press, New York, 1978, chap. 3.
42. **Scheer, H. and Inhoffen, H. H.**, *The Porphyrins*, Vol. 2, Dolphin, D., Ed., Academic Press, New York, 1978.
43. **Dörnemann, D. and Senger, H.**, *Photochem. Photobiol.*, 43, 573, 1986.
44. **Scheer, H., Gross, E., Nitsche, B., Cmiel, E., Schneider, S., Schäfer, W., Schiebel, H.-M., Schulten, H. R.**, *Photochem. Photobiol.*, 43, 559, 1986.
45. **Struck, A. and Scheer, H.**, unpublished.
46. **Gottstein, J. and Scheer, H.**, *Proc. Natl. Acad. Sci. U.S.A.*, 80, 2231, 1983.
47. **Scheer, H. and Parson, W. W.**, *Biochim. Biophys. Acta*, 766, 653 and 666, 1984.

Table III.4
ESTIMATION OF CHLOROPHYLLS BY SPECTROSCOPIC TECHNIQUES

Compound	Method	Solvent	Ref.
Chlorophylls *a,b*	Absorption	Acetone or 80% methanol	1
Chlorophylls *a,b*, pheophytins *a,b*	Absorption	80% Acetone	2
	Absorption	96% Ethanol	3
	Absorption	Ether, 80% acetone	4
Chlorophylls, pheophytins *a,b*	Absorption, carotenoid interference	90% Acetone or methanol	5
Chlorophylls, chlorophyllides, pheophytins, pheophorbides *a,b*	Absorption (+ demetalation and extraction)	Ether	6
Chlorophyll *a* (in the presence of chlorophyllide(s) and pheophorbide(s)	Absorption (+ demetalation and extraction)	Hexane (acetone, ether)	7
Chlorophylls *a,b* + pheophytins *a,b*	Absorption	90% Acetone	8
Chlorophylls (a,b	Absorption, nomograms	80% or pure acetone	9
	Absorption	80% Acetone, ether	10
Protochlorophyll + protopheophytin (chlorophylls, pheophytins *a,b*)	Absorption (interferences evaluated)	Dimethyl formamide	11 12
Chlorophyll *c*	Absorption (+ extraction)	80% Acetone	13
Chlorophylls *a,b,c*	Absorption (trichromatic)	90% Acetone	14—16
Bacteriochlorophyll *a* (in chromatophores)	Absorption	Water	17
Chlorophylls *a,b*	Derivative absorption	Ether or 80% acetone	18
Chlorophylls, pheophytins *a,b*	Reflection, satellite measurements at 443 and 550 nm	Oceanic water	19, 20
Chlorophylls, pheophytins, and Co-pheophytins *a,b*	Absorption fluorescence	Ether	21
Chlorophylls, pheophytins, and Zn-pheophytins *a,b*	Absorption fluorescence	80% Acetone	22
Chlorophylls *a,b*, pheophytins *a,b*	Fluorescence	(a) 90% Acetone, (b) 90% + HCl	23—28
Chlorophylls *a,b,c* + pheophytins *a,b,c*	Fluorimetry with selective excitation	90% Acetone	29
N-Methyltetraphenylporphine	Stable fluorescence standard for chlorophyll compounds	Various organic + aqueous organic solvents	30
Chlorophyll *c*	Delayed fluorescence	Adsorbed on paper	31
Algal chlorophylls	Gc phytol	—	32
Chlorophyll (+ ide) *a*, pheophytin (+ ide) *b*, chlorophyll *c*	Adsorption + extraction	(a) Hexane, (b) acetone	33

REFERENCES

1. **Hoffmann, P. and Werner, D.,** *Jena Rev.,* 11, 300, 1966.
2. **Vernon, L. P.,** *Anal. Chem.,* 32, 1144, 1960.
3. **Wintermans, J. F. G. M. and DeMots, A.,** *Biochim. Biophys. Acta,* 109, 448, 1965.
4. **French, C. S.,** *Handbuch der Pflanzenphysiologie,* Vol. V(1), Ruhland, W., Ed., Springer-Verlag, Berlin, 1960, 252.
5. **Riemann, B.,** *Limnol. Oceanogr.,* 23, 1059, 1978.
6. **White, R. C., Jones, I. D., and Gibbs, E.,** *J. Food Sci.,* 28, 431, 1963.
7. **Whitney, D. E. and Darley, W. M.,** *Limnol. Oceanogr.,* 24, 183, 1979.
7a. **Owens, T. S.,** *Phytochemistry,* 21, 979, 1982.
8. **Delaporte, N. and Laval-Martin, D.,** *Anal. Chim. Acta,* 55, 415, 1971.
9. **Šestak, Z.,** *Plant Photosynthetic Production. Manual of Methods,* Šestak, Z., Catsky, J., and Jarvis, P. G., Eds., Dr. W. Junk Publishing, The Hague, 1971, 672.

10. **Ziegler, R. and Egle, K.,** *Beitr. Biol. Pflanz.,* 41, 11, 1965.
11. **Moran, R.,** *Plant Physiol.,* 69, 1376, 1982.
12. **Moran, R. and Porath, D.,** *Plant Physiol.,* 65, 478, 1980.
13. **Parsons, T. R.,** *J. Mar. Res.,* 21, 164, 1963.
14. **Strickland, J. D. H. and Parsons, T. R.,** *Practical Handbook of Sea Water Analysis,* Fisheries Research Board of Canada, Ottawa, 1968.
15. **Jeffrey, S. W. and Humphrey, G. F.,** *Biochem. Physiol. Pflanz.,* 167, 191, 1975.
16. SCOR-UNESCO Rep. Working Group 17, Monographs on Oceanic Methodology, UNESCO, Paris, 1966, 69.
17. **Neufang, H.,** *Biochim. Biophys. Acta,* 681, 327, 1982.
18. **Navarro, S., Almela, L., and Garcia, A. L.,** *Photosynthetica,* 16, 134, 1982.
19. **Smith, R. C. and Baker, K. S.,** *Mar. Biol.,* 66, 269, 1982.
20. **Smith, R. C., Eppley, R. W., and Baker, K. S.,** *Mar. Biol.,* 66, 281, 1982.
21. **White, R. C., Jones, I. D., Gibbs, E., and Butler, L. S.,** *J. Agric. Food Chem.,* 25, 143, 1977.
22. **Jones, I. D., White, R. C., Gibbs, E., and Butler, L. S.,** *J. Agric. Food Chem.,* 25, 146, 1977.
23. **Holm-Hansen, O. C., Lorenzen, C. J., Holmes, R. W., and Strickland, J. D. H.,** *J. Cons. Perm. Int. Explor. Mer.,* 30, 3, 1965.
24. **Lorenzen, C. J.,** *Limnol. Oceanogr.,* 12, 343, 1967.
25. **Platt, T. and Conover, R. J.,** *Mar. Biol.,* 10, 2065, 1971.
26. **Berman, T. and Rohde, W.,** *Mitt. Int. Ver. Limnol.,* 19, 266, 1971.
27. **Yentsch, C. S. and Menzel, O. W.,** *Deep Sea Res.,* 10, 221, 1963.
28. **Loftus, M. E. and Carpenter, J. H.,** *J. Mar. Res.,* 29, 319, 1971.
29. **Boto, K. G. and Bunt, J. S.,** *Anal. Chem.,* 50, 392, 1978.
30. **Lavallee, D. K., MacDonough, T. J., Jr., and Cioffi, L.,** *Appl. Spectrosc.,* 36, 430, 1982.
31. **Onue, Y., Kotani, M., Hiraki, K., Shigematsu, T., and Nishikawa, Y.,** *Bunseki Kag.,* 31, E 45, 1982.
32. **Wun, C. K. and Litsky, W.,** *Environ. Sci. Technol.,* 16, 335, 1982.
33. **Owens, T. G. and Falkowski, P. G.,** *Phytochemistry,* 21, 979, 1982.

Table III.5

DISCRIMINATORY DETECTION WAVELENGTHS FOR CHLOROPHYLLS

Chlorophyll	Detection wavelength[a] (nm)	Discrimination against	No discrimination against[b]
Chl *a*	667	Chl *b*, carotenoids	Phe *a*
	435	Phe *a*	Carotenoids[c] Chl *b*, Phe *b*
Phe *a*[d]	500, 410	Chlorophylls	Carotenoids[c]
	667	Carotenoids	Chl a
Chl *b*	650	Carotenoids, (Chl *a*)	Phe *b*
Phe *b*	435	Phe *a*	Carotenoids, Chl *a*
	655	Carotenoids	Chl *a,b*, Phe *a*
Chl *c*	—[e]		
Bchl *a*	770	Plant chlorophylls, carotenoids (Bphe *a*), 3-acetyl-Chl *a*	Bchl *b*, Bphe *b*
	605	Bphe *a*, *b*, carotenoids, (3-acetylchlorophylls) chl *a*	Bchl *b* (plant chlorophylls)
Bphe *a*	750	Carotenoids, plant chlorophylls	Bchl *a,b*, Bphe *b*
	525	Bchl *a*, *b*, plant chlorophylls	Bphe *b*, (carotenoid plant pheophytins, 3-acetyl-Phe *a*
Bchl *b*	795	(Bchl *a*), Bphe *a*, *b;* plant chlorophylls, carotenoids	
	625	Bphe *a*, *b*, (Bchl *a*), (plant chlorophylls), 3-acetyl-Chl *a*, carotenoids	
Bphe *b*	750	Plant chlorophylls, 3-acetyl-Chl *a*, carotenoids	Bchl *a*, *b*, Bphe *a*
	530	Bchl a, b	Bphe *b,a* 3-acetyl-phe *a*,[f] (carotenoids)

[a] The values are given for RP-phases and methanol-water mixtures. They may vary by a few nanometers, depending on the percentage of the mixture, and more for other separation systems.

[b] As a rule, most of the common degradation products absorb at wavelengths similar to the parent compound and cannot be discriminated easily by the detection wavelength.

[c] 435 + 667 discriminates against carotenoids.

[d] 500 + 667 discriminates against carotenoids.

[e] No discrimination with single-wavelength detection possible. In dual-wavelength detection, characteristic high ratio E^{441}_{626}.

[f] Discrimination against 3-acetyl-Chl *a* and -Phe *a* (as likely alteration products of especially Bchl *b* and Bphe *b*, respectively) by additional setting at 680 nm.

Table III.6
DEGRADATION PRODUCTS OF CHLOROPHYLLS

Pigment	Conditions of formation	Ref.
13^2-OH-Chlorophyll a (2 epimers)	Dry tetrahybrofuran or methanol, lithium bromide, air (catalyzed by trace metals)	1, 2, 24
13^2-OCH$_3$-13^1A-homooxa-chlorophyll a (= "13^2-methoxylactone")	Dry methanol, lithium bromide, air (catalyzed by formation of cation radical)	1
13^2-OH-Bacteriochlorophyll a	Dry tetrahydrofuran or methanol, lithium bromide, air (catalyzed by trace metals)	1
Several "changed chlorophylls"	Chromatography on silica and cellulose	3
Several chlorophyll a alteration products	Marine phytoplankton, formed during senescence and/or zooplankton grazing	4
Pheophytins	Canned spinach, frozen spinach	5, 23, 24
	Frozen or blanched spinach	5
Prime-chlorophylls	Nucleophiles like pyridine, imidazole	6, 7
20-Cl-Chlorophylls	Oxidants which generate Cl$^+$	8—12
Oxidation of macrocyclic single bonds	High-potential quinones or strong reductants followed by oxygen	13, 14, 15
Hydrolysis of phytylester	Chlorophyllase action, other degradative processes	2, 16, 17
Other central metals	Transmetalation, acid, and heat catalyzed	18, 19
Oxidation products of 8-ethylidene group in bacteriochlorophylls b and g	Photo- and dark oxidations; acetone good promoter	20, 21
Vinyl group modifications	Oxidative and/or hydrolytic reactions	14, 15, 22
Pyrochlorophylls	Enzymatic reactions(?), heat	5, 23, 24

REFERENCES

1. **Schaber, P. M., Hunt, J. E., Fries, R., and Katz, J. J.,** *J. Chromatogr.,* 316, 25, 1984.
2. **Endo, H., Hosoya, H., Koyama, T., and Ichioka, M.,** *Agric. Biol. Chem.,* 46, 2183, 1982.
3. **Bacon, M. F.,** *Biochem. J.,* 101, 34c, 1966.
4. **Hallegraeff, G. M. and Jeffrey, S. W.,** *Deep-Sea Research,* 32, 697, 1985.
5. **Schwartz, S. J., Woo, S. L., and von Elbe, J. H.,** *J. Agric. Food Chem.,* 29, 533, 1981.
6. **Watanabe, T., Nakazato, M., and Honda, K.,** *Chem. Lett.,* p. 253, 1986.
7. **Watanabe, T., Kobayashi, M., Hongu, A., Nakazato, M., Hiyama, T., and Murata, N.,** *FEBS Lett.,* 191, 252, 1985.
8. **Woodward, R. B. and Skaric, V.,** *J. Am. Chem. Soc.,* 83, 4676, 1961.
9. **Hynninen, P. H. and Lötjönen, S.,** *Tetrahedron Lett.,* 22, 1845, 1981.
10. **Scheer, H., Gross, E., Nitsche, B., Cmiel, E., Schneider, S., Schäfer, W., Schiebel, H.-M., and Schulten, H.-R.,** *Photochem. Photobiol.,* 43, 559, 1986.
11. **Dörnemann, D. and Senger, H.,** *Photochem. Photobiol.,* 43, 572, 1986.
12. **Hynninen, P. H. and Sievers, G.,** *Z. Naturforsch.,* B 36, 1000, 1981.
13. **Smith, J. R. L. and Calvin, M.,** *J. Am. Chem. Soc.,* 88, 4500, 1966.
14. **Scheer, H. and Inhoffen, H. H.,** in *The Porphyrins,* Vol. 2, Dolphin, D., Ed., Academic Press, New York, 1978, 49.
15. **Fischer, H. and Orth, H.,** *Die Chemie des Pyrrols,* Vol. 2, 2nd half, Akademische Verlagsges., Leipzig, 1940, 154.
16. **Gassmann, J., Strell, I., Brandl, F., Sturm, M., and Hoppe, W.,** *Tetrahedron Lett.,* p. 4609, 1971.
17. **Jeffrey, S. W. and Hallegraeff, G. M.,** *Mar. Ecol. Physiol.,* 35, 293, 1987.
18. **Schwartz, S. J.,** *J. Liq. Chromatogr.,* 7, 1673, 1984.
19. **Buchler, J. W.,** in *The Porphyrins,* Vol. 1, Dolphin, D., Ed., Academic Press, New York, 1978, 389.
20. **Steiner, R., Cmiel, E., and Scheer, H.,** *Z. Naturforsch.,* 38c, 748, 1983.
21. **Brockmann, H., Jr. and Lipinski, A.,** *Arch. Microbiol.,* 136, 17, 1983.

22. **Scheer, H.,** Ph.D. thesis, Universität Braunschweig, Braunschweig, W. Germany, 1970.
23. **Schoch, S., Scheer, H., Schiff, J. A., Rüdiger, W., and Siegelmann, H. W.,** *Z. Naturforsch.*, 35c, 827, 1981.
24. **Haidl, H., Knödlmayer, K., Rüdiger, W., Scheer, H., Schoch, S., and Ullrich, J.,** *Z. Naturforsch.*, 40c, 685, 1985.

PAPER CHROMATOGRAPHY

Table III. PC 1
SURVEY OF CHROMATOGRAPHY SYSTEMS

Geometry	Paper	Developing agent	Ref.
One-way	Cellulose (Whatman No. 1)	Ether + petroleum ether (1:1) + 1% *n*-propanol	1
		Carbon disulfide	2
		Petroleum ether + 0.5—2% *n*-propanol	3, 4
		Benzene + *i*-propanol	5
		Petroleum ether + chloroform (3:1)	1
		Benzene-petroleum ether-acetone (10:2.5:2)	6
		Petroleum ether-benzene-chloroform-acetone-2-propanol (50:35:10:5:0.17)	7
		Hexane + ether + *n*-propanol (70:30:0.5)	8
		Hexane + diisopropyl-ether + *n*-propanol (70:30)	8
		Petroleum ether + acetone + isopropanol (90:7.5:0)	8
	Ion-exchange	Petroleum ether + 0.5—1% *n*-propanol	9
	Cellulose + Wesson® oil	Methanol-acetate-water (20:4:3)	7
	Cellulose + Vaseline®	80% methanol	10
	Cellulose + glycerine	Petroleum ether + 0.5% *n*-propanol	10
	Cellulose + 80% MeOH	Petroleum ether	10
	Cellulose + 22% SiO_2	Petroleum ether + acetone (7:3)	11
	Glass	Petroleum ether	10
Two-way	Cellulose	Petroleum ether (60—80°C) + 4% *n*-propanol	12
		Petroleum ether + 30% chloroform	
	Cellulose	Petroleum ether + 1% *n*-propanol	7, 13
		Petroleum ether + chloroform (3:1)	
Radial	Cellulose	Carbon disulfide	14
		Petroleum ether + 1.5% *n*-propanol	15
		Petroleum ether + 2% acetone	15

REFERENCES

1. **Sherma, J.,** *Anal. Lett.,* 3, 35, 1970.
2. **Strain, H. H. and Sherma, J.,** *J. Chromatogr.,* 73, 371, 1972.
3. **Strain, H. H. and Svec, W. A.,** *Adv. Chromatogr.,* 8, 119, 1969.
4. **Strain, H. H., Sherma, J., Benton, F. L., and Katz, J. J.,** *Biochim. Biophys. Acta,* 109, 1, 1965.
5. **Uspenskaya, V. E., Kondrat'eva, E. N., and Akulovich, N. K.,** *Dokl. Akad. Nauk SSSR,* 167, 702, 1966.
6. **Michel-Wolwertz, M. R. and Sironval, C.,** *Biochim. Biophys. Acta,* 94, 330, 1965.
7. **Strain, H. H. and Sherma, J.,** *J. Chem. Educ.,* 46, 476, 1969.
8. **Burke, S. and Aronoff, S.,** *Anal. Biochem.,* 101, 103, 1980.
9. **Sherma, J. and Strain, H. H.,** *Anal. Chim. Acta,* 40, 155, 1968.
10. **Strain, H. H.,** *J. Phys. Chem.,* 57, 638, 1953.
11. **Sherma, J.,** *J. Chromatogr.,* 61, 202, 1971.
12. **Jeffrey, S. W. and Allen, M. B.,** *Limnol. Oceanogr.,* 12, 533, 1967.
13. **Strain, H. H., Sherma, J., Benton, F. L., and Katz, J. J.,** *Biochim. Biophys. Acta,* 109, 16, 1965.
14. **Strain, H. H. and Svec, W. A.,** *Chromatography,* 3rd ed., Heftmann, E., Ed., Van Nostrand-Reinhold, Princeton, N.J., 1975, 744.
15. **Strain, H. H., Sherma, J., Benton, F. L., and Katz, J. J.,** *Biochim. Biophys. Acta,* 109, 23, 1965.

THIN-LAYER CHROMATOGRAPHY

Table III. TLC 1
SURVEY OF DIFFERENT ADSORBENTS USED FOR CHLOROPHYLLS

Geometry	Adsorbent	Developing agent	Ref.
One-way	Sucrose	Petroleum ether + 5—10% acetone	1
		Petroleum ether + 2% methanol	2
	Cellulose	Petroleum ether-benzene-chloroform-acetone-*i*-propanol (50:35:10:5:0.17)	3
		n-Hexane + 3% *n*-propanol	4
		Methanol-dichloromethane-water (100:18:20)	5
		Carbon tetrachloride + 0.25% *n*-propanol	3
		Petroleum ether-acetone-*n*-propanol (90:10:0.45)	6
		Petroleum ether + pyridine (9:1)	7
		Heptane + pyridine (7:3)	7
	Silica gel G	Benzene + 0—10% acetone	8
		Petroleum ether-benzene-ethanol (40:15:3.6)	9
		Petroleum ether + ethanol (16:1)	2
		iso-Octane + acetone + diethyl ether (3:1:1)	10
	Silica gel H	Benzene-ethyl acetate-ethanol (80:20:5—30)	11
		Same with toluene	21
	Silica gel + Wesson® oil	Methanol-acetone-water (20:4:3)	12
	Cellulose + unsaturated triglycerides	Methanol-acetone-water (30:10:3)	2
		Acetonitrile + tetrahydrofuran + carbon tetrachloride + water (70:15:10:5)	13
	Polyethylene	90% Acetone	14
	C-18 bonded silica	Methanol + acetone + water (20:4:3)	15
		Butanol-methanol (8:7)	15
Two-way	Cellulose	Petroleum ether + 1% *n*-propanol	16
		Petroleum ether + chloroform (3:1)	
	Silica gel G	Benzene-petroleum ether-acetone (10:2.5:2)	17
		Benzene-petroleum ether-acetone-methanol (10:2.5:1:0.25)	
	Sucrose	Petroleum ether (60—80°C) + 0.8% *n*-propanol	18
		Petroleum ether + 20% chloroform	
	Silica gel G	Petroleum ether-benzene-acetone (2.5:10:2)	19
		Petroleum ether-acetone-propanol (9:1:0.45)	
	Cellulose	Ligroin + 2.5% *n*-propanol	20
		Chloroform-ligroin-acetone (25:75:0.25)	

REFERENCES

1. **Nutting, M. D., Voet, M., and Becker, R.,** *Anal. Chem.,* 37, 445, 1965.
2. **Strain, H. H. and Svec, W. A.,** *Adv. Chromatogr.,* 8, 119, 1969.
3. **Sherma, J.,** *Anal. Lett.,* 3, 35, 1970.
4. **Mattox, K. R. and Williams, J. P.,** *J. Phycol.,* 1, 191, 1965.
5. **Schneider, H. A. W.,** *J. Chromatogr.,* 21, 448, 1966.
6. **Bacon, M. F.,** *J. Chromatogr.,* 17, 322, 1965.
7. **Sivers, G. and Hynninen, P.,** *J. Chromatogr.,* 134, 359, 1977.
8. **Sherma, J.,** *J. Chromatogr.,* 52, 177, 1970.
9. **Schaltegger, K.,** *J. Chromatogr.,* 19, 75, 1965.
10. **Strain, H. H. and Sherma, J.,** *J. Chem. Educ.,* 46, 476, 1969.
11. **Seliskar, C. J.,** *Anal. Biochem.,* 17, 174, 1966.
12. **Jones, I. D., Butler, L. S., Gibbs, E., and White, R.C.,** *J. Chromatogr.,* 70, 87, 1973.

13. **Scholz, B., Willaschek, D., Müller, H., and Ballschmiter, K.,** *J. Chromatogr.,* 208, 156, 1981.
14. **Jeffrey, S. W.,** *Nature (London),* 220, 1032, 1968.
15. **Sherma, J. and Latta, M.,** *J. Chromatogr.,* 154, 73, 1978.
16. **Strain, H. H., Sherma, J., Benton, F. L., and Katz, J. J.,** *Biochim. Biophys. Acta,* 109, 16, 1965.
17. **Schanderl, S. H. and Lynn, D. Y. C.,** *J. Food Sci.,* 31, 141, 1966.
18. **Jeffrey, S. W.,** *Biochim. Biophys. Acta,* 162, 271, 1968.
19. **Lynn, D. Y. C. and Schanderl, S. H.,** *J. Chromatogr.,* 26, 442, 1967.
20. **Jeffrey, W.,** *Limnol. Oceanogr.,* 26, 191, 1981.
21. **Rebeiz, C. A., Matheis, J. R., Smith, B. B., Rebeiz, C. C., and Dyton, D. F.,** *Arch. Biochem. Biophys.,* 171, 549, 1975.

Table III. TLC 2
TLC OF TOTAL PIGMENT EXTRACTS

	L1	L1	L2	L3	L4	L5	L6	L7
Layer	L1	L1	L2	L3	L4	L5	L6	L7
Solvent	S1	S2	S1	S3	S3	S4	S5	S6
Flow rate	—	—	—	—	—	—	—	—
Temperature	a	a	a	a	a	a	a	a
Detection	D1	D1	D1	D1	D1	D1	D1	D1
Literature	1	1	1	1	1	1	2	3
Compound	R_f^a	R_f^a	R_f^a	R_f^a	R_f^a	R_f^a	R_f	R_f
Chlorophyll *a*	4	2	3	3b	3b	1b	0.27	0.29
Chlorophyll *a'*	n.d.	n.d.	n.d.	n.d.	n.d.	n.d.	n.d.	0.25
Chlorophyll *b*	2b	1a	2b	2b	2b	1a	0.10	0.40
Chlorophyll *b'*	n.d.	n.d.	n.d.	n.d.	n.d.	n.d.	n.d.	0.35
Chlorophyll *c*[b]	—	—	—	—	—	—	0.00	—
Carotenes[c]	5	5	5	4	4	5	0.96	0.09
Lutein	3	4	4	3a	3a	3	0.73	0.58
Neoxanthin	1	1b	1	1	1	2	0.05	0.89
Violaxanthin	2a	3	2a	2a	2a	4	0.48	0.77
Pheophytins	n.d.	n.d.	n.d.	n.d.	n.d.	n.d.	n.d.	0.14
Fucoxanthin[d]	—	—	—	—	—	—	0.28	—
Neofucoxanthin (A,B)	—	—	—	—	—	—	0.08	—
Diadinoxanthin	—	—	—	—	—	—	0.46	—
Diatoxanthin	—	—	—	—	—	—	0.55	—

[a] Only the sequence of the pigments given. Letters a and b refer to zones not fully separated.
[b] c_1 and c_2 unresolved.
[c] Unresolved mixture of "carotenes".
[d] Unresolved mixture of "pheophytins".

Packing	P1 = powdered sugar
	P2 = starch
	P3 = powdered cellulose
	P4 = paper
	P5 = magnesia/celite 545 = 1:1
	P6 = paper (Whatman 3 MM)
	P7 = cellulose impregnated with olive oil (Dante, Genoa, Italy)
Solvent	S1 = petroleum ether-propanol = 99.5:0.5
	S2 = benzene
	S3 = petroleum ether-propanol = 99:1
	S4 = petroleum ether-acetone = 70:30
	S5 = petroleum ether-chloroform = 70:30
	S6 = acetonitrile-tetrahydrofuran-carbon tetrachloride-water = 70:15:10:5
Detection	D1 = visual inspection and subsequent UV-vis specrum

REFERENCES

1. **Strain, H. H., Sherma, J., and Gandolfo, M.**, *Anal. Biochem.*, 24, 54, 1968.
2. **Eskins, K., Scholfield, C. R., and Dutton, H. J.**, *J. Chromatogr.*, 135, 217, 1977.
3. **Scholz, B., Willaschek, K., Müller, H., and Ballschmiter, K.**, *J. Chromatogr.*, 208, 156, 1981.

Table III. TLC 3
TLC OF NONESTERIFIED PIGMENTS

Packing	P1	P2	P3	P4
Solvent	S1	S2	S3	S4
Temperature	Ambient	Ambient	Ambient	Ambient
Detection	Visual	Visual	Visual	Visual
Literature	1	1	2[a]	3

Compound		R_f values		
Chlorophyll C_1	0.45	0.38[d]	—	—
Chlorophyll C_2	0.45	0.28[d]	—	—
Chlorophyll C_3[c]	0.59	0.32[d]	—	—
Mg-2,4-divinylpheoporphyrin a_5-monomethylester	0.48	0.49	—	—
Pheophorbide *a*	—	—	0.44[c]	—
Pyropheophorbide *a*	—	—	0.43	—
13-OH-Pheophorbide *a*	—	—	0.39	—
Protochlorophyllide *a*	—	—	—	0.32
Chlorophyllide *b*	—	—	—	n.a.
Methyl-protochlorophyllide *a*	—	—	—	0.51

[a]   Other pigments (unknown structures) well separated.
[b]   Polyethylene plates.
[c]   Structure unknown.
[d]   Tailing.

Packing	P1	= C-8 bonded silica (Merck HPTLC)
	P2	= Polyethylene (PRX-1025, Polysciences Inc.)
	P3	= Silica gel G (Merck)
	P4	= Silica gel type 60 (EM laboratories)
Solvent	S1	= methanol-water = 9:1 (v/v)
	S2	= acetone
	S3	= benzene-ethyl acetate-ethanol-*n*-propanol = 16:4:1:1 (v/v)
	S4	= toluene-ethyl acetate-ethanol-hexane = 80:40:20:1 (v/v)

REFERENCES

1. **Jeffrey, S. W. and Wright, S. W.**, Private communication.
2. **Endo, H., Hosoya, H., Koyama, T., and Ichioka, M.**, *Agric. Biol. Chem.*, 46, 2183, 1982.
3. **Bednarik, D. P. and Hoober, K. J..**, *Science*, 230, 450, 1985.

LIQUID CHROMATOGRAPHY

Table III. LC 1
SURVEY OF DIFFERENT ADSORBENTS USED FOR CHLOROPHYLLS

Adsorbent	Developing agent	Ref.
Sucrose	Petroleum ether + 0.5—2% n-propanol	1—4
	Petroleum ether + 15% diethyl ether	3
	Petroleum ether + benzene	3
	Iso-octane + 0.5% n-propanol	5
Starch	Petroleum ether + 0.5—2% n-propanol	3,6
	Petroleum ether + 1—3% acetone	7
Cellulose	Petroleum ether + 0.5—2% n-propanol	3,7
	n-Hexane + 3% n-propanol	8
	Pyridine + water (8:1)	23
	Chloroform-cyclohexane-nitromethane	9
	Petroleum ether + acetone + propanol	22
Cellulose + dimethylformamide	Petroleum ether	6
DEAE-cellulose	Chloroform + methanol	20
Celite	Petroleum ether + 5% acetone	10
Celite + 7% Wesson® oil	Methanol-acetone-water (20:4:3)	6
Silica gel	Benzene + 0—10% acetone	3
	Iso-octane + acetone + ether (3:1:1)	10
Alumina		19
Sephadex LH-20	Formamide + i-propanol (3:1)	3
	Chloroform + methanol	11
	Formamide + 2-propanol (1000:8)	18
	Petroleum ether + diethyl ether (80:20)	14
	Hexane + 2-propanol + ether (gradient)	16
	Hexane + diethyl ether + acetone (gradient)	17
Sephasorb HP ultrafine	Cyclohexane-tetrahydrofuran (1000:8)	13
Polyethylene	Acetone + 30—15% water	5,7
	Methanol + 25—10% water	7
	Water + acetone (gradient)	15
Polypropylene	Acetone + 30—15% water	7
	Methanol + 25—10% water	7
Kel-F 300	Acetone + 30—15% water	7
	Methanol + 25—10% water	12
Polyamide	Methanol	12
Celite + alumina (1:1)		21

Note: For HPLC normal phase and HPLC reverse phase, see HPLC tables.

REFERENCES

1. **Strain, H. H. and Svec, W. A.**, *Chromatography,* Heftmann, E., Ed., Van Nostrand-Reinhold, Princeton, N.J., 1977, 744.
2. **Strain, H. H., Cope, B. T., and Svec, W. A.**, *Methods in Enzymology,* Vol. 23, San Pietro, A., Ed., Academic Press, New York, 1971, 452.
3. **Strain, H. H. and Svec, W. A.**, *Adv. Chromatogr.,* 8, 119, 1969.
4. **Strain, H. H. and Svec, W. A.**, *The Chlorophylls,* Vernon, L. P. and Seely, G. R., Eds., Academic Press, New York, 1966, 21.
5. **Anderson, A. P. H. and Calvin, M.**, *Nature (London),* 194, 285, 1962.
6. **Strain, H. H. and Sherma, J.**, *J. Chem. Educ.,* 46, 476, 1969.
7. **Strain, H. H., Sherma, J., and Gandolfo, M.**, *Anal. Biochem.,* 24, 54, 1968.
8. **Mattox, K. R. and Williams, J. P.**, *J. Phycol.,* 1, 191, 1965.
9. **Smith. W. F., Jr. and Eddy, K. L.**, *J. Chromatogr.,* 22, 296, 1966.
10. **Strain, H. H., Sherma, J., and Gandolfo, M.**, *Anal. Chem.,* 39, 926, 1967.
11. **Shimizu, S.**, *J. Chromatogr.,* 59, 440, 1971.

Table III. LC 1 (continued)
SURVEY OF DIFFERENT ADSORBENTS USED FOR CHLOROPHYLLS

12. **Fric, F. and Haspel-Horavtovic, E.,** *J. Chromatogr.,* 68, 264, 1972.
13. **Scholz, B. and Ballschmiter, K.,** *J. Chromatogr.,* 208, 148, 1981.
14. **Downey, W. K., Murphy, R. F., and Keogh, M. K.,** *J. Chromatogr.,* 46, 120, 1970.
15. **Chow, H.-O., Caple, M. B., and Strouse, C. E.,** *J. Chromatogr.,* 151, 357, 1978.
16. **Iryama, K. and Yoshiura, M.,** *J. Chromatogr.,* 177, 154, 1979.
17. **Iryama, K., Yoshiura, M., and Shiraki, M.,** *J. Chem. Soc. Chem. Commun.,* p. 40, 1979.
18. **Schenk, J. and Dässler, H. G.,** *Pharmazie,* 24, 419, 1969.
19. **Verma, M. R. and Rai, J.,** *Indian Stand. Instr. Bull.,* 20, 495, 1968.
20. **Sato, N. and Murata, N.,** *Biochim. Biophys. Acta,* 501, 103, 1978.
21. **Buckle, K. A. and Rahmann, F. M. M.,** *J. Chromatogr.,* 171, 385, 1979.
22. **Bazzaz, M. B. and Rebeiz, C. A.,** *Photochem. Photobiol.,* 30, 709, 1979.
23. **Risch, N.,** *J. Chem. Res.,* 3, 116, 1981.

Table III. LC 2
LC OF TOTAL PIGMENT EXTRACTS

Packing	P1	P1	P2	P3	P4	P5[a]	P6	P7	P8
Solvent	S1	S2	S1	S3	S4	S5	S6	S7	S8
Temperature	a	a	a	a	a	a	a	a	a
Literature	1	1	1	1	1	2	3	3	3
Compound[b]	**EO**	**EO**	**EO**	**EO**	**EO**	**EO**	**EO**	**EO**	**EO**
Chl *a*	2	4a	3	2b	5a	5[a]	3	2a	3
Chl *b*	4b	4b	4a	3b	5b	4	4	2b	3
Carotenes	1	1	1	1	1	6	1	3	1[c]
Lutein	3	2	2	2a	3	3	5a	1a	3
Neoxanthin	5	4c	5	4	4	1	7	1b	3
Violaxanthin	4a	3	4b	3a	2	2	6	1c	3
Cryptoxanthin	n.d.	n.d.	n.d.	n.d.	n.d.	n.d.	2	1d	3
Zeaxanthin	n.d.	n.d.	n.d.	n.d.	n.d.	n.d.	5b	13	3

[a]   HPLC, for details see table.
[b]   The numbers give the elution order (EO); letters a, b, and c refer to zones.
[c]   α-Carotene runs in front of β-carotene.

Packing P1 = powdered sugar
        P2 = starch
        P3 = powdered cellulose
        P4 = magnesia-Celite = 1:1
        P5 = Sil 60 RP 18 (Riedel de Haen, Hannover, FRG)
        P6 = calcium carbonate
        P7 = polyethylene
        P8 = magnesia
Solvent S1 = petroleum ether-propanol = 99.5:0.5
        S2 = benzene
        S3 = petroleum ether-propanol = 99:1
        S4 = petroleum ether-acetone = 40:60
        S5 = A: methanol-acetonitrile = 25:75; B: water, linear gradient 75 →
              100% A in 20 min, then isocratic, 1.5 mℓ/min
        S6 = petroleum ether-acetone = 80:20
        S7 = acetone-water = 70:30
        S8 = petroleum ether-acetone = 95:5

REFERENCES

1. **Strain, H. H., Sherma, J., and Gandolfo, M.,** *Anal. Biochem.,* 24, 54, 1968.
2. **Braumann, T., and Grimme, H.,** *Biochim. Biophys. Acta,* 637, 8, 1981.
3. **Svec, W.,** *The Porphyrins,* Vol. 5, Dolphin, D., Ed., Academic Press, New York, 1978, 34.

Table III. LC 3
PREPARATIVE LC OF TOTAL PIGMENT EXTRACTS

Packing	P1	P1	P1	P1	P2	P3	P4[a]	P5[a]
Column								
length	n.a.	n.a.	n.a.	n.a.	n.a.	n.a.	n.a.	n.a.
diameter	n.a.	n.a.	n.a.	n.a.	n.a.	n.a.	n.a.	n.a.
material	n.a.	n.a.	n.a.	n.a.	n.a.	n.a.	n.a.	n.a.
Solvent	S1	S1	S1	S1	S2	S3	S4	S5
Flow rate	F1	F1	F1	F1	F2	F3	F4	F5
Temperature	Ambient	Ambient	Ambient	Ambient	Ambient	Ambient	Ambient	Ambient
Detection	D1	D1	D1	D1	D1	D1	D1	D1
Literature	1	1	1	1	2	3	4	4
Compound[b]	**EO**[c]	**EO**[d]	**EO**[e]	**EO**[f]	**EO**	**EO**	**EO**	**EO**
Carotenes[g]	1	1	1	1a	1[h]	5	1	—
Cryptoxanthin	2	—	—	—	—	—	1	—
Chlorophyll *a*	4	3	3	—	2[i]	3[i]	2	1
Chlorophyll *a'*	3	2	2	—	2[i]	3[i]	2	1
Lutein	5a	—	4a	—	1[h]	1[j]	1	—
Zeaxanthin	5b	—	4b	—	1[h]	1[j]	1	—
Chlorophyll *b*	7	—	—	—	3[i]	2[i]	2	2
Chlorophyll *b'*	6	—	—	—	3[i]	2[i]	2	2
Violaxanthin	8	4	—	—	1[h]	1[j]	—	—
Neoxanthin	9	—	—	—	1[h]	1[j]	1	—
Fucoxanthin	—	5	—	—	—	—	—	—
Neofucoxanthin B	—	6	—	—	—	—	—	—
Neofucoxanthin A	—	7	—	—	—	—	—	—
Chlorophylls *c*[k]	—	8	—	—	—	—	—	—
Chlorophyll *d*	—	—	5	—	—	—	—	—
Spirilloxanthin	—	—	—	1b	—	—	—	—
Bacteriochlorophyll *a*	—	—	—	2	—	—	—	—
Pheophytin *a*	—	—	—	—	1[h]	4[i]	1	—
Pheophytin *b*	—	—	—	—	1[h]	4[i]	1	—

[a] The two columns (P4,P5) are used in series.
[b] The numbers correspond to the elution order (EO).
[c] Green plant extract.
[d] Brown algal extract.
[e] Red algal extract.
[f] Purple bacterial extract.
[g] Unresolved mixture of "carotenes".
[h] Unresolved mixture of carotenoids, xanthophylls, and pheophytins.
[i] Unresolved.
[j] Unresolved mixture of "xanthophylls".
[k] Unresolved mixture of chlorophylls c_1 and c_2.

Packing
P1 = powdered sugar
P2 = Sephasorb HP ultrafine (Pharmacia, Uppsala, Sweden)
P3 = powdered polyethylene, MI < 2 (Dow, U.S.)
P4 = DEAE-Sepharose CL-6B (Pharmacia, Uppsala, Sweden)
P5 = Sepharose CL-6B (Pharmacia, Uppsala, Sweden)

Solvent
S1 = petroleum ether with *n*-propanol, increasing from 0.5 to 2%
S2 = cyclohexane-tetrahydrofuran = 1000:8
S3 = step gradient acetone-water = 7:1, increased to 85:15 after elution of xanthopylls; the chlorophylls are separated on sugar in a second step
S4 = acetone, after elution of carotenoids and pheophytins, changed to acetone-methanol = 10:3; chromatography should be done as fast as possible
S5 = hexane-2-propanol = 20:1, after elution of chlorophyll *a* changed to 10:1

Flow rate
F1 = columns run with suction from aspirator, reduced to approx. 0.5 atm
F2 = 25 mℓ/hr, slight pressure
F3 = gravity
F4 = 2 mℓ/min
F5 = 1.4 mℓ/min

Detection D1 = naked eye

Table III. LC 3 (continued)
PREPARATIVE LC OF TOTAL PIGMENT EXTRACTS

REFERENCES

1. **Svec, W. A.**, in *The Porphyrins*, Vol. 5, Dolphin, D., Ed., Academic Press, New York, 1978, 341.
2. **Scholz, B. and Ballschmiter, K.**, *J. Chromatogr.*, 208, 148, 1981.
3. **Anderson, A. F. H. and Calvin, M.**, *Nature*, 194, 285, 1962.
4. **Omata, T. and Murata, N.**, *Photochem. Photobiol.*, 31, 183, 1980.

Table III. LC 4
LC OF REDUCTION PRODUCTS OF CHLOROPHYLLS
a AND *b*

Packing	P1
Column	
length	500
diameter	30
material	Glass
Solvent	S1
Flow rate	n.a.
Temperature	4°C
Detection	D1
Literature	1

Compound[a]	r_v[b]
Chlorophyll *a*	355[c]
Chlorophyll *a'*	246[c]
13^1-Deoxo-13^1-hydroxychlorophyll *a*	
(13^{1}R, 13^{2}R)[d]	451[c]
(13^{1}R, 13^{2}S)[d]	440
(13^{1}S, 13^{2}R)[d]	502[c]
(13^{1}S, 13^{2}S)[d]	383[c]
Pheophytin *a'*	214
Pheophytin *a*	227
13^1-Deoxo-13^1-hydroxypheophytin *a* (mixture)	287
7-Demethyl-7-hydroxymethylchlorophyll *a*	423
7-Demethyl-7-hydroxymethylchlorophyll *a*	280
7-Demethyl-7-hydroxymethyl-13^1-deoxo-13^1-hydroxychlorophyll *a*	
(13^{1}R, 13^{2}R)[d]	553
(13^{1}S, 13^{2}R)[d]	647
(13^{1}S, 13^{2}S)[d]	465

[a] For formula, see Table III.1.
[b] Retention volume in milliliters.
[c] Mean value of several runs.
[d] Tentative assignment of absolute configuration at C-13^1 and C-13^2.

Packing	P1	icing sugar
Solvent	S1	petroleum ether (60—80°C b.p.)-*n*-propanol = 99.5:0.5
Flow rate	n.a.	
Detection	D1	absorption

REFERENCE

1. **Hynninen, H.**, *J. Chromatogr.*, 175, 75, 1979.

Table III. LC 5
LC OF BACTERIOCHLOROPHYLL *e* DERIVATIVES

	P1	P2
Packing		
Column		
length	310	n.a.
diameter	25	n.a.
material	Glass	n.a.
Solvent	S1	S1
Flow rate	n.a.	n.a.
Temperature	Ambient	Ambient
Detection	n.a.	n.a.
Literature	1	1

Compounds[a]	r[b]	r
Bacteriomethylpheophorbide *e* (substituents at C-8 and C-10 indicated)		
i-But-Et	1.9	3.3
n-Pr-Et	2.9	6.5
Et-Et	4.1	10.0

[a] For formula see Table III. 1.
[b] Only relative retention times are available.

Packing P1 LiChrosorb Si 100 (10μm)
P2 Lichrosorb Si 100 (5 μm)
Solvent S1 hexane-acetone = 6:1

REFERENCE

1. **Risch, N., Kemmer, T., and Brockmann, H., Jr.,** *Liebigs Ann. Chem.,* 1978, 585, 1978.

Table III. LC 6
LIQUID-LIQUID PARTITION OF CHLOROPHYLL DERIVATIVES

	100-Tube countercurrent machine	
Column	S1	S2
Solvent	1	2
Literature		

Compound	Partition coefficient	
β-Carotene	—	—
Pheophytin *a*	42.1	—
Chlorophyll *a*	2.83	1.06
Mg-purpurin 7-lactone-alkylether-methylphytylester	0.75	—
10-Hydroxy-chl *a*	0.697	—
Chlorophyll *b*	0.482	0.59
Mg-*b*-purpurin 7-lactone-alkylether-methylphytylester	0.250	—
Lutein	0.238	—
10-Hydroxy-chl *b*	0.140	—
Violaxanthin	0.094	—
Neoxanthin	0.015	—
Chlorophyllides	0	—

Solvent S1 = petroleum ether-benzene-methanol-formamide
S2 = hexane-90% aqueous ethanol
Detection visual inspection and subsequent UV-vis spectra

REFERENCES

1. **Hynninen, P.H. and Ellfolk, N.,** *Acta Chem. Scand.,* 27, 1463, 1973.
2. **Lancaster, C. R., Lancaster, E. B., and Dutton, H. J.,** *Am. J. Oil Chem. Soc.,* 27, 386, 1950.

HIGH PERFORMANCE LIQUID CHROMATOGRAPHY

Table III. HPLC 1
HPLC OF TOTAL EXTRACTS

Packing	P1	P2	P3	P4	P4	P5
Column						
length	300	300	65	1220	1220	250
diameter	3	3	0.5	7	7	3
material	Glass	Glass	PTFE	SS	SS	SS
Solvent	S1	S2	S3	S4	S5	S6
Flow rate	1.5	2.0	0.016	4.0	2.5	1.0
Temperature	Ambient	Ambient	Ambient	18°C	28°C	Ambient
Detection	445	445	380	440	440	445
Literature	1	2	3	4	4	5
Compound	**r**[a]	**r**[a]	**r**[a]	**r**	**r**	**r**[b]
Neoxanthin	10.6	10.9	68.0	2.3	—	12.5
cis-Neoxanthin	9.5	8.6	n.d.	n.d.	—	n.d.
Trihydroxy-2-carotene	11.6	n.d.	n.d.	n.d.	—	n.d.
Violaxanthin	13.5	15.4	56.3	30	—	6.8
cis-Violaxanthin	12.2	n.d.	n.d.	n.d.	—	n.d.
Violaxanthin	n.d.	21.6	n.d.	n.d.	—	n.d.
Lutein-5,6-epoxide	15.4	25.2	n.d.	n.d.	—	n.d.
Antheraxanthin	16.1	12.9	n.d.	n.d.	—	5.5
Lutein	18.9	24.3	43.7	53	—	4.3
cis-Lutein	17.6	n.d.	n.d.	n.d.	—	n.d.
Chlorophyll *b*	21.7	29.6	38.6	41	—	3.2
Chlorophyll *b'*	n.d.	n.d.	c	n.d.	—	n.d.
Chlorophyll *a*-derivative[d]	22.9	n.d.	n.d.	n.d.	—	n.d.
Chlorophyll *a*	23.7	31.8	15.4	72	192.5	2.4
Chlorophyll *a'*	25.0	30.4[e]	c	n.d.	n.d.	n.d.
α-Carotene	26.4	n.d.	2.15[f]	96[f]	265[f]	n.d.
cis-α-Carotene	27.9[g]	n.d.	2.15[f]	96[f]	265[f]	n.d.
β-Carotene	27.4	38.3	2.15	96[f]	265[f]	1.5
cis-β-Carotene	28.4[g]	n.d.	2.15[f]	96[f]	265[f]	n.d.
Pheophytin *b*	30.1	37.7	c	92[h]	274[h]	n.d.
Pheophytin *a*	32.4	37.7	10.1	92[h]	274[h]	n.d.
Chlorophyll *c*	—	—	—	—	58.5	—
Tucoxanthin	—	—	—	—	79.5	—
Neofucoxanthin (A,B)	—	—	—	—	93.2	—
Diadinoxanthin	—	—	—	—	135.2	—
Diatoxanthin	—	—	—	—	157.2	—

[a]   Retention times taken from figures.
[b]   Calibration factors are given for quantitative analysis.
[c]   Identified in separate runs under different conditions; for details, see original reference.
[d]   Spectral characteristics like chlorophyll *a*, structure unknown.
[e]   From comparison with other HPLC data, this is likely to be a chlorophyll *a* derivative of unknown structure (see L1).
[f]   Unresolved peak for "carotenes".
[g]   Shoulder.
[h]   Unresolved peak for "pheophytins".

Table III. HPLC 1 (continued)
HPLC OF TOTAL EXTRACTS

Packing   P1 = Sil 60-RP 18 (Riedel de Haen, Hannover, F.R.G.)

            P2 = Sorb-Sil 60-D 10 C_{18} (Macherey & Nagel, Düren, F.R.G.)

            P3 = Silica gel SS05 (Japan Spectroscopic, Tokyo, Japan)

            P4 = 37—75 μm Bondapak C_{18}-Porasil B (Waters, U.S.)

            P5 = Nucleosil 50-5 Silica (Macherey & Nagel, Düren, F.R.G.)

Solvent   S1 = A: methanol-acetonitrile = 25:75; B: water, gradient 75 → 100% B in 20 min, then isocratic

            S2 = A: methanol; B: water; C: methanol-ethanol = 1:1, step gradient; A-B = 85:15 for 17.5 min, A-B = 95:5 for 9.5 min, 100% A for 6 min, then C

            S3 = isopropanol in hexane, step gradient: 1% for 20 min, 2% for 30 min, 5% for 12 min, then 10%

            S4 = step gradient: methanol-water = 98:2 for 77 min, then methanol-water = 1:1

            S5 = step gradient: A: methanol; B: water; C: ether; A-B = 80:20 for 20 min, 90:10 for 45 min, 95:5 for 45 min, 97.5:2.5 for 65 min, 100% A for 40 min; A-C = 90:10 for 30 min, 50:50 for 25 min, then 25:75

            S6 = iso-Octane-98% ethanol = 9:1.

REFERENCES

1. **Braumann, T. and Grimme, L. H.,** *Biochim. Biophys. Acta,* 637, 8, 1981.
2. **Braumann, T. and Grimme, L. H.,** *J. Chromatogr.,* 170, 264, 1979.
3. **Iriyama, K., Yoshiura, M., and Shiraki, M.,** *J. Chromatogr.,* 154, 302, 1978.
4. **Eskins, K., Scholfield, C. R., and Dutton, H. J.,** *J. Chromatogr.,* 135, 217, 1977.
5. **Stransky, H.,** *Z. Naturforsch.,* 33c, 836, 1978.

Table III. HPLC 2
HPLC OF CHLOROPHYLLS AND DERIVATIVES

	P1	P2	P2	P2	P2	P2	P2	P2	P3	P4	P5
Packing											
Column											
length (mm)	250	250	250	250	250	250	250	250	120	250	1800
diameter (mm) (OD)	4.6	4.6	4.6	4.6	4.6	4.6	4.6	4.6	4.5	4.5	3
material	SS	SS	SS	SS	SS	SS	SS	SS	SS	SS	SS
Solvent	S1	S2	S2	S3	S4	S2	S2	S2	S5	S2	S6
Flow rate (mℓ/min)	1.5	1.5	1.5	1.5	1.5	1.5	1.5	1.5	3	4	0.7
Temperature (°C)	Ambient	Ambient	Ambient	Ambient	Ambient	Ambient	Ambient	Ambient	Ambient	Ambient	Ambient
Detection	D1	D1	D2	D3	D4	D4	D3	D3	D5	D5	D6
Literature	1	2	2	3	3	4	5	5	7	8	9
Compound[a,b]							t_R' (corrected)[c] (min)				
Pheophytin a_{CG}	4.6	7.4	—	—	—	—	—	—	—	—	—
Pheophytin a_{DHGG}	5.6	9.4	—	—	—	—	—	—	—	—	—
Pheophytin $a_{THGG\text{-}1}$	6.7	11.5	—	—	—	—	—	—	—	—	—
Pheophytin a_P	8.0	13.6	—	—	—	—	—	—	—	—	—
Pheophytin a'_P	9.4	15.2	—	—	—	—	—	—	—	—	5.5
Pheophytin b_{GG}	—	—	5.0	—	—	—	—	—	—	—	—
Pheophytin b_{DHGG}	—	—	6.1	—	—	—	—	—	—	—	—
Pheophytin b_{THGG}	—	—	8.2	—	—	—	—	—	—	—	—
Pheophytin b_P	—	—	10.2	—	—	—	—	—	—	—	10
3-Acetyl-3-devinylpheophytin a_{CG}	—	—	3.9	—	—	—	—	—	—	—	—
Pyropheophytin a_{CG}	—	7.7[d]	—	—	—	—	—	—	—	—	—
Pyropheophytin a_P	—	8.0[e]	—	—	—	—	—	—	—	—	—
Methylpheophorbide a	—	1.3[c]	—	—	—	—	—	—	—	—	—
Pyromethylpheophorbide a	—	2.1[e]	—	—	—	—	—	—	—	—	—
Chlorophyll a_P	—	7.2	—	—	—	—	—	—	34	22	—
Chlorophyll a_P	—	—	—	—	—	—	—	—	44	—	—
Chlorophyll b_P	—	—	—	—	—	—	—	—	19	10	—
Chlorophyll b_P'	—	—	—	—	—	—	—	—	27	—	—
Pyrochlorophyll a_P	—	8.3	—	—	—	—	—	—	—	—	—
Pheophytin a_F	—	4.5[f]	—	—	—	—	—	—	—	—	—

Compound			
Bacteriochlorophyll a_P	12.1	—	3.0
Bacteriochlorophyll a_{GG}	—	—	1.5
Bacteriopheophytin a_P	—	6.0	—
Bacteriopheophytin a_{GG}	—	3.9	—
Bacteriochlorophyll b_P	22.5	—	—
Bacteriochlorophyll $b_{THGG\text{-}II}$	17.1	4.7	—
Bacteriopheophytin b_P	19.5[g]	3.5	—
Bacteriopheophytin $b_{THGG\text{-}II}$	11.5[g]	—	—
3-Acetyl-3-devinyl-8-hydroxyethylchlorophyll a_P	13.3[g]	—	—
3-Acetyl-3-devinyl-8-hydroxyethyl-8-deethylchlorophyll $a_{THGG\text{-}II}$	10.1[g]	—	—
Oxidation products of bacteriochlorophyll b_P (Structures unknown)	15.3[g]	—	—
Oxidation products of bacteriochlorophyll $b_{THGG\text{-}II}$ (Structures unknown)	14.7[g]	—	—

[a] The subscripts GG, DHGG, THGG-I, THGG-II, P, and F stand for the pigments esterified with geranyl-geraniol (= Δ2,6,10,14-phytatetraenol), dihydrogeranyl geraniol (= Δ2,10,14-phytatrienol), tetrahydrogeranyl geraniol I (= Δ2,14-phytadienol), tetrahydrogeranyl geraniol II (= Δ2,10-phytadienol), phytol (= Δ2-phytaenol), and farnesol, respectively. See Formula in Table III.1 for phytanol.

[b] For the structure of these compounds see Formula .

[c] t_R' = retention time corrected for dead time.

[d] Shoulder of pheophytin a_{GG}.

[e] Reference 5.

[f] Reference 6.

[g] Best detection wavelength for degradation products of bchl b is 680 nm.

[h] Columns filled with material from Merck, Darmstadt, F.R.G.

Packing   P1 = μBondapak C_{18}, 10 μm (Waters, Königstein)
P2 = Lichrosorb C_8, 10 μm (Knauer, Oberursel)[b]
P3 = Lichrosorb RP-18, 5 μm (Knauer, Oberursel)[h]
P4 = Partisil PX S 1025 ODS 2 (Whatman, U.S.A.)
P5 = Corasil II

Solvent   S1 = methanol-acetone (90:10 v/v)
S2 = methanol-water (95:5 v/v)
S3 = methanol-1% sodium ascorbate in water (89:11 v/v)
S4 = methanol-1% sodium ascorbate in water (95:5 v/v)
S5 = acetonitrile-water (94:6)
S6 = ethylacetate-petroleum ether (20:80)

Table III. HPLC 2 (continued)
HPLC OF CHLOROPHYLLS AND DERIVATIVES

Detection

D1	=	absorption at 667 nm
D2	=	absorption at 655 nm
D3	=	absorption at 600 nm
D4	=	absorption at 525 nm
D5	=	absorption at 436 nm
D6	=	absorption at 412 and 434 nm

REFERENCES

1. **Schoch, S., Lempert, U., Wieschhoff, H., and Scheer, H.,** *J. Chromatogr.,* 157, 357, 1978.
2. **Benz, J. and Rüdiger, W.,** *Z. Naturforsch.,* 36C, 51, 1981.
3. **Steiner, R.,** Zulassung, University of Munich, Munich, 1980.
4. **Scheer, H.,** unpublished.
5. **Schoch, S., Wieschhoff, H., and Scheer, H.,** unpublished.
6. **Benz, J.,** Dissertation, University of Munich, Munich, 1980.
7. **Scholz, B. and Ballschmiter, K.,** *J. Chromatogr.,* 208, 148, 1981.
8. **Shoaf, W. T.,** *J. Chromatogr.,* 152, 247, 1978.
9. **Evans, N., Games, D. E., Jackson, A. H., and Matlin, S. A.,** *J. Chromatogr.,* 115, 325, 1975.

Table III. HPLC 3
HPLC OF CHLOROPHYLLOUS PIGMENTS INCLUDING FREE ACIDS

Compound[a]	P1	P1	P2	P3	P3	P3	P3	P4	P4	P4
Packing										
Column length (mm)	n.a.	250	150	250	250	250	250	n.a.	n.a.	300
diameter (mm)	n.a.	4.6	n.a.	4.6	4.6	4.6	4.6	n.a.	n.a.	4
material	SS	SS	SS	SS	SS	SS	SS	n.a.	n.a.	SS
form	st	st	st	st	st	st	st	n.a.	n.a.	st
Solvent	S1	S2	S3	S4	S5	S6	S7	S8	S9	S10
Flow rate (mℓ/min)	2.0	1.5	1.5	1.5	1.5	2.0	2.0	4.0	4.0	1.0
Temperature (°C)	Ambient	40	Ambient	Ambient	Ambient	Ambient	Ambient	29	25	Ambient
Detection	D1	D2	D2	D3	D4	D5	D5	D6	D6	D7
Literature	1	2	3	4	4	5	5	6	6	7
t_R (min)	b					b,c	b,c	b	b	b,c,d
Chlorophyll a_p	16	13.0	7.1	—	30.8	18.2	18.5	—	—	29.4
Chlorophyll a'_p	16.5	13.8	—	—	—	—	—	—	—	—
Chlorophyllide a	6.6	—	2.2	—	3.0	5.5	—	16.5–18[c]	—	8
Pheophorbide b	7.6	—	—	—	—	—	—	—	—	—
Pyropheophytin b	19.5	—	—	—	—	—	—	28	8.5,13	—
Chlorophyll c[f]	—	—	1.5	—	—	8.6	7	—	—	—
Chlorophyll a_{THGG}	11.0	—	—	—	—	—	—	—	—	—
Chlorophyll a_{DHGG}	9.3	—	—	—	—	—	—	—	—	—
Chlorophyll a_{GG}	8.0	—	—	—	—	—	—	—	—	—
Chlorophyll a_F	5.3	—	—	—	—	—	—	—	—	—
Chlorophyll a_G	3.7	—	—	—	—	—	—	—	—	—
Ethyl chlorophyllide a	2.52	—	—	—	—	—	—	—	—	—
Methyl chlorophyllide a	2.46	—	—	—	—	—	—	—	—	—
13^2-Hydroxychlorophyll a_p	10.6	—	—	—	—	—	—	—	—	—
Pyrochlorophyll a_p	17.8	—	—	—	—	—	—	—	—	—
Pheophytin a_p	22	53.9	14.7	—	33.7	—	—	—	—	—
Pheophytin a'_p	22.5	—	—	—	—	—	—	—	—	—
Pheophorbide a	8.7	—	3.0	—	7.0	—	—	33–35[c]	13–15[c]	—
Pyropheophytin a_p	24	—	—	—	—	—	—	—	—	—

Table III. HPLC 3 (continued)
HPLC OF CHLOROPHYLLOUS PIGMENTS INCLUDING FREE ACIDS

Compound								
Chlorophyll b_p	14	7.2	5.6	—	—	—	—	—
Chlorophyll b'_p	14.5	8.1	—	—	27.2	—	—	—
Chlorophyllide b	3.8	—	—	—	—	6,10.5[c]	—	26.3
Pheophytin b_p	18	34.8	10.8	—	—	—	—	—
Pheophytin b'_p	19	—	—	—	—	—	—	—
Protochlorophyll$_p$[g,k]	—	19.7	—	—	—	—	—	30.8
Protochlorophyll$_{THGG}$[g]	—	16.6	—	—	—	—	—	—
Protochlorophyll$_{DPHGG}$[g]	—	14.1	—	—	—	—	—	—
Protochlorophyll$_{GG}$[g]	—	12.1	—	—	—	—	—	—
Protochlorophyll$_{c-18}$[g,h]	—	10.0	—	—	—	—	—	—
Protochlorophyll$_F$[g]	—	7.0	—	—	—	—	—	—
Protochlorophyll$_{c-13}$[g]	—	5.8	—	—	—	—	—	—
Protochlorophyll x	—	2.5—11.2	—	—	—	17	—	12.7
Protochlorophyllide	—	—	—	30.5	—	—	—	—
Protopheophytin$_p$	—	58.0	—	—	—	—	—	—
Bacteriochlorophyll a_p	—	6.4	—	—	—	—	—	—
Bacteriochlorophyll a'_p	—	7.3	—	—	—	—	—	—
Bacteriochlorophyll a_{THGG}[i]	5.6	—	—	—	—	—	—	—
Bacteriochlorophyll a_{DHGG}[i]	4.9	—	—	—	—	—	—	—
Bacteriochlorophyll a_{GG}	—	4.4	—	—	—	—	—	—
Bacteriomethyl chlorophyllide a	—	—	—	—	—	—	—	—
Bacteriochlorophyllide a	—	—	—	—	—	—	—	—
Bacteriopheophytin a	—	27.5	—	34.3	—	—	—	—
Bacteriomethylpheophorbide a	—	—	—	22.0	—	—	—	—
Bacteriopheophorbide a	—	—	—	3.5	—	—	—	—
Bacteriopyropheophytin a	—	—	—	35.4	—	—	—	—
Bacteriopyromethalpheophorbide a	—	—	—	25.6	—	—	—	—
Bacteriopyropheophorbide a	—	—	—	12.2	—	—	—	—

a   The index indicates the esterifying alcohol: P = Δ^2-phytaenol (phytol); THGG = Δ2,14-phytatrienol (tetrahydrogeranyl-geraniol); DHGG = Δ2,10,14-phytatrienol (dihydrogeranyl-geraniol); GG = Δ2,6,10,14-phytatrienol (geranyl-geraniol); F = Δ2,6,10-farnatrienol (farnesol); G = Δ2,6-geraniadienol (geraniol). The prime pigments (e.g., a') are the 13²-epimers.

b   Retention times estimated from figures.

c   The system has been used for the simultaneous separation of carotenoids.

d   Some peaks show evidence of containing more than one compound.

e   More than one peak; smaller peak probably due to the "prime" pigments.

f   No separation of C_1 and C_2.

g   No difference in retention times for protochlorophyll and 8-desethyl-8-vinylprotochlorophyll (= "mono-" and "divinyl-protochlorophyll", respectively.) Differentiation is possible from fluorescence excitation spectra.

h   Alcohol chain length estimated from t_R-diagram.

i   Alcohols not rigorously identified with respect to the position of double bonds; pigments isolated from *Rhodopseudomonas palustris*.

j   Accompanying peak at 23.6 (= a'?).

k   10 Further protochlorophylls probably differing in their esterifying alcohols have been identified in *Cucumis moschato* seedcoats.

Packing   P1   =   Sorbax ODS (C_{18}, DuPont)

P2   = C_8 — Rp

P3   =   Partisil 10 ODS (Waters)

P4   =   P × S 1025 ODO-2 (Whatman)

Solvent   S1   =   A: methanol-water = 75:25; B: ethyl acetate; convex gradient (Waters No. 7) 0—50% B in 10 min, the n isocratic

S2   =   methanol

S3   =   aqueous methanol, step gradient: 4 min 90%, 12 min 98% methanol

S4   =   A: methanol-water = 80:20; B: ethyl acetate, convex gradient (Waters No. 8) 0—50% B in 30 min

S5   =   same as S4, except A: methanol-water = 91:9

S6   =   A: methanol; B: water, 70—95% B

S7   =   same as S6, but water containing 5 mM tetrabutyl ammonium phosphate (Waters Pic A); this improves the resolution of accompanying carotenoids and retards free acids

S8   =   gradient: MeOH-H_2O = 85:15 (0—25 min), 95:5 (25—60 min)

S9   =   MeOH-H_2O = 95:5

S10   =   gradient: A: methanol-water = 8:2. B: ethyl acetate, 0—50% B in 20 min

Detection   D1   =   absorption at 654 nm

D2   =   fluorescence with selected excitation and emission wavelengths

D3   =   absorption at 525 nm (605 for bacteriochlorophyll)

D4   =   absorption at 667 nm

D5   =   absorption at 440 nm

D6   =   absorption at 445 nm

D7   =   absorption at 436 nm

REFERENCES

1. **Schwartz, S. J., Woo, S. L., and von Elbe, J. H.,** *J. Agric. Food Chem.,* 29, 533, 1981.
2. **Shioi, Y., Fukae, R., and Sasa, T.,** *Biochim. Biophys. Acta,* 722, 72, 1983; **Shioi, Y. and Sasa, I.,** *Plant Cell Physiol.,* 23, 24, 835, 1983.
3. **Falkowski, P. G. and Sucher, J.,** *J. Chromatogr.,* 213, 349, 1981.
4. **Schoch, S.,** private communication.
5. **Davies, D. and Holdsworth, E. S.,** *J. Liq. Chromatogr.,* 3, 123, 1980.
6. **Burke, S. and Aronoff, S.,** *Anal. Biochem.,* 114, 367, 1981.
7. **Eskins, K. and Harris, L.,** *Photochem. Photobiol.,* 33, 131, 1981.

Table III. HPLC 4
HPLC OF BACTERIOPHEOPHYTINS *a*

Packing	P1	P1	P1	P2
Column				
length	240	240	240	240
diameter	7	7	7	9
material	Glass	Glass	Glass	Glass
Solvent	S1	S2	S3	S4
Flow rate	F1	F2	F3	F4
Temperature	T1	T1	T1	T1
Detection	D1	D1	D1	D1
Literature	1	1	1	1
Compound[a]	t_R	t_R	t_R	t_R (min)
Bacteriopheophytin a'_p	18.2	21.8	18.8	17.9
Bacteriopheophytin a_p	15.8	20.3	16.0	17.0
Bacteriopheophytin a'_{GG}	21.5	26.2	24.0	—
Bacteriopheophytin a_{GG}	18.4	24.2	19.7	—

[a] Formulas, see Table III.1.
[b] Dimethyl ether
[c] Hexamethyl phosphor triamide.
[d] Methylcyclohexane.

Packing	P1	Partisil 5 (Whatman)
	P2	Lichrosorb S; 60 5 μm (Merck)
Solvent	S1	pentane-ether-water saturated with ether = 16:15:5
	S2	pentane-benzene-DME[b] = 40:40:3
	S3	pentane-MC[c]-DME[b]-HMPA[c] = 100:20:4:0.3
	S4	pentane-MC[d]-acetonitrile = 50:10:3
Flow rate	F1	1.5 mℓ/min
	F2	1.42 mℓ/min
	F3	1.08 mℓ/min
	F4	2.67 mℓ/min
Detection		absorption 254, 260 nm

REFERENCE

1. **Walter, E., Schreiber, J., Zass, E., and Eschenmoser, A.,** *Helv. Chim. Acta,* 62, 899, 1979.

Table III. HPLC 5
HPLC OF CHLOROPHYLLS *a*, *b*, *c*, AND SOME UNKNOWNS

Packing	P1	P2	P3	P4
Column				
length	65	250	250	250
diameter	0.5	n.a.	4.6	4.6
material	PTFE	SS	SS	SS
Solvent	S1	S2	S3	S4
Flow rate	F1	F2	F3	F3
Temperature	Ambient	24°C	n.a.	n.a.
Detection	D1	D2	D3	D3
Literature	1	2	3	3
Compound	t_R	t_R	t_R	t_R(min)
Chlorophyll *a*	15.2	—	18.3	21.4
Chlorophyllide *a*	—	—	5.2	—
Chlorophyll *b*	38.5	48.6	—	—
Chlorophyll $c_1 + c^2$	—	—	8.3	8.1
Pheophytins *a* of unknown structure				
1	—	45.8	—	—
2	—	46.2	—	—
3	—	46.6	—	—
4	—	47.2	—	—
Chlorophyll *a* (E 446, F 674)[a]	—	47.7	—	—
Chlorophyll *a* (E 443, E 672)[a]	—	51.2	—	—
Chlorophyll *a* (E 432, F 662)[a]	—	61.8	—	—
Chlorophyll *a* (E 436, F 670)[a]	—	73.2	—	—

[a] Mixture of several compounds of unknown structure; indicated are the maximum wavelengths for fluorescence excitation (E) and emission (F) of the major component as identified by fluorescence analysis.

Packing	P1	silica gel SS-05 (Japan Spectroscopic)
	P2	Spherisorb (5 μm)
	P3	Partisil 10 ODS
Solvent	S1	step gradient of isopropanol in hexane (0—20 min: 1%; 20—50 min: 2%; 50—60 min: 5%; 60—75 min: 10%
	S2	step gradient: 0—39 min, benzene-hexane = 1:1; then acetone-benzene: 39—53 min, 6:94; 53—65 min, 8:92, then 16:84
	S3	gradient methanol-water = 70:30—95:5
	S4	S3 containing 5 m*M* tetrabutyl ammonium phosphate
Flow rate	F1	16 μℓ/min
	F2	n.a.
	F3	2 mℓ/min
Detection	D1	absorption 380 nm
	D2	fluorescence excited at 425 nm
	D3	absorption 440 or 650 nm

REFERENCES

1. **Iriyama, K., Shiraki, M., and Yoshikura, M.,** *J. Liq. Chromatogr.*, 2, 255, 1979.
2. **Rebeiz, C. A., Belanger, F. C., Freyssinet, G., and Saab, D. S.,** *Biochim. Biophys. Acta*, 590, 234, 1980.
3. **Davies, D. and Holdsworth, E. S.,** *J. Liq. Chromatogr.*, 3, 123, 1980.

Table III. HPLC 6
HPLC OF DERIVATIVES OF
BACTERIOCHLOROPHYLLS c

Packing	P1	P1	P1
Column			
length	300	300	300
diameter	7.8	7.8	7.8
material	SS	SS	SS
Solvent	S2	S3	S1
Flow rate	F1	F2	F1
Temperature	T1	T1	T1
Detection	D1	D1	D1
Literature	1	2	1, 3

Compound[a]		t_R (min)	
Bacteriomethylpheophorbide c			
Fraction 1(2^1-R)	18	46	
Fraction 1(2^1-S)	—	50	
Fraction 2(2^1-R)	23	58	
Fraction 2(2^1-S)	—	63	
Fraction 3(2^1-R)	31	75	
Fraction 3(2^1-S)	33	79	
Fraction 4(2^1-R)	—	91	
Fraction 4(2^1-S)	43	98	
Bacteriomesomethylpheophorbide c			
Fraction 1			12.7
Fraction 2			14.5
Fraction 3			17.6
Fraction 4			20.7

[a]   For formulas, see Table III. 1.

Packing	P1	=	μ-Bondapak C-18
Solvent	S1	=	acetonitrile-water = 90:10
	S2	=	acetonitrile-water = 67:33
	S3	=	methanol-water = 85:15
Flow rate	F1	=	2 mℓ/min
	F2	=	3 mℓ/min
Detection	D1	=	absorption 405 nm
Temperature	T1	=	ambient

REFERENCES

1. **Smith, K. M., Bushell, M. J., and Rimmer, J.**, *J. Am. Chem. Soc.*, 102, 2437, 1980.
2. **Smith, K M., Kehres, L. A., and Tabba, H. D.**, *J. Am. Chem. Soc.*, 102, 7149, 1980.
3. **Smith, K. M., Bisset, G. M. F., and Bushell, M. J.**, *J. Org. Chem.*, 45, 2218, 1980.

Table III. HPLC 7
HPLC OF PRIME-CHLOROPHYLLS

Packing	P1	P1	P2
Column			
length	250	150	150
diameter	30	4.6	4.6
material	SS	SS	SS
Solvent	S1	S2	S3
Flow rate (mℓ/min)	10	1	1
Temperature	Ambient	Ambient	7°C
Detection	D1	D2	D2˙
Literature	1	1	2

Compound[a]	t_R (min)		
Chlorophyll *a*	21	9.6	6.0
Chlorophyll *a'*	16	5.6	3.8
Chlorophyll *b*	37	16.4	21.1
Chlorophyll *b'*	26	15.0	n.a.
Pheophytin *a*	—[b]	5.8	4.1
Pheophytin *a'*	—[b]	4.5	3.4
Pheophytin *b*	—[b]	12.6	n.a.
Pheophytin *b'*	—[b]	10.7	n.a.

[a]   Retention times measured from figures.
[b]   Data not available, but separation achieved under similar conditions.

Packing	P1	= Nucleosil 50-5 Macherey & Nagel
	P2	= Silica Senshupack 50-5

Solvent	S1	= hexane/2-propanol = 97:3
	S2	= hexane/2-propanol = 98.6: 1.4
	S3	= hexane/2-propanol = 98.4: 1.6

Detection	D1	= absorption
	D2	= absorption, 430 nm

REFERENCES

1. **Watanabe, T., Hongu, A., Honda, K., Nakazato, M., Konno, M., and Saitoh, S.,** *Anal. Chem.,* 56, 251, 1984.
2. **Watanabe, T., Nakazato, M., Mazaki, H., Hongu, A., Konno, M., Saitoh, S., and Honda, K.,** *Biochim. Biophys. Acta,* 807, 110, 1985.

Table III. HPLC 8
HPLC OF CHLOROPHYLL DERIVATES

Packing	P1	P1
Column		
length	250	250
diameter	4.6	4.6
material	SS	SS
Solvent	S1	S1
Flow rate	1.5	1.5
Temperature	40°C	40°C
Detection	D1	D1
Literature	1	2

Compound[a,b]	t_R (min)	
Methylchlorophyllide *a*	2.46	—
Ethylchlorophyllide *a*	2.52	—
Chlorophyll a_G	3.74	—
Chlorophyll a_F	5.27	—
Chlorophyll a_{GG}	8.04	—
Chlorophyll a_{DHGG}	9.29	—
Chlorophyll a_{THGG}	11.03	—
Chlorophyll a_P	13.03	—
Chlorophyll a'_P	15.80	—
Pheophytin a_P	53.87	—
10-Hydroxychlorophyll a_P	10.62	—
Pyrochlorophyll a_P	17.85	—
Chlorophyll b_P	7.25	—
Chlorophyll b'_P	8.14	—
Pheophytin b_P	34.85	—
Protochlorophyll$_{GG}$	12.12	—
Protochlorophyll $_{DHGG}$	14.10	—
Protochlorophyll $_{THGG}$	16.62	—
Protochlorophyll$_P$	19.69	—
Protopheophytin$_P$	58.05	—
Bacteriochlorophyll a_{GG}	4.36	—
Bacteriochlorophyll a_{DHGG}	4.93	—
Bacteriochlorophyll a_{THGG}	5.61	—
Bacteriochlorophyll a_P	6.45	—
Bacteriochlorophyll a'_P	7.28	—
Bacteriopheophytin a_P	27.52	—
Chlorophyll b_{GG}	—	4.72
Chlorophyll b_{DHGG}	—	5.36
Chlorophyll b_{THGG}	—	6.19
Chlorophyll b_P	—	7.22

[a] Subscripts denote the esterifying alcohol. P = phytol, GG = geranyl-geraniol, DHGG = dihydrogeranylgeraniol, THGG = tetrahydrogeranylgeraniol, F = farnesol.

[b] Sensitivity enhancement can be achieved by proper excitation and emission wavelengths. See original reference for details.

Packing	P1 = C18-silica, Sorbax-ODS (DuPont)
Solvent	S1 = Methanol
Detection	D1 = Fluorimetry + Integrator

REFERENCES

1. **Shioi, Y., Fukae, R., and Sasa, T.,** *Biochim. Biophys. Acta,* 722, 72, 1983.
2. **Shioi, Y. and Sasa, T.,** *Biochim. Biophys. Acta,* 756, 127, 1983.

Table III. HPLC 9
HPLC OF BACTERIOMETHYL-
PHEOPHORBIDES *d*

Packing	P1
Column	
length	250
diameter	7.8
material	SS
Solvent	S1
Flow rate	5mℓ/min
Temperature	Ambient
Detection	D1
Literature	1

Compound[a,b,c,]			Retention volume (mℓ)
R	Et	Et	74
R	Pr^n	Et	124
S	Bu^i	Et	182
S	NP	Et	232
R	Et	Me	49
R	Pr^n	Me	85
S	Bu^i	Me	136
S	NP	Me	177

[a]   The isomers are characterized (in this order) by the stereochemistry at C-3^1 (R or S), the substituent at C-8 (Et = ethyl, Pr^n = *n*-propyl, Bu^i = *iso*-butyl, NP = neopentyl), and the substituent at C-12 (Me or Et).

[b]   Elution volumes are given in mℓ.

[c]   R and S isomers with identical substituents at C-8 and C-12 can be separated.

Packing	P1	= μ-Bondapak C-18 (Waters)
Solvent	S1	= methanol-water = 85:15
Detection	D1	= absorption

REFERENCE

1. **Smith, K. M. and Goff, J.** *J. Chem. Soc. Perkin I*, p. 1099 1985.

Table III. HPLC 10
HPLC OF NONESTERIFIED CHLOROPHYLL PIGMENTS

Packing	P1	P2	P3	P4	P4	P5
Column						
length	a	20	150	250	250	250
diameter	a	10	4	4.6	4.6	4.6
material	a	Glass	SS	SS	SS	n.a.
Solvent	S1	S2	S3	S4	S5	S6
Flow rate	1	n.a.	1	1	1	0.2
Temperature	Amb.	Amb.	25°C	40°C	40°C	20°C
Detection	D1	Visual	D2	D3	D3	D4
Literature	1	2	3	4[d]	4	5[e]

Compound			t_R (min)			
Protochlorophyllide *a*	12.3	—	—	6.4	—	22(65)
8-Deethyl-8-vinyl-protochlorophyllide *a*	—	—	—	—	—	34(65)
Protopheophorbide *a*	—	—	—	14.4	—	—
Protochlorophyll *a*$_p$	—	—	—	—	—	—
Chlorophyllide *a*	7.9	33[b]	5.6	4.4	—	17(50)
Chlorophyllide *a*′	—	—	—	—	—	—
8-Deethyl-8-vinyl-chlorophyllide *a*	—	—	—	—	—	21(50)
Pheophorbide *a*	—	21.4[b]	11.8	13.5	—	—
Pheophorbide *a*′	—	—	—	16.6	—	—
Methylchlorophyllide *a*	—	—	9.7	—	—	—
Methylpheophorbide *a*	—	—	16.5	—	—	—
Chlorophyll *a*	—	3.8[b]	20.3	—	—	18(80)
Chlorophyll *a*′	—	—	20.8	—	—	—
Pheophytin *a*	—	3.2[b]	27.7	—	—	41(80)
Pheophytin *a*′	—	—	29.0	—	—	—
Chlorophyllide *b*	—	—	2.6	—	7.9	—
Chlorophyllide *b*′	—	—	—	—	9.1	—
Pheophorbide *b*	—	—	9.1	—	30.6	—
Pheophorbide *b*′	—	—	—	—	35.8	—
Methylchlorophyllide *b*	—	—	5.1	—	—	—
Methylpheophorbide *b*	—	—	13.9	—	—	—
Chlorophyll *b*	—	—	18.9	—	—	14.4 (80)
Chlorophyll *b*′	—	—	19.4	—	—	—
Pheophytin *b*	—	—	24.4	—	—	30 (80)
Pheophytin *b*′	—	—	25.2	—	—	—
Chlorophyll c_1	—	—	8.6	4.7	—	16.6 (67)
Chlorophyll c_2	—	—	8.6	7.3	—	24.5 (67)
Methylchlorophyllide c_1	—	—	11	—	—	—
Methylchlorophyllide c_2	—	—	11	—	—	—

[a]   Analytical column, no dimensions given.
[b]   Elution volumes.
[c]   AA = ammonium acetate.
[d]   This system also separates chlorophylls esterified with different alcohols.
[e]   This system also separates chlorophylls and bacteriochlorophylls esterified with different alcohols.

Table III. HPLC 10 (continued)
HPLC OF NONESTERIFIED CHLOROPHYLL PIGMENTS

Packing	P1	=	μ-Bondapak C-18 (Waters)
	P2	=	DEAE-Sepharose CL-6B (Pharmacia)
	P3	=	5 μ Spherisorb ODS-2 (Phase Sep)
	P4	=	Partisil-10 ODS-2 (Whatman) or Sorbax-ODS (DuPont)
	P5	=	Polyethylene, RP-HPLC grade (Polysciences)
Solvents	S1	=	A: methanol-water = 80:20; B: ethyl-acetate, gradient linear A to 50% B.
	S2[c]	=	A: acetone; B: acetone-methanol = 10:1; C: acetone-methanol-AA = 70:30:0.3; D: acetone-water-AA = 80:20:1, step gradient.
	S3[c]	=	A: methanol-AA (1 M) in water = 8:2; B: methanol-acetone = 8:2; linear gradient A to 100% B in 15 min, then isocratic B
	S4	=	methanol-water = 95:5, containing 13 m M acetic acid (pH = 4.2)
	S5	=	methanol-water = 85:5, containing 13 m M acetic acid (pH = 4.2)
	S6	=	acetone-water mixtures. The acetone concentration (v/v) is given in brackets with the retention times.
Detection	D1	=	absorption 436 nm
	D2	=	absorption 650 nm
	D3	=	fluorimetry and densitometry
	D4	=	fluorimetry at optimum wavelengths.

REFERENCES

1. **Eskins, K. and Harris, L.,** *Photochem. Photobiol.,* 33, 131, 1986.
2. **Araki, S., Oohusa, T., Owata, T. and Murata, N.,** *Plant Cell Physiol.,* 25, 841, 1984.
3. **Zapata, M., Ayala, A. M., Franco, J. M., and Garrido, J. L.,** *Chromatographia,* 23, 26, 1987.
4. **Shioi, Y., Doi, M. and Sasa, T.,** *J. Chromatogr.,* 238, 141, 1984.
5. **Shioi, Y. and Beale, S. I.,** *Anal. Biochem.,* 162, 493, 1987.

Index

INDEX

Milton Keynes UK
Ingram Content Group UK Ltd.
UKHW051932141024
449569UK00027B/1453